The Optics of Rays,
Wavefronts, and Caustics

This is Volume 38 in
PURE AND APPLIED PHYSICS
A Series of Monographs and Textbooks
Consulting Editors: H. S. W. MASSEY AND KEITH A. BRUECKNER
A complete list of titles in this series appears at the end of this volume.

The Optics of Rays, Wavefronts, and Caustics

O. N. STAVROUDIS

Optical Sciences Center
University of Arizona
Tucson, Arizona

ACADEMIC PRESS New York and London 1972

ACADEMIC PRESS, INC.
111 Fifth Avenue, New York, New York 10003

United Kingdom Edition published by
ACADEMIC PRESS, INC. (LONDON) LTD.
24/28 Oval Road, London NW1

LIBRARY OF CONGRESS CATALOG CARD NUMBER: 72-84280

PRINTED IN THE UNITED STATES OF AMERICA

Respectfully dedicated to the Department of Defense Project

THEMIS

which supported a portion of the research reported in this book. Defense support for this kind of research may appear strange, but there is a precedent:

...I am very well acquainted, too, with matters mathematical,
I understand equations, both the simple and quadratical,
About binomial theorem I'm teaming with a lot o' news—
With many cheerful facts about the square of the hypotenuse.

I'm very good at integral and differential calculus,
I know the scientific names of beings animalculous;
In short, in matters vegetable, animal and mineral,
I am the very model of a modern Major-General.

—W. S. Gilbert
H. M. S. Pinafore

Contents

Preface

...je ne prétends ni n'ai jamais prétendu être de la confidence secrète de la Nature. Elle a des voies obscures et cachées que je n'ai jamais entrepris de pénétrer; je lui avois seulement offert un petit secours de géométrie au sujet de la réfraction, si elle en eût eu besoin.

—FERMAT
Oeuvres, 1891, p. 483

Geometrical optics can be described as seventeenth and eighteenth century physics camouflaged behind nineteenth and twentieth century mathematics. Consider, for example, the corpuscular theory, in which light is assumed to consist of a flow of tiny particles, with mass but without charge, which are supposed to obey the usual laws of mechanics. The path taken by each corpuscle is determined by Fermat's principle of least time—that is, out of all possible curves connecting two points, the path taken by a corpuscle is that one for which the time of transit is an extremum.

So naive a concept seems out of place in this age of quantum electronics. Nevertheless, whether we approve of it or not, geometrical optics maintains a unique position in modern technology. It remains the only convenient means by which the gross properties of an optical system can be described in terms of its design parameters. Except for those rare lenses in which the geometrical aberrations are small relative to the diffraction pattern produced by its aperture, the description of the image of a point formed by an optical system in terms of its geometrical aberrations is remarkably faithful to what is seen of such an image through a microscope.

The process of designing an optical system depends almost exclusively on ray tracing, and the assumptions on which ray tracing is based are purely geometrical. Rays are, after all, the paths along which our presumed corpuscles travel, and they can be visualized as narrow beams of light passing through the lens. However, if we try to isolate a single ray by making the beam narrower and narrower, by means of a physical stop with an aperture whose diameter we can control, we find that, beyond some point, the beam broadens and becomes divergent. Diffraction rears its ugly head and our purpose is defeated. A ray therefore exists only as a geometrical abstraction and has no observable physical counterpart.

So we are faced with a paradox. On the one hand the procedures, the formulas used so successfully in optical design, are tailored out of the fabric of geometrical optics. On the other hand geometrical optics is, from the most generous point of view, an absurdly naive statement of the physics of the propagation of light. To some extent this paradox was resolved by R. K. Luneburg in 1964 and by M. Kline and I. W. Kay in 1965, who showed that geometrical optics can be thought of as an approximation to the electromagnetic wave theory of light. Yet the lack of dependence of the conclusions of geometrical optics upon the Maxwell equations, their universal applicability to important problems of a practical nature as well as their simplicity and elegance, to me justify the application of Occam's razor to the real physics of the problem and the treatment of the subject as a quasi-physical, quasi-mathematical entity. It is in this spirit that the subject will be approached in this volume.

There are certainly many good books on geometrical optics. What most fail to provide, however, is a concrete link with the underlying mathematical foundation on which geometrical optics has been erected. We see the bricks and the mortar, the shingles and the eaves, the stove and the kitchen sink; for the most part this is all one needs to know in order to live in the house. But when a remodeling job is contemplated, then a knowledge of the foundation becomes vital. That information is what this book is meant to provide: which partitions can bear the weight of an additional story and which can be knocked down and eliminated.

The plan that I follow is to go from the general to the particular, to begin with the underlying general mathematical facts, which, when applied in an optical context, lead to those particular theorems that constitute the substance of, and that are peculiar to, geometrical optics. I am convinced that the usefulness of this approach transcends being merely heuristic and pedagogical. In what follows we will see some new ideas and perhaps a few new results. More important, we will see geometrical optics as a vibrant and vital field that is far from the dry and dead subject that it has for so long been depicted to be.

It is certain that in time the optical designer will have in hand practical tools that will enable him to do his job better and more efficiently and that, moreover, will be firmly based on real physics. When that day comes these pages will be obsolete. However, until that day comes we are stuck with geometrical optics. So let's make the best of it.

Acknowledgments

The two direct quotations of Pierre de Fermat are from "Oeuvres," Vol. 2, Gauthier-Villars, Paris. The quotation from "H. M. S. Pinafore" was taken from "Plays and Poems of W. S. Gilbert," Random House, New York. Thanks are due to both publishers for their kind permission to quote from their works.

It is a great pleasure to thank the many hands whose help have made this labor light. Martha Stockton and her staff, Judi Tafoya, Marti Porterfield, and Kathy Seeley, as well as Janet Rowe, all contributed to the preparation of the manuscript. They were also my arbiters on questions of style and taste and tolerantly and tactfully modified my imaginative spelling.

The illustrations were drawn by Christopher Stavroudis who also, along with Gregory Stavroudis, assisted in the preparation of the index. Dorle Stavroudis, whose valuable comments and criticisms comprise an integral part of this work, also cheerfully relinquished, for weeks on end, her dining room table and allowed the strangest litter to accumulate on top of her piano.

My thanks go to the Optical Sciences Center, University of Arizona, for patiently and unstintingly supporting my efforts and, in particular, to its director and founder, Aden B. Meinel, whose genius created the environment (and whose projects provided the fertilizer) in which these ideas were nourished, grew, and finally bore fruit.

I

Introduction

Authors of the classical period always began their tales by invoking one or more of the muses. Perhaps that is why scientific works frequently contain brief historical synopses that often are far too brief to be more than just barely accurate. Their purpose seems to be only to conjure up the ghosts of dead and famous men and beseech them for the right kind of inspiration.

To do more than this, however, might be a little awkward if not outright dangerous. As E. T. Bell once put it, mathematicians rush in where historians fear to tread. History is a Pandora's box. Opened by the unskilled hand, it may unleash all sorts of philosophical dragons to stalk the unwary and uncertain traveler. This hazard I gladly risk. Geometrical optics, this peculiar science, this utter anachronism, is best understood in a historical perspective. I shall attempt to provide this here. So, with apologies in advance for my errors and the omissions that I must make, I shall begin.

I could start with Aristotle and his contemporaries, who concerned themselves with the nature of light and the mechanics of vision.

Empedocles, said Aristotle, regarded light as a substance ejected from a luminous object, with a finite velocity, and causing an alteration of the properties of the transparent medium through which it passed. Aristotle himself took an opposing view. He regarded light as only the alteration of the transparent medium; it was neither caused by nor accompanied by any emission from a luminous object. This state or quality was acquired by all parts of the illuminated transparent medium at the same time just as an entire body of water may under the right conditions be turned to ice simultaneously. In our terminology, Aristotle regarded the velocity of light as being infinite (Sabra, 1967, p. 46). These two points of view represent the vanguard of what we now call, respectively, the corpuscular theory and the field theory of light.

But the above constitutes only a remark and not a proper beginning for this tale. I could also begin with Hero of Alexandria and the law of reflection and with Willebrord Snell and the law of refraction. But these technologically indispensible, empirically determined laws, although crucial in the development of a theory describing light and its properties, fail to lie at the center of the development of such a theory. The greatest value of these laws, from our rather parochial point of view, is as a touchstone to provide verification of a proposed theory. Or rather, to state their function more precisely, any theory of light that fails to agree with these laws of reflection and refraction must be abandoned.

Clues to our proper starting point are the words *experimental* and *verification*. The systematic use of experimental data to verify or disprove a theory is the hallmark of the modern scientific method. Although there were earlier experimenters (Roger Bacon comes to mind), Descartes was the first philosopher to break with tradition and acknowledge the value of experimentally obtained information (Sabra, 1967, Chapter 1). Modern science, and therefore modern physics and modern optics, begins with Descartes. His life and times is then the proper beginning of this tale and his contemporaries, Newton and Fermat in particular, are the proper *dramatis personae*.

THE SEVENTEENTH CENTURY

The 1600s were marked as a period of extraordinary stress. The plague and religious wars tramped back and forth across Europe. By our standards, it was a period marked by abject poverty, abysmal ignorance, and rampant superstition. The seventeenth century and the men who inhabited it are far more alien to the twentieth century mind than, say, the classical eras of Greece and Rome of fifteen centuries earlier. We raise our eyebrows at the young Isaac Newton, who, in 1660, the year of his matriculation at Cambridge, when asked what he wished to study, replied "mathematics, because I wish to test judicial astrology" (Dampier, 1961, p. 148). We can wonder what horrendous crime it was that caused Pierre de Fermat, the jurist of Toulouse, to condemn a priest to the stake (Bell, 1961, p. 302). And we are amazed at the Spirit of Truth who, on the tenth of November 1619, the eve of St. Martin, revealed to René Descartes a "wonderful discovery," "a new science," and who planned once and for all the course of his future investigations (Mahaffy, 1880, p. 28; Haldane, 1905, pp. 51–54). Yet, in spite of the divineness of this revelation, just to be on the safe side, Descartes delayed publication of this new science until the storm over Galileo's books blew over (Sabra, 1967, p. 18).

These three incidents perhaps overstress the negative side of three of the most crucial lives in scientific history. From our remote point of view one was befuddled with superstitious mysticism; another was marked with a callousness and brutality quite consistent with his times; the third was perhaps a little mad. How would these admittedly great men have fared in our enlightened age? Newton's strange ideas may have barred his admission to Cambridge. Fermat could have been hanged at Nuremburg. It is easy to picture Descartes as a jet set outpatient.

DESCARTES

Descartes agreed with Aristotle that light was propagated at an infinite speed. From those who disagreed with him he demanded experimental evidence, conceding that if such verification were available his whole theory of physics would topple. This uncharacteristic particle of honesty reveals the depth and breadth of his break with tradition. Descartes' reliance on experimental evidence appears again in the discussion of his second law of motion—when two bodies meet, one loses as much motion as is gained by the other—which he asserted is affirmed by experiment. Nevertheless, he felt that he must buttress the statement, arguing that, because God is immutable, the amount of motion deposited in the universe remains constant for all time. This seventeenth century conservation law therefore has a sound theological basis. On the other hand, he was so sure that light could not have a finite velocity that he did not bother to invoke theology to support his thesis (Sabra,1967,pp.50–53).

Descartes' physical reasoning—and his explanations—occupied two levels of sophistication. The higher level was pure physics with perhaps a touch of applied theology to make things reasonable. Here facts were jostled into position to reveal ultimate truth. The goal was a theory of the universe on the grand scale. The lower level was where details were taken care of. Less care was taken with logical rigor—the emphasis was perhaps intellectual vigor. It was on this level that he, like Jacob, wrestled with the angel. This is where his self doubts and struggles and failures are really revealed, where his mind at work is best seen. Here, he blusters, he is not revealing ultimate truth but only explaining how things work. Here he is at his mechanistic best playing with the wheels and cogs of his cosmological *Weltenuhr*. Here theology is less conspicuous. This is the domain of the schoolboy explanation where the biggest schoolboy is the schoolmaster himself.

Descartes knew that rays existed (there was never any argument about these), that these rays were associated with light's propagation, which took place at an infinite speed. He was committed to a mechanical universe—events were described in terms of moving particles, collisions, pressures, and the like. He deftly created a hydrostatic analog for the propagation of light that is mechanistic and at the same time consistent with his notion of infinite velocity. Light is a pressure applied to the transparent medium by a luminous object. Rays are in the directon of the force. The pressure is applied throughout the medium at exactly the same time. The hydraulic fluid, a substance that he called the *second element*, which was inelastic and incompressible, would be the kind of medium that could transmit a pressure instantaneously (Sabra, 1967, pp. 51 *et seq.*). This was the serious physics of the higher level of sophistication.

For an explanation of reflection and refraction Descartes descended to his second level (Sabra, 1967, Chapter 3). Here the mechanical analog is more obvious and more concrete. He harks back to a rather old idea, due originally to Hero of Alexandria, who explained reflection as a process of balls bouncing off a plane surface. For both a reflecting beam of light and a bouncing ball, the angle of incidence always equals the angle of reflection or of rebound. Hero observed that the path of a ray between two fixed points that included a reflection off a mirror was such that the path length was a minimum. This *minimum-path* principle was true, according to Hero, because nature does nothing in vain.

But how did Descartes extricate himself from the position that light is only a pressure that is exerted instantaneously everywhere and yet does not flow? Deftly! Being a pressure, light is a *tendency* to motion, not motion itself, and is propagated instantly. And a tendency to motion must obey the laws of motion just as though motion had indeed taken place. The bouncing ball then becomes a legitimate explanation of reflection.

Descartes' derivation of the law of refraction also uses bouncing balls rather than the upper-level static pressure theory. The first step in his argument is that light penetrates more easily in denser than in rarer media. Why? Because light can be generated only in a material medium. Where there is more material, light can be generated and propagated more easily. Light therefore bends toward the normal in the denser medium and away from the normal in the rarer medium. This explanation is more or less consistent with his upper-level theories. However, when he becomes quantitative he reverts to the bouncing-ball model. Now the bouncing ball must travel faster in the denser medium and slower in the rarer. And how is this accomplished? By a tennis racket that adds motion to the ball entering the denser medium and by a net that slows it down on leaving it. After all, if Maxwell can have his demon, should not Descartes be allowed his tennis player?

From an argument like this he derived Snell's law, that the sine of the angle of incidence i is proportional to the sine of the angle of refraction r:

$$\sin i = k \sin r. \tag{I-1}$$

As far as Snell had been concerned, k was a number that was characteristic of the medium. When light enters glass, for example, k is greater than unity. When it exits, k is less than *one*. In Descartes' derivation, k became a ratio of velocities

$$k = v_{\mathrm{r}}/v_{\mathrm{i}}\,, \tag{I-2}$$

and since k is greater than one when light enters a denser medium, its velocity must be greater in that medium.

This law, stated in Eq. (I-1), was obtained empirically by Willebrord Snell prior to 1621. He died before it was published, and his results were uncovered among his papers not before 1632. A point of controversy is whether Descartes derived this law independently or whether he had seen Snell's papers and was influenced by them. This derivation appeared in Descartes' *Le Monde ou Traité de la Lumière* (Sabra, 1967, p. 18), which was completed by the middle of 1633 but not published until some years later when the storm that Galileo raised had blown over. A charge of plagiarism was made (Sabra, 1967, p. 104). Fermat, and much later, Huygens and Leibnitz, rejected the proof for logical and mathematical reasons, and asserted that Descartes "cooked up" the proof to fit Snell's results which he already knew. (For a conflicting point of view, see Boegehold, 1919.)

NEWTON AND FERMAT

Newton seemed to waver between the two positions taken by Descartes. He did talk about corpuscles (these were shrunken tennis balls) and followed Descartes' proof although he endowed it with greater physical plausibility. In this way refraction remained a special case of particle dynamics. He was nevertheless uncomfortable in this position and seems to have rewritten one of his early papers to appear to be less of a corpusculist (Sabra, 1967, p. 244).

Fermat was highly critical of Descartes. He was convinced that there was no merit at all in Descartes' proof of the law of refraction and ultimately came to regard him as something of a fraud. It was a point of honor that spurred Fermat on to undertake his own derivation of the refraction law. He was convinced that Descartes' was wrong—correct enough for experimental work but not correct in the analytical, deductive sense that would appeal to a mathematician. He would have none of the bouncing balls, and the idea of a static pressure that resulted in no motion left him cold (Sabra, 1967, pp. 142–143). Fermat embarked himself on a course of research that lasted several years.

Fermat committed himself to a light that propagated at a finite velocity. Moreover, in formulating a law of motion for light, he reached back to Hero of Alexandria—not to the bouncing balls but to the notion of *minimum path*. Although correct for reflection, this obviously was not the proper postulate for refraction. What was required was a formula that would work for both cases and that would also reduce to the minimum-path postulate in the special case of reflection. Fermat arrived at a *least time* principle. The path of a light ray connecting two fixed points is the one for which the time of transit, not the length, is a minimum (Sabra, 1967, Chapter 5). His justification was mystical: Nature is essentially lazy:

> *Je reconnois premièrement ... la vérité de ce principe, que la nature agit toujours par les voies les plus courtes* [*Oeuvres*, 1891, p. 354].

These *shortest courses* he interpreted as *paths of least time*—almost *paths of least resistance*—but not quite (Sabra, 1967, pp. 138, 155).

Imagine Fermat's disappointment when, after several years of hard labor, he arrived at, exactly, Snell's law of refraction, Eq. (I-1). His results agreed perfectly with Descartes' up to the constant of proportionality k. Beyond that there was a difference. Descartes' findings indicated that k was the ratio of the velocity of the refracted ray to the velocity of the incident ray, as shown in Eq. (I-2). Fermat's results were exactly opposite, that

$$k = v_{\mathrm{i}}/v_{\mathrm{r}} , \tag{I-3}$$

and that therefore light must travel *more slowly* in the denser medium.

The fact that both Descartes and Fermat agreed at least on Snell's law might have been the occasion of some sort of reconciliation between these two giants. Fermat did make some overtures to a few of Descartes' partisans. They were, however, unwilling to compromise. They argued that, although Fermat's results were correct, his basic assumptions, his principle in fact, was faulty. It was only geometry, not physics. Fermat's last word, a letter to Clerselier, one of Descartes' partisans, was frosty and biting (Sabra, 1967, p. 154):

> I do not pretend nor have I ever pretended to be Nature's private confidant. She has obscure and secret ways that I have never attempted to penetrate; I have only offered her a little geometrical assistance on the subject of refraction if it should be needed. But since you assure me, Monsieur, that her affairs are in order without it and that she is content to follow the course that M. Descartes has prescribed, I surrender to you my supposed physical conquest and content myself with the purely geometrical problem...
>
> [*Oeuvres*, 1891, p. 483; author's translation].

HUYGENS

But Descartes, Newton, and Fermat were not the only players in this game. There were many others. For our purposes we mention only Huygens, who proposed a theory upon which much of modern twentieth-century physical optics is based. It is paradoxial that, in the context of geometrical optics, Christian Huygens' contribution is puny when compared with those of Descartes, Fermat, and Newton. Huygens, like Fermat, espoused the cause of the finite velocity of light. Light, in his explanation, was transmitted in longitudinal waves in very much the same way as ripples in a pool of water. A luminous object was the origin of such a wavefront. The wavefronts progressed in a train emanating from the luminous object. But he went considerably further than this. Each point on a wavefront was regarded as a source of a special kind of wavelet. These are not seen, but their envelope constitutes a subsequent wave in the train. This is the kind of mechanism Huygens visualized for light propagation. Essentially he regarded it as a longitudinal wave phenomenon with an underlying structure of wavelets (Sabra, 1967, Chapter 8).

The wavelets are the secret of success of the theory's application to refraction and reflection. The argument is so familiar it does not bear repeating (Luneburg, 1964, pp. 34–36). The important consequence is that Huygens' wave theory again is in agreement with Snell's law. However, the constant k agrees with Fermat's conclusion, Eq. (I-3), and not with Descartes', Eq. (I-2).

Thus were the battle lines drawn, Fermat and Huygens favoring a slower speed of light in a denser medium and Descartes and Newton disagreeing. Diffraction effects appeared to be in accord with Huygens' wave theory, yet Newton's corpuscular theory for a time appeared to provide a better explanation of color.

THE NINETEENTH CENTURY

At any rate the constant k turned out to be the open sesame of both optics and mechanics. In both, the minimization of an integral was involved,

$$I = \int k\, ds, \tag{I-4}$$

where ds is an element of arc length. If Fermat's rule holds, then k is inversely proportional to the speed of light. If Newton and Descartes are right, then k is directly proportional, which is correct also for the path of a material body in mechanics.

Fermat's principle, expressed as a problem of minimizing the integral in Eq. (I-1), with k being proportional to v, evolved into Maupertuis' *principle of least action*. This also was based on the postulate that Nature was essentially lazy and moved her objects in such a manner that "action," the integral in Eq. (I-4), was minimized. "Action" ultimately evolved into momentum, and Maupertuis' principle became a statement of the conservation of energy.

Hamilton's application of the calculus of variations to Eq. (I-4) led ultimately to applications in both mechanics and optics. But that is a story for a different chapter.

At any rate the integral in Eq. (I-4) is the starting point for any kind of optics. For light optics we write it as

$$I = \int n \, ds$$

where n is the *index of refraction* or the *refractive index* of the medium through which light passes and is inversely proportional to the reciprocal of the speed of light in that medium.

For electron optics the purely mechanical integral is valid (and let us stick to nonrelativistic electrons and eschew magnetic fields). The index of refraction is directly proportional to the electron's velocity and can be regarded as a potential function (Luneburg, 1964, p. 84).

THE BRACHISTOCHRONE

It is appropriate then to begin the study of geometrical optics with a problem in mechanics as an introduction to the variational principles underlying both fields. Let us consider the problem of the brachistochrone. We will follow Bliss (1925). Suppose there is a wire connecting two points, on which there slides a bead. We assume that the bead moves without friction and that, for simplicity, the entire wire lies in a vertical plane. The only force acting on the bead is that of gravity. Let T represent the total time required for the bead to slide from the uppermost endpoint, A, of the wire to the lower, B. Our problem is to find that particular shape for the wire for which T is a minimum. (See Fig. I-1.)

Let some shape for the wire be given by the curve $y = y(x)$. If the coordinates of A are (a_1, a_2) and those of B are (b_1, b_2), then necessarily, for any such curve, $a_2 = y(a_1)$ and $b_2 = y(b_1)$. Let t represent time and let s denote arc length along the curve $y = y(x)$. Then $ds^2 = dx^2 + dy^2 = (1 + y'^2)\, dx^2$, where y' denotes the derivative of y with respect to x. The velocity of the bead at any point is given by $v = ds/dt$, and its acceleration is $a = d^2s/dt^2$, so that the force f acting on the bead is

$$f = ma = m\, d^2s/dt^2,$$

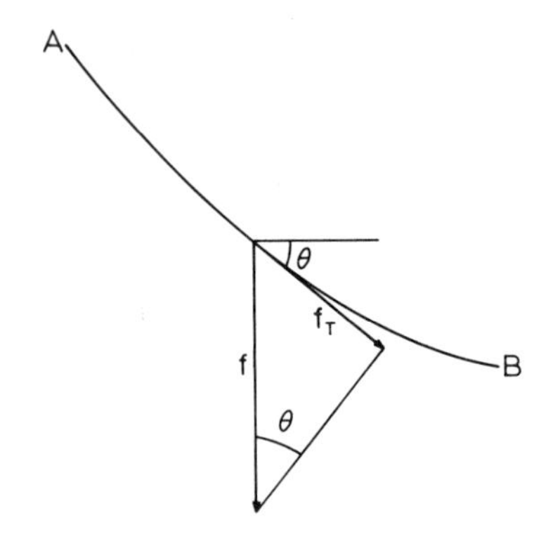

FIG. I-1. The brachistochrone. The forces acting on the sliding bead.

where m is its mass. The total time T is of course given by

$$T = \int_A^B dt.$$

Gravity exerts a force in a downward direction on the bead, $f = mg$, where g is the gravitational constant. However, the motion of the bead is constrained by the wire so that only that component of the force of gravity in the direction of the tangent to the wire has an effect on the motion of the bead. From Fig. I-1, it can be seen that the tangential component of the force is given by

$$f_T = mg \sin \theta.$$

But $\tan \theta$ in the slope of the curve $y = y(x)$ and is therefore equal to y'. This relationship yields an expression for $\sin \theta$ in terms of y' and, upon substitution, we obtain

$$f_T = mgy'/(1 + y'^2)^{1/2} = mg\, dy/(dx^2 + dy^2)^{1/2} = mg\, dy/ds,$$

an expression involving the derivative of y with respect to the arc length parameter s.

We have thus obtained two expressions for the force acting on the bead. Equating these and dropping the common factor m results in

$$d^2s/dt^2 = g\, dy/ds.$$

Multiplying by $2\, ds/dt$,

$$2(ds/dt)(d^2s/dt^2) = 2g(dy/ds)(ds/dt) = 2g\, dy/dt.$$

This is readily integrable,

$$(ds/dt)^2 = 2gy + k,$$

where k is a constant of integration. Note that ds/dt is the bead's velocity. If we now assume that the initial velocity of the bead is v_A, the constant of integration is determined and

$$ds/dt = [2g(y - a_2) + v_A{}^2]^{1/2}.$$

Solving for dt and recalling that $ds^2 = (1 + y'^2)\,dx^2$, we obtain

$$dt = \left(\frac{1 + y'^2}{2g(y - a_2) + v_A{}^2}\right)^{1/2} dx.$$

The time of descent is therefore

$$T = \int_A^B dt = \int_{a_1}^{a_2} \left(\frac{1 + y'^2}{2g(y - a_2) + v_A{}^2}\right)^{1/2} dx. \tag{I-5}$$

Now we can state the problem of the brachistochrone with a little more precision. Associated with each curve connecting A and B is a value of T obtained from the above integral. Out of all such curves we seek the one for which T is a minimum. We defer until later a discussion of techniques for obtaining such a curve, which provides the required shape for our wire. At this point it is sufficient to say that the solution is an arc of a cycloid. A wire in this shape connecting A and B has the property that, if it is slightly distorted in any way, the time of descent of the bead will be increased.

We have found that the velocity of the bead at any point is given by

$$v = ds/dt = [2g(y - a_2) + v_A{}^2]^{1/2} \tag{I-6}$$

and depends only on the vertical distance the bead has dropped from the initial point A. Now we change the problem. Suppose we remove the wire completely, repeal the law of gravity, and in their stead *assume* that the velocity of the bead conforms to the above law. In the expression for the time of descent T nothing changes. Any path the free-falling bead follows in traveling from A to B takes exactly the same time as it would sliding along a wire under the influence of gravity. Moreover, if the path taken is the arc of a cycloid, then that time of descent is a minimum.

Turning the problem around still further, we can ask under what conditions the true path of the free-falling bead is the arc of a cycloid. The answer is "always" if Fermat's principle is true. Fermat's principle states that a particle, acted on by any system of forces, travels along that path for which its time of descent is a minimum.

This is exactly what happens in a transparent medium if the velocity of light in that medium is given by the velocity function in Eq. (I-6). In other words, Fermat's principle states that, if v is any velocity law obeyed by light moving between two points A and B, the path taken by a light "corpuscle" is that for which its time of flight,

$$T = \int_A^B (1/v)\, dx, \tag{I-7}$$

is a minimum. The path that it takes is called a *ray*. To sum up, if the bead is a light corpuscle in a medium with a velocity law given by Eq. (I-5), then its path will be a cycloid. The details of this will be covered in Chapter IV.

But what of material particles such as electrons in an electron microscope—can they be treated in the same way? The answer is in the affirmative if in the integral in Eq. (I-7) we replace the velocity by its reciprocal,

$$I = \int v\, dx.$$

The velocity law for an electron beam whose ray paths would be cycloids then would be, from Eq. (I-5),

$$v = [2g(y - a_2) + v_A^{-2}]^{-1/2},$$

where v_A is the initial velocity of a particular electron. An electrostatic field that resulted in a velocity function like this would produce cycloidal ray paths.

Finally, in dealing with optical problems, we rarely refer to velocity but refer instead to the index of refraction n, which is inversely proportional to the velocity of light in a medium. The constant of proportionality is c, the velocity of light *in vacuo*. The index of refraction is therefore dimensionless.

With this in mind we multiply the above integral by c, converting it to

$$S = \int n\, dx. \tag{I-8}$$

A minimum for S defines a ray between A and B. The corresponding value of S has the dimensions of length and is called the *optical path length* between A and B along the ray.

REFERENCES

Bell, E. T. (1961). "The Last Problem." Simon and Shuster, New York.

Bliss, G. A. (1925). "Calculus of Variations." Open Court Publ., LaSalle, Illinois.

Boegehold, H. (1919). Einiges aus der Geschichte des Brechungsgesetzes, *Cent. Ztg. Opt. U. Mech.* **40**, 94–124.

Dampier, W. C. (Sir) (1961). "A History of Science and Its Relations with Philosophy and Religion," 4th ed. Cambridge Univ. Press, London and New York.

Fermat, P. de (1891). "Oeuvres" (P. Tannery and C. Henry, eds.), Vol. 2. Paris.

Haldane, E. S. (1905). "Descartes: His Life and Times." Murray, London.

Luneburg, L. K. (1964). "Mathematical Theory of Optics." Univ. of California Press, Berkeley.

Mahaffy, J. P. (1880). "Descartes." Blackwood, Edinburgh and London.

Sabra, A. I. (1967). "Theories of Light from Descartes to Newton." Oldbourne, London.

II

The Calculus of Variations

In the final decade of the seventeenth century, according to Bliss (1925), an extraordinary rivalry developed between the brothers Bernoulli, in which each attempted to outdo the other in displays of mathematical virtuosity. This rivalry played perhaps a crucial role in the early development of the calculus of variations. In 1696 Jean Bernoulli, the younger of the two brothers, publicly proposed the problem of the brachistochrone and then exhorted the mathematical community to present solutions. A prize of 50 ducats was offered. In the following year both he and his brother Jacques published their results. Jean's solution depended on the analogy between mechanics and optics, which we have already discussed. He noticed that the problem is identical to that of finding the path of a ray of light in a medium in which the velocity of light is, as we have already determined,

$$v^2 = 2g(y - a_2) + v_1^2.$$

Using modern notation, the index of refraction of this medium is $n = c/v$, where c is the velocity of light *in vacuo*. The initial velocity v_1

can be interpreted as being proportional to the reciprocal of the index of refraction at the boundary. We have already noted that v, and therefore n, depends on $y - a_2$, the vertical distance the bead has dropped or the thickness of the optical medium. Jean approximated this medium with a stack of thin, plane laminae, each with a constant refractive index, the refractive index of each layer being determined by the above formula. The path of a ray in this medium was then determined by repeated applications of Snell's law and what we would call today "ray tracing." In the limit, as the number of laminae incresed and as their thickness approached zero the ray path so obtained approached the ray path for the continuous medium and therefore the solution of the brachistochrone.

Jacques' solution, on the other hand, dealt more directly with the problem and was stated in more general terms so as to be applicable to a more general class of problems. Although he failed to win the prize his brother offered, his work marks the beginning of what we now know as the calculus of variations.

To be completely candid one must add that this account of the brothers Bernoulli and the brachistochrone may be mere legend. Woodhouse (1810) tells a somewhat different tale. Nevertheless, accurate or not, this story provides an appropriate setting for the main theme of this chapter, the calculus of variations and its relation to optics.

THE STATEMENT OF THE PROBLEM

A more general type of problem treated in the calculus of variations is the following. Given the function $f(\alpha, \beta)$, continuous and possessing continuous partial derivatives of at least the first order, we form the integral

$$I = \int_a^b f(y, y')\, dx, \tag{II-1}$$

where y is an unspecified function of the variable x defined and continuous over the interval (a, b). The $y(x)$ need not possess a continuous derivative over (a, b), but its derivative y' may have a finite number of discontinuities. A final stipulation is that all functions $y(x)$ under consideration will assume the same values at the endpoints of (a, b). In other words, we are considering those curves with, at most, a finite number of "kinks" or corners and for which the integral I is defined.

Out of all such functions we seek one $\bar{y}(x)$ which renders I a minimum.†

For our purposes, this is stated far too strongly. It is sufficient for problems in geometrical optics to require only that I be an extremum, and the results to be developed here will be directed toward that end. However, it should be mentioned that the requirement that I be a minimum (or maximum) demands much stronger conditions than those that we will see here.

The function $\bar{y}(x)$, the presumed solution to the problem, for which I is an extremum, is called, quite appropriately, an *extremal*. Let us assume for the moment that such a function is known. We next select a second function $m(x)$ entirely arbitrary except that it must vanish at the endpoints of (a, b) and that it must possess a derivative m' over that interval. The function

$$y = \bar{y} + \alpha m,$$

where α is constant, satisfies all the requirements of the class of functions under consideration. This is illustrated in Fig. II-1. Substituting this

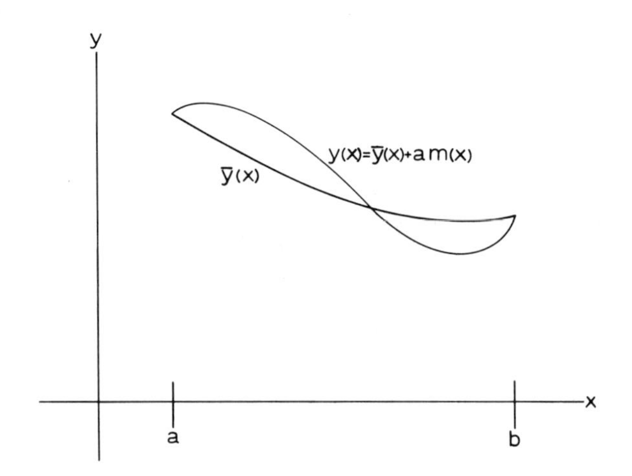

FIG. II-1. The extremal.
The heavy line represents the presumed extremal arc.

function into the integral yields

$$I(\alpha) = \int_a^b f(\bar{y} + \alpha m, \bar{y}' + \alpha m)\, dx.$$

† The standard reference in English to the calculus of variations is Bliss (1946). A more modern and more readable text is Clegg (1968).

Clearly, when $\alpha = 0$, then $I(\alpha)$ is at an extremum. Therefore the derivative of I with respect to α must vanish when $\alpha = 0$. Carrying out this calculation yields

$$\frac{dI}{d\alpha} = \int_a^b \left\{\frac{\partial f}{\partial y} m + \frac{\partial f}{\partial y'} m'\right\} dx = \int_a^b m \frac{\partial f}{\partial y} dx + \int_a^b m' \frac{\partial f}{\partial y'} dx = 0. \quad \text{(II-2)}$$

Applying integration by parts to the first integral in the above expression yields

$$\int_a^b m \frac{\partial f}{\partial y} dx = \left[m \int_a^x \frac{\partial f}{\partial y} dx\right]_a^b - \int_a^b \left[\int_a^x \frac{\partial f}{\partial y} dx\right] m' \, dx.$$

Since m is zero at the endpoints of (a, b), the first term vanishes and we are left with

$$\frac{dI}{d\alpha} = -\int_a^b m' \left[\int_a^x \frac{\partial f}{\partial y} dx - \frac{\partial f}{\partial y'}\right] dx = 0.$$

This is zero only if the quantity in brackets, which we now denote by θ, is constant. To see this, note first of all that if θ is indeed constant, then

$$\int_a^b m'\theta \, dx = \theta \int_a^b m' \, dx = \theta[m]_a^b = 0$$

since m vanishes at the endpoints of (a, b). Next, note that m is an arbitrary function which can be chosen at will and that, moreover, θ does not depend on m since we have set $\alpha = 0$. Now if θ were not constant, it would be possible to select a function m so that the integral would not be zero. But the integral must be zero for all choices of m. It follows that $\theta = \text{const}$. From this we obtain

$$\frac{\partial f}{\partial y'} = \int_a^x \frac{\partial f}{\partial y} dx + c. \quad \text{(II-3)}$$

One rather important point must be made here. We did not exclude the possibility that the first derivative of the extremal $\bar{y}$ be discontinuous. Nevertheless, the quantity c in Eq. (II-3) is constant over the entire interval (a, b) and is blind to the presence or absence of "kinks" or corners in the extremal curve.

If f has second partial derivatives, then, between the discontinuities in $\bar{y}$,

$$\frac{d}{dx} \frac{\partial f}{\partial y'} = \frac{\partial f}{\partial y}. \quad \text{(II-4)}$$

This property of extremals is known as the Euler equation. One needs to emphasize that Eq. (II-3) is valid over the entire range (a, b), whereas Eq. (II-4) is valid only between the corners in the extremal curve.

If the extremal can be assumed to have second derivatives, the differentiation indicated in Eq. (II-4) can be carried out, resulting in

$$\frac{\partial^2 f}{\partial y'^2} y'' + \frac{\partial^2 f}{\partial y' \partial y} y' = \frac{\partial f}{\partial y}. \tag{II-5}$$

By multiplying this by y', it is possible to reduce the above expression to

$$\frac{d}{dx}\left[\frac{\partial f}{\partial y'} y' - f\right] = 0, \tag{II-6}$$

whence

$$\frac{\partial f}{\partial y'} y' - f = \text{const.} \tag{II-7}$$

Strictly speaking, Eqs. (II-5) and (II-6) are valid only between corners those points at which the first derivative of $\bar{y}$ is discontinuous, whereas Eq. (II-7) is valid over the entire interval (a, b). Thus the pair of equations (II-3) and (II-7) are valid, notwithstanding the presence or absence of corners, over the whole interval (a, b) whereas the remaining equations hold only between corners. We will return to this point later.

THE PROBLEM IN THREE DIMENSIONS

The problem can be generalized to any number of dimensions. In three dimensions it is expressed by the following integral:

$$I = \int_a^b f(x, y, z, y', z')\, dx,$$

where we seek as an extremal the space curve $y(x)$, $z(x)$ for which I is an extremum. Here the independent variable x also appears as an argument in the integrand function f in the variational integral. As before, we assume that $\bar{y}(x)$ and $\bar{z}(x)$ are the required functions representing the extremal, and we form the functions

$$y = \bar{y} + \alpha m, \qquad z = \bar{z} + \beta n,$$

where $m(x)$ and $n(x)$ are two functions that are differentiable over (a, b)

and vanish at the endpoints and where α and β are constants. The variational integral I is now a function of the two parameters α and β. Upon differentiating I with respect to these, following the procedures of the earlier case we obtain the analog of Eq. (II-3),

$$\frac{\partial f}{\partial y'} = \int_a^x \frac{\partial f}{\partial y}\,dx + c, \qquad \frac{\partial f}{\partial z'} = \int_a^x \frac{\partial f}{\partial z}\,dx + d. \tag{II-8}$$

These equations, like those in Eq. (II-3), are valid over the entire interval (a, b).

Differentiation of Eq. (II-8) leads to the Euler equations,

$$\frac{d}{dx}\frac{\partial f}{\partial y'} = \frac{\partial f}{\partial y}, \qquad \frac{d}{dx}\frac{\partial f}{\partial z'} = \frac{\partial f}{\partial z}. \tag{II-9}$$

Here, of course, it is assumed that f possesses second partial derivatives. Like their analog, Eq. (II-4), these are valid only between corners on the extremal, those points, if any, at which the first derivatives of either y or z are discontinuous.

In a region in which both y and z possess second derivatives, it is possible to derive an analog to Eq. (II-6). Multiplying the first equation of Eq. (II-9) by y', the second by z', and adding, by making use of the fact that

$$\frac{df}{dx} = \frac{\partial f}{\partial x} + \frac{\partial f}{\partial y}y' + \frac{\partial f}{\partial z}z' + \frac{\partial f}{\partial y'}y'' + \frac{\partial f}{\partial z'}z'',$$

we get

$$\frac{d}{dx}\left[f - \frac{\partial f}{\partial y'}y' - \frac{\partial f}{\partial z'}z'\right] = -\frac{\partial f}{\partial x}. \tag{II-10}$$

Equation (II-10) is valid between corners. By integrating Eq. (II-10) we get the analog of Eq. (II-7),

$$f - \frac{\partial f}{\partial y'}y' - \frac{\partial f}{\partial z'}z' = -\int_a^x \frac{\partial f}{\partial x}\,dx + \text{const}, \tag{II-11}$$

valid over the entire interval (a, b).

Now Eqs. (II-8) and (II-11) are valid over the whole interval (a, b); in particular, they hold at the corners where either y' or z' is discontinuous. The right members of the two equations of Eq. (II-8) and the right member of Eq. (II-11), since they are the sum of a constant and an integral, are continuous. It follows that at a corner of the extremal curve the right-hand limits of the three left members must equal their left-hand limits.

This is a statement of the corner condition. At a corner of an extremal curve, the right- and left-hand limits of the following three quantities must be equal:

$$\frac{\partial f}{\partial y'}, \qquad \frac{\partial f}{\partial z'}, \qquad f - \frac{\partial f}{\partial y'} y' - \frac{\partial f}{\partial z'} z'.$$

THE SHORTEST DISTANCE BETWEEN TWO POINTS

A rather interesting, albeit trivial, illustration of the application of the calculus of variations to a problem is the proof that the shortest distance between a pair of points is a straight line. Suppose we start with the chosen pair of points and imagine that they are connected by a curve $y(x)$, $z(x)$. The differential of distance along this curve is, of course,

$$ds = (1 + y'^2 + z'^2)^{1/2}\, dx,$$

and therefore the distance between the two points along this curve is given by

$$D = \int (1 + y'^2 + z'^2)^{1/2}\, dx,$$

where the limits of the integral are determined by the coordinates of the two points.

We now ask for that curve for which D is a minimum. The integral is now a variational integral, in which

$$f = (1 + y'^2 + z'^2)^{1/2}.$$

The derivatives of f are

$$\partial f/\partial x = 0, \qquad \partial f/\partial y = 0, \qquad \partial f/\partial z = 0,$$

$$\frac{\partial f}{\partial y'} = \frac{y'}{(1 + y'^2 + z'^2)^{1/2}}, \qquad \frac{\partial f}{\partial z'} = \frac{z'}{(1 + y'^2 + z'^2)^{1/2}}.$$

Applying these to Eq. (II-9), we get the Euler equations for this problem,

$$\frac{d}{dx} \frac{y'}{(1 + y'^2 + z'^2)^{1/2}} = 0, \qquad \frac{d}{dx} \frac{z'}{(1 + y'^2 + z'^2)^{1/2}} = 0.$$

Integration yields

$$\frac{y'}{(1 + y'^2 + z'^2)^{1/2}} = a, \qquad \frac{z'}{(1 + y'^2 + z'^2)^{1/2}} = b$$

(where a and b are constants of integration), a pair of simultaneous equations that, when solved, yield

$$y' = \frac{a}{(1 - a^2 - b^2)^{1/2}}, \qquad z' = \frac{b}{(1 - a^2 - b^2)^{1/2}}.$$

When these are integrated, we obtain a four-parameter family of straight lines,

$$y = \frac{ax + c}{(1 - a^2 - b^2)^{1/2}}, \qquad z = \frac{bx + d}{(1 - a^2 - b^2)^{1/2}}.$$

Our proof is not exactly legitimate. We have shown that a straight line is an extremal for this problem; that for a straight line the variational integral is an extremum. That is far from saying that the integral is a minimum. However, to show this with the appropriate rigor and vigor goes a bit beyond the scope of this book.

However, when we apply this ritual to an optical problem, we can get a positive, rather than simply a nonnegative, result. Suppose we ask for the ray paths in a medium in which the refractive index is a constant n. Now,

$$f = n(1 + y'^2 + z'^2)^{1/2};$$

however, the Euler equations and their solutions remain the same, a four-parameter family of straight lines. This result is indeed legitimate. Ray paths need be only extremals; a minimum property is not called for.

THE PROBLEM IN PARAMETRIC FORM

Now we consider variational problems in which an extremal is to be expressed in parametric form, $x(t)$, $y(t)$, $z(t)$, where t is some convenient parameter, the choice of which is not related in any way to the shape of the resulting curve. By changing t, say, by transforming it into a new parameter τ, by $t = \beta(\tau)$, $x(\tau)$, $y(\tau)$, $z(\tau)$ would represent exactly the same curve.

For variational problems of the type we have been examining, we require that the endpoints of all curves considered as candidates for extremals be fixed. This is done for the parametric case by stipulating that the initial point be designated by, say, $x(t_1)$, $y(t_1)$, $z(t_1)$ and the final point by $x(t_2)$, $y(t_2)$, $z(t_2)$. The parameter t is then restricted to the interval (t_1, t_2). The variational integral takes the form

$$I = \int_{t_1}^{t_2} f(x, y, z, x_t, y_t, z_t)\, dt, \tag{II-12}$$

where $x_t = dx/dt$, etc.

One could well ask why t does not appear as an explicit argument in the variational integrand f. However, the numerical value of I is an intrinsic property of the space curve $x(t)$, $y(t)$, $z(t)$ and therefore must be independent of the choice of parameters. If t appeared explicitly in f, a change of parameters would result in a change of the value of f and therefore of I, whereas, as we have seen above, no change would have taken place in the form of the given space curve. This state of affairs is not permissible. We conclude then that the parameter t cannot occur as an explicit argument of f.

The invariance of the variational integral I under transformations of the space curve parameter leads to a second property of f. Suppose we perform a transformation on t: $t = \beta(\tau)$. Then I becomes

$$I = \int_{\tau_1}^{\tau_2} f(x, y, z, x_\tau/\beta', y_\tau/\beta', z_\tau/\beta')\beta'\, d\tau,$$

which must lead to an identical value of I for all choices of β. Moreover, this independence of I on the choice of the parameter transform β, extends to any choice of initial and final points of the extremal curve we are seeking. Therefore, f itself must be invariant under such a transformation and as a result,

$$f(x, y, z, x_\tau/\beta', y_\tau/\beta', z_\tau/\beta')\beta' = f(x, y, z, x_t, y_t, z_t),$$

and we are led to the conclusion that f must be a homogeneous function of the first order in x_t, y_t, z_t. For any k,

$$f(x, y, z, kx_t, ky_t, kz_t) = kf(x, y, z, x_t, y_t, z_t).$$

Differentiating this with respect to k leads to the well-known relation for homogeneous functions,

$$\frac{\partial f}{\partial x_t} x_t + \frac{\partial f}{\partial y_t} y_t + \frac{\partial f}{\partial z_t} z_t = f. \tag{II-13}$$

Following exactly the same sort of argument that led to Eqs. (II-3) and (II-8), we obtain

$$\frac{\partial f}{\partial x_t} = \int_{t_1}^{t} \frac{\partial f}{\partial x}\, dt + a,$$

$$\frac{\partial f}{\partial y_t} = \int_{t_1}^{t} \frac{\partial f}{\partial y}\, dt + b, \tag{II-14}$$

$$\frac{\partial f}{\partial z_t} = \int_{t_1}^{t} \frac{\partial f}{\partial z}\, dt + c,$$

where a, b, and c are constants. As in the case of Eqs. (II-3) and (II-8), Eqs. (II-14) are valid over the entire interval (t_1, t_2).

Differentiating Eqs. (II-14) with respect to t leads to the Euler equations,

$$\frac{d}{dt}\frac{\partial f}{\partial x_t} = \frac{\partial f}{\partial x}, \qquad \frac{d}{dt}\frac{\partial f}{\partial y_t} = \frac{\partial f}{\partial y}, \qquad \frac{d}{dt}\frac{\partial f}{\partial z_t} = \frac{\partial f}{\partial z}. \tag{II-15}$$

These are, of course, analogs of Eqs. (II-4) and (II-9).

The analog of Eq. (II-10) is trivial by virtue of Eq. (II-13) and the fact that f is not explicitly dependent on t. The statement of the corner condition is simply that the three quantities

$$\partial f/\partial x_t, \qquad \partial f/\partial y_t, \qquad \partial f/\partial z_t$$

have the property that their left-hand limits and their right-hand limits are equal at a corner of an extremal curve.

It is convenient at this point to make a change of notation. Let $\mathbf{P}$ be a vector function of t,

$$\mathbf{P}(t) = (x(t), y(t), z(t)),$$

which describes the space curve we are considering, and let its derivative be denoted by

$$\mathbf{P}_t(t) = (x_t, y_t, z_t).$$

Then the integrand function f can be regarded as a scalar function of these two vectors, $f = f(\mathbf{P}, \mathbf{P}_t)$.

We introduce the familiar del operator,

$$\nabla = (\partial/\partial x, \partial/\partial y, \partial/\partial z),$$

and another operator with a slightly different form,

$$\nabla_t = (\partial/\partial x_t, \partial/\partial y_t, \partial/\partial z_t).$$

Now f is a homogeneous function of $\mathbf{P}_t$ and Eq. (II-13) can be written as a scalar product,

$$\nabla_t f \cdot \mathbf{P}_t = f. \tag{II-16}$$

Moreover, Eq. (II-14) can be expressed in the form

$$\nabla_t f = \int_{t_1}^{t} \nabla f \, dt + \mathbf{A}, \tag{II-17}$$

where $\mathbf{A}$ is a constant vector. Finally, we are able to write the Euler equations, Eq. (II-15), in a particularly neat and compact form,

$$d \nabla_t f/dt = \nabla f. \tag{II-18}$$

THE APPLICATION TO GEOMETRICAL OPTICS

We have seen that the path of a ray in a medium having index of refraction n is that curve for which the integral, from Eq. (I-8), $S = \int n \, ds$, where s, which is arc length, is an extremum. This is the same thing as saying that the ray path is an extremal for this integral. Consider an inhomogeneous but isotropic medium in which the refractive index depends on position,

$$n = n(x, y, z) = n(\mathbf{P}),$$

and we seek the ray path connecting two points in this medium, $\mathbf{P}_1$ and $\mathbf{P}_2$. It would appear natural to look for an extremal curve expressed in terms of an arc length parameter s. However, we have no *a priori* knowledge of what the arc length of any curve connecting $\mathbf{P}_1$ and $\mathbf{P}_2$ might be. We cannot determine the two values of s corresponding to the initial and terminal points and therefore we cannot, in terms of s, fix the limits of the variational integral. Since our whole development has been contingent upon these limits being fixed, we must introduce a parametrization for all curves connecting $\mathbf{P}_1$ and $\mathbf{P}_2$ that is not dependent on arc length and that results in a variational integral with fixed endpoints. See Fig. II-2.

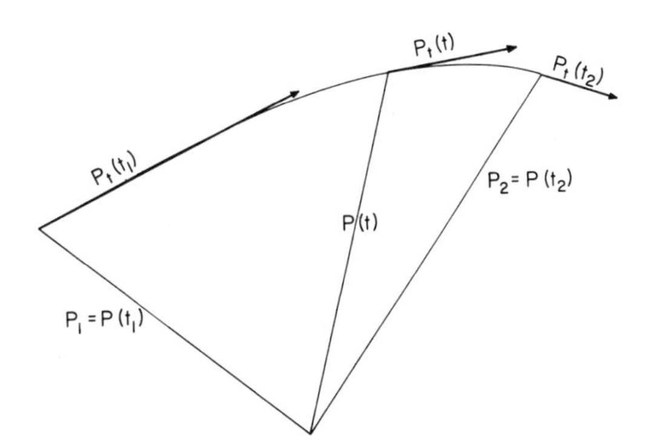

FIG. II-2. The ray path.

$\mathbf{P}_1$ is the starting point, $\mathbf{P}_2$ the end point. Note that the derivatives are tangent vectors.

This presents no great problem. We need only specify that every curve under consideration, every candidate for the ray path, be expressed as a function of a parameter t, $\mathbf{P} = \mathbf{P}(t)$, and that, moreover, each such curve has the property that

$$\mathbf{P}_1 = \mathbf{P}(t_1), \qquad \mathbf{P}_2 = \mathbf{P}(t_2).$$

Moreover, $\mathbf{P}(t)$ must be defined on the interval (t_1, t_2) over which its derivatives can have at most a finite number of discontinuities.

On any such curve the following identities hold:

$$ds = (x_t^2 + y_t^2 + z_t^2)^{1/2}\, dt = (\mathbf{P}_t^2)^{1/2}\, dt.$$

Our variational integral therefore becomes

$$S = \int_{t_1}^{t_2} n(\mathbf{P})(\mathbf{P}_t^2)^{1/2}\, dt$$

and the integrand function in Eq. (II-12) becomes

$$f(\mathbf{P}, \mathbf{P}_t) = n(\mathbf{P})(\mathbf{P}_t^2)^{1/2}.$$

By applying the two del operators to f we get

$$\nabla_t f = n(\mathbf{P})\, \mathbf{P}_t(\mathbf{P}_t^2)^{1/2}, \qquad \nabla f = (\mathbf{P}_t^2)^{1/2}\, \nabla n,$$

where we can interpret ∇n as the gradient of the refractive index. Substituting these expressions into the Euler equation (II-18), we obtain

$$\frac{d}{dt}\left[n(\mathbf{P})\, \frac{\mathbf{P}_t}{(\mathbf{P}_t^2)^{1/2}}\right] = (\mathbf{P}_t^2)^{1/2}\, \nabla n.$$

But $(\mathbf{P}_t^2)^{1/2}$ is exactly and precisely ds/dt, which leads to

$$\frac{d}{dt}\left[n\,\frac{d\mathbf{P}/dt}{ds/dt}\right] = \frac{ds}{dt}\,\nabla n;$$

this, in turn, leads with no difficulty to

$$\frac{d}{ds}\left[n\,\frac{d\mathbf{P}}{ds}\right] = \nabla n.$$

We can see that we can replace our parameter t with the arc length parameter s and in these terms we have obtained a differential equation for a general ray in an inhomogeneous medium,

$$d(n\mathbf{P}')/ds = \nabla n, \qquad \text{(II-19)}$$

where $\mathbf{P}' = d\mathbf{P}/ds$. Equation (II-19) is one of our main conclusions. It is generally known as the ray equation and was known to Hamilton (see Synge, 1937, Chapter 5).

We have returned to the point of view of representing the ray path as a function of the arc length parameter s. As we will see in the next chapter the advantages of this formulation are legion. For the present note that

$$\mathbf{P}' = \frac{d\mathbf{P}}{ds} = \frac{d\mathbf{P}/dt}{ds/dt} = \frac{\mathbf{P}_t}{(\mathbf{P}_t^2)^{1/2}},$$

whence $\mathbf{P}'^2 = 1$. The vector $\mathbf{P}'$ is therefore a unit vector in the direction of propagation.

We will frequently use an expanded version of (II-19). Noting that $(d/ds)\,n(\mathbf{P}) = \nabla n \cdot \mathbf{P}'$, we can write the ray equation in the form

$$n\mathbf{P}'' + (\nabla n \cdot \mathbf{P}')\,\mathbf{P}' = \nabla n. \qquad \text{(II-20)}$$

The Euler equations are in general a pair of second-order ordinary differential equations. Under certain conditions there exists a general solution that depends on four arbitrary constants. Thus, we are led to a four-parameter family of extremals associated with every properly behaved variational integrand f. However, when we specify a particular set of end points, what we are doing is specifying a set of boundary conditions for the Euler differential equations. Thus we can say that a solution to a particular fixed end point problem is embedded in a four-parameter family of extremals.

To translate this into an optical context, we need change only a few words. In general, any medium with a continuous refractive index has associated with it a four-parameter family of ray paths constituting the general solution of the ray equation. A ray connecting any two points in that medium must be a member of that family.

This is intuitively obvious. A ray can be completely determined by its point of intersection with some arbitrary plane and its direction at that point. This requires the specification of two coordinates on a plane and two direction cosines, exactly four parameters in all.

We have assumed from the start that the refractive index n is a continuously differentiable function of position, thus precluding the consideration of the very practical problem of determining the ray path through a lens. In general terms, a lens consists of an arrangement of pieces of glass, each with a constant index of refraction, either in contact or separated by air. A lens can therefore be considered as a medium with a discontinuous refractive index, a medium in which the glass surfaces are the surfaces on which the refractive index is discontinuous. A ray passing through a lens consists of straight line segments with corners at these surfaces of discontinuity. It is tempting to conclude that the corner condition applies in this case. An underlying assumption on which the corner condition is based is thet the variational integral f, and therefore the refractive index n are continuous. The case where the refractive index is discontinuous requires additional analytic tools and must be deferred to a subsequent chapter.

REFERENCES

Bliss, G. A. (1925). "Calculus of Variations." Open Court Publ., La Salle, Illinois.

Bliss, G. A. (1946). "Lectures on the Calculus of Variations." Univ. of Chicago Press, Chicago.

Clegg, John C. (1968). "Calculus of Variations." Wiley, New York.

Synge, J. L. (1937). "Geometrical Optics. An Introduction to Hamilton's Method." Cambridge Univ. Press, London and New York.

Woodhouse, R. (1810). "A Treatise on Isoperimetrical Problems and the Calculus of Variations." Cambridge Univ. Press, London and New York. (Reprinted as "A History of the Calculus of Variations in the Eighteenth Century." Chelsea, Bronx, New York, 1964.)

III

Space Curves

We have found that ray paths in a general inhomogeneous medium, in which the refractive index is a continuously differentiable function of position, can be described as a vector function of a single parameter. We obtained a second-order differential equation for such ray paths that showed that the most convenient choice of parameter is arc length. It is appropriate at this point to examine space curves in general and to determine what geometric properties distinguish ray paths.

The most direct route is by means of differential geometry, in particular, the differential geometry of space curves, the origins of which go back over a century and a half and are associated with Gaspard Monge. He, incidentally, has the additional distinction of having been the French state's head on the day that Louis XVI lost his. Among his pupils was Etienne L. Malus, the author of a crucial theorem in geometrical optics that time has eroded into a rather trivial and obvious statement. Later Hamilton not only contributed to the development of differential geometry but relied on certain of its results in his work in optics and mechanics. However, the real power of differential geometry was not realized until it was cast in a form that made use of the vector notation. In fact, in this form it appears tailor-made for application to geometrical optics. It is this form that will be developed here. Our development will be brief and cursory. For a detailed account of this material the reader is referred to the many excellent texts that are available. Among these the most definitive of the classical treatments is that of Blaschke (1930). Its English counterpart is the text of Struik (1961), and a recent modern treatment is that of Stoker (1969).

Let $\mathbf{P}(t) = [x(t), y(t), z(t)]$ represent a vector from some arbitrary coordinate origin to a point on a space curve. Each value of t, over some interval (t_1, t_2), corresponds to a particular point on the space curve. We assume, moreover, that $\mathbf{P}(t)$ is differentiable over (t_1, t_2). We single out a point on the curve and designate it as $\mathbf{P}(t_0)$. A nearby point is then $\mathbf{P}(t_0 + \delta)$ if δ is small. This is illustrated in Fig. III-1. The vector

$$[\mathbf{P}(t_0 + \delta) - \mathbf{P}(t_0)]/\delta$$

is then a chord connecting the two points. As $\delta \to 0$, the second point

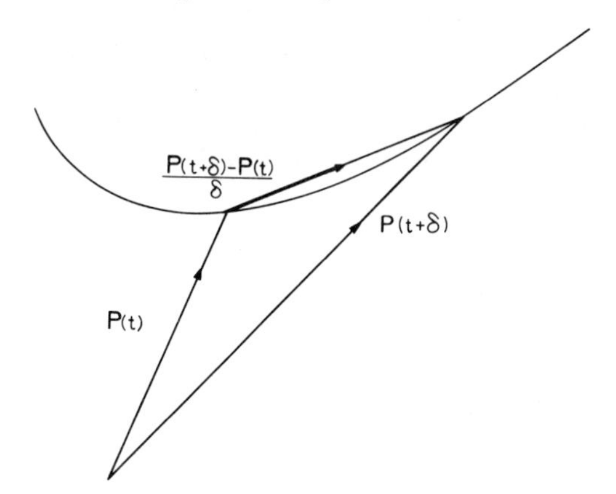

FIG. III-1. The difference quotient for a vector function.

approaches the first and the limiting position of the chord is a vector tangent to the curve at the point $\mathbf{P}(t_0)$. At the same time the limiting value of the vector quantity representing the chord is the derivative of the vector $\mathbf{P}(t)$, denoted by $\mathbf{P}_t(t)$, evaluated at t_0. Therefore, $\mathbf{P}_t(t)$ is a vector in the direction of the tangent to the curve. Its length is $(\mathbf{P}_t^2)^{1/2}$, which we have seen is equal to ds/dt, where s is arc length. The unit tangent vector is therefore

$$\frac{\mathbf{P}_t}{(\mathbf{P}_t^2)^{1/2}} = \frac{d\mathbf{P}/dt}{ds/dt} = \frac{d\mathbf{P}}{ds}.$$

We conclude that if we select arc length as the parameter, then the derivative of **P** with respect to s is the *unit* tangent vector to the curve. The choice of arc length as the parameter for a general space curve is therefore natural and convenient and, in what follows, we will assume that a vector function **P**, representing any space curve, is given in terms of the arc length parameter s. We will designate its derivative by $\mathbf{P}'$ and the unit tangent vector that it represents by **t**.

Since $\mathbf{P}'$ is a unit vector, $\mathbf{P}'^2 = 1$ and its derivative is $\mathbf{P}' \cdot \mathbf{P}'' = 0$. The vectors $\mathbf{P}'$ and $\mathbf{P}''$ are therefore perpendicular and, since $\mathbf{P}' = \mathbf{t}$ is the tangent vector to the curve, $\mathbf{P}''$ must be a normal. However, unlike $\mathbf{P}'$, $\mathbf{P}''$ is not automatically a unit vector. Define *curvature* $(1/\rho)$ and *radius of curvature* ρ by the formula

$$1/\rho = (\mathbf{P}''^2)^{1/2}. \tag{III-1}$$

The vector $\mathbf{n} = \rho\mathbf{P}''$ is called the *unit normal vector* to the curve.

We now have two orthogonal unit vectors, **t** and **n**, associated with each point of a space curve. A third, called the *binormal vector*, is defined as $\mathbf{b} = \mathbf{t} \times \mathbf{n}$, which, since it is perpendicular to **t**, is also normal to the curve. Since it is the vector product of two orthogonal unit vectors, it is also a unit vector.

Thus we have a system of three orthogonal unit vectors, often referred to as the *trihedron*, at each point of a space curve governed by the following relations:

$$\begin{aligned} &\text{tangent} && \mathbf{t} = \mathbf{P}', \\ &\text{normal} && \mathbf{n} = \rho\mathbf{P}'', \\ &\text{binormal} && \mathbf{b} = \mathbf{t} \times \mathbf{n}; \end{aligned} \tag{III-2}$$

$$\begin{aligned} \mathbf{t} \times \mathbf{n} &= \mathbf{b}, & \mathbf{t} \cdot \mathbf{n} &= 0, \\ \mathbf{n} \times \mathbf{b} &= \mathbf{t}, & \mathbf{n} \cdot \mathbf{b} &= 0, \\ \mathbf{b} \times \mathbf{t} &= \mathbf{n}, & \mathbf{b} \cdot \mathbf{t} &= 0. \end{aligned} \tag{III-3}$$

This is illustrated in Fig. III-2.

Two points regarding these vectors must be made. Since they are defined at each point **P** on a space curve given by the vector function

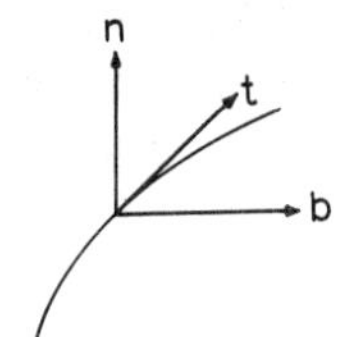

FIG. III-2. The trihedron.
Tangent, normal, and binormal vectors associated with a space curve.

$\mathbf{P}(s)$, any other vector, in particular any vector $\mathbf{V}$ passing through $\mathbf{P}$, can be expressed as a linear combination of $\mathbf{t}$, $\mathbf{n}$, and $\mathbf{b}$:

$$\mathbf{V} = a\mathbf{t} + b\mathbf{n} + c\mathbf{b}.$$

They, the tangent, normal, and binormal vectors, can be thought of as comprising a local coordinate system with its origin at $\mathbf{P}$.

Since $\mathbf{t}$, $\mathbf{n}$, and $\mathbf{b}$ are defined at every point on the space curve $\mathbf{P}(s)$, they may be regarded as vector functions of the arc length s. It is convenient to visualize this kinematically by imagining these three orthogonal unit vectors in motion along the curve as s increases, each maintaining its proper orientation with the curve. We now ask how these vectors behave as s varies.

THE FRENET EQUATIONS

The Frenet equations are nothing more than formulas for the derivatives of each of the three orthogonal unit vectors. If we visualize these as a triad of vectors sliding along a space curve, the Frenet equations tell us how $\mathbf{t}$, $\mathbf{n}$, and $\mathbf{b}$ vary with s. We have already the derivative of $\mathbf{t}$:

$$\mathbf{t}' = d\mathbf{P}'/ds = \mathbf{P}'' = (1/\rho)\mathbf{n}.$$

Next, consider the normal vector. Note first that we may write its derivative as a linear combination of $\mathbf{t}$, $\mathbf{n}$, and $\mathbf{b}$:

$$\mathbf{n}' = (\mathbf{n}' \cdot \mathbf{t})\mathbf{t} + (\mathbf{n}' \cdot \mathbf{n})\mathbf{n} + (\mathbf{n}' \cdot \mathbf{b})\mathbf{b}.$$

Since $\mathbf{n}^2 = 1$, by differentiating we get $\mathbf{n} \cdot \mathbf{n}' = 0$, which tells us that $\mathbf{n}'$ lies in a plane perpendicular to $\mathbf{n}$ and containing $\mathbf{t}$ and $\mathbf{b}$. (This plane is called the *rectifying plane.*) The second term in the above expression for $\mathbf{n}'$ is therefore zero. Since $\mathbf{n} \cdot \mathbf{t} = 0$, by differentiating we discover that

$$\mathbf{n}' \cdot \mathbf{t} + \mathbf{n} \cdot \mathbf{t}' = 0,$$

from which it follows that $\mathbf{t} \cdot \mathbf{n}' = 1/\rho$. For the third term, we must actually calculate the derivative of $\mathbf{n}$:

$$\mathbf{n}' = d(\rho\mathbf{P}'')/ds = \rho'\mathbf{P}'' + \rho\mathbf{P}'''.$$

Therefore,

$$\begin{aligned}\mathbf{n}' \cdot \mathbf{b} &= (\rho'\mathbf{P}'' + \rho\mathbf{P}''') \cdot (\mathbf{t} \times \mathbf{n}) \\ &= \rho(\rho'\mathbf{P}'' + \rho\mathbf{P}''')(\mathbf{P}' \times \mathbf{P}'') \\ &= \rho^2(\mathbf{P}' \times \mathbf{P}'') \cdot \mathbf{P}'''.\end{aligned}$$

We define this last quantity as the *torsion* of the curve:

$$\frac{1}{\tau} = \frac{(\mathbf{P}' \times \mathbf{P}'') \cdot \mathbf{P}'''}{\mathbf{P}''^2}. \tag{III-4}$$

We now have

$$\mathbf{n}' = -\frac{1}{\rho}\mathbf{t} + \frac{1}{\tau}\mathbf{b}.$$

The derivative of the binormal is now simple:

$$\begin{aligned}\mathbf{b}' &= \frac{d(\mathbf{t} \times \mathbf{n})}{ds} = \mathbf{t}' \times \mathbf{n} + \mathbf{t} \times \mathbf{n}' \\ &= \left(\frac{1}{\rho}\mathbf{n}\right) \times \mathbf{n} + \mathbf{t} \times \left(-\frac{1}{\rho}\mathbf{t} + \frac{1}{\tau}\mathbf{b}\right) \\ &= \frac{1}{\tau}\mathbf{t} \times \mathbf{b} = -\frac{1}{\tau}\mathbf{n}.\end{aligned}$$

We now have the Frenet formulas:

$$\mathbf{t}' = \frac{1}{\rho}\mathbf{n}, \qquad \mathbf{n}' = -\frac{1}{\rho}\mathbf{t} + \frac{1}{\tau}\mathbf{b}, \qquad \mathbf{b}' = -\frac{1}{\tau}\mathbf{n}. \tag{III-5}$$

Now to interpret them, once again we consider two neighboring points, $\mathbf{P}(s + \delta)$ and $\mathbf{P}(s)$. Expanding $\mathbf{t}(s + \delta)$ to the first order, we get

$$\mathbf{t}(s + \delta) = \mathbf{t}(s) + \delta\mathbf{t}'(s) = \mathbf{t} + \frac{\delta}{\rho}\mathbf{n},$$

which tells us that the motion of the tangent vector as s increases is in the direction of the normal vector. The rate of change of the tangent vector $1/\rho$ is, naturally, the curvature, and its reciprocal ρ is the radius of curvature of the curve at the given point. It must be borne in mind that these quantities are not constant but depend on s. We now treat the binormal in the same way:

$$\mathbf{b}(s + \delta) = \mathbf{b}(s) + \delta\mathbf{b}'(s) = \mathbf{b} - \frac{\delta}{\tau}\mathbf{n},$$

which tells us that the motion of **b** is also in the direction of the normal vector **n**. The rate of change of the binormal vector is $1/\tau$, the torsion. These relations are shown in Fig. III-3.

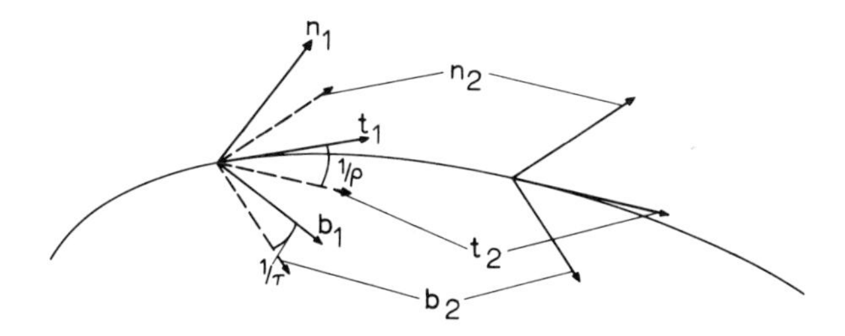

FIG. III-3. The Frenet formulas.

Curvature $1/\rho$ and torsion $1/\tau$ as rates of change of the tangent vector **t** and of the binormal vector **n**, respectively.

We will soon need the third derivative of **P**. Differentiating the expression for **P**″,

$$\mathbf{P}'' = \frac{1}{\rho}\mathbf{n},$$

and using the Frenet equation for the derivative of **n**, Eq. (III-5), we get

$$\mathbf{P}''' = -\frac{1}{\rho^2}\mathbf{t} - \frac{\rho'}{\rho^2}\mathbf{n} + \frac{1}{\rho\tau}\mathbf{b}. \qquad \text{(III-6)}$$

THE CYLINDRICAL HELIX

The oldest chestnut used to describe the nature of curvature and torsion is the cylindrical helix, the shape assumed by a coil spring or the threads of a bolt (see Fig. III-4). It can be best written in terms of a vector formula

$$\mathbf{P}(t) = (r\cos t, r\sin t, kt),$$

where r is the radius of the cylinder on which the helix is inscribed and k is a measure of the pitch of the threads. However, we should by now be accustomed to the use of the arc length parameter s rather than the parameter t in the above expression. Since

$$ds = (\mathbf{P}_t^2)^{1/2}\,dt = (r^2 + k^2)^{1/2}\,dt,$$

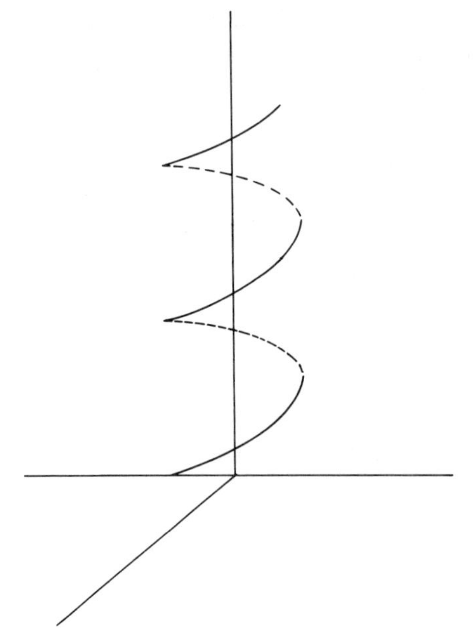

FIG. III-4. The cylindrical helix.

and since both r and k are constants, we may write

$$t = \frac{s}{(r^2 + k^2)^{1/2}},$$

so that the equation for the helix becomes

$$\mathbf{P}(s) = \left(r \cos \frac{s}{(r^2 + k^2)^{1/2}}, r \sin \frac{s}{(r^2 + k^2)^{1/2}}, \frac{ks}{(r^2 + k^2)^{1/2}}\right).$$

Its derivatives are

$$\mathbf{P}' = \frac{1}{(r^2 + k^2)^{1/2}} \left(-r \sin \frac{s}{(r^2 + k^2)^{1/2}}, r \cos \frac{s}{(r^2 + k^2)^{1/2}}, k\right),$$

$$\mathbf{P}'' = \frac{1}{r^2 + k^2} \left(-r \cos \frac{s}{(r^2 + k^2)^{1/2}}, -r \sin \frac{s}{(r^2 + k^2)^{1/2}}, 0\right),$$

$$\mathbf{P}''' = \frac{1}{(r^2 + k^2)^{3/2}} \left(r \sin \frac{s}{(r^2 + k^2)^{1/2}}, -r \cos \frac{s}{(r^2 + k^2)^{1/2}}, 0\right).$$

From Eq. (III-2), its curvature is

$$1/\rho = r/(r^2 + k^2),$$

and the torsion, from Eq. (II-6), is

$$1/\tau = k/(r^2 + k^2).$$

The torsion of the cylindrical helix is therefore proportional to k, which in turn is a measure of the pitch of the threads on the bolt. The expression for curvature, however, appears a little strange; one would have expected ρ to be equal to r, the radius of the cylinder.

THE OSCULATING SPHERE

Consider next two nearby points on a curve, $\mathbf{P}(s+\delta)$ and $\mathbf{P}(s)$. Let $\mathbf{W}$ represent the chord joining these points,

$$\mathbf{W} = \mathbf{P}(s+\delta) - \mathbf{P}(s),$$

which, when expanded in a power series, becomes

$$\begin{aligned}\mathbf{W} &= \delta\mathbf{P}' + \frac{1}{2}\delta^2\mathbf{P}'' + \frac{1}{6}\delta^3\mathbf{P}''' + \cdots \\ &= \delta\mathbf{t} + \frac{1}{2\rho}\delta^2\mathbf{n} + \frac{1}{6}\delta^3\mathbf{P}''' + \cdots.\end{aligned}$$

Thus the chord represented by the vector $\mathbf{W}$ lies on the plane determined by the tangent and normal vectors provided that we disregard terms of degree greater than *two*. This plane, termed the osculating plane, has the property that of all possible planes through $\mathbf{P}$ it fits the curve best.

Next we construct a sphere of radius r that passes through the point $\mathbf{P}$. Now, let $\mathbf{C}$ be any vector of length r; that is $\mathbf{C}^2 = r^2$. Take $\mathbf{P}+\mathbf{C}$ as the sphere's center. Then any vector $\mathbf{Y}$, such that $(\mathbf{P}+\mathbf{C}-\mathbf{Y})^2 = r^2$, lies on the surface of the sphere. If $\mathbf{Z}$ is any vector whatsoever, then the distance between the endpoint of $\mathbf{Z}$ and the sphere's surface along an extended radius of the sphere is given by F, where

$$F^2 = (\mathbf{P}+\mathbf{C}-\mathbf{Z})^2 - r^2.$$

Now let $\mathbf{Z} = \mathbf{P}(s+\delta)$ and expand the expression for F:

$$\begin{aligned}F^2 &= [\mathbf{P}(s) + \mathbf{C} - \mathbf{P}(s+\delta)]^2 - r^2 \\ &= \left(\mathbf{C} - \delta\mathbf{P}' - \frac{1}{2}\delta^2\mathbf{P}'' - \frac{1}{6}\delta^3\mathbf{P}''' - \cdots\right)^2 - r^2.\end{aligned}$$

Rejecting powers of δ greater than *three*, and making use of the properties of the derivatives of $\mathbf{P}$, we get

$$F^2 = -2\delta\mathbf{P}'\cdot\mathbf{C} + \delta^2(1 - \mathbf{P}''\cdot\mathbf{C}) - \frac{1}{3}\delta^3\mathbf{P}'''\cdot\mathbf{C}.$$

Using the Frenet equations, Eqs. (II-5) as well as Eq. (II-6), this becomes

$$F^2 = -2\delta\mathbf{t}\cdot\mathbf{C} + \delta^2\Big(1 - \frac{1}{\rho}\mathbf{n}\cdot\mathbf{C}\Big) - \frac{1}{3}\delta^3\Big(-\frac{1}{\rho^2}\mathbf{t} - \frac{\rho'}{\rho^2}\mathbf{n} + \frac{1}{\rho\tau}\mathbf{b}\Big)\cdot\mathbf{C}.$$

The quantity F is the distance from the point $\mathbf{P}(s + \delta)$ to the sphere whose center is at $\mathbf{P}(s) + \mathbf{C}$. Its formula contains powers of δ. We can define the degree of contact between the sphere and the curve $\mathbf{P}(s)$ as the lowest power of δ appearing in the expression for F. Thus, as it now stands, the sphere and the curve have a degree of contact equal to *one*. What we want to do is adjust the sphere by changing its center and radius until we find the sphere that has the highest degree of contact with the given space curve.

Note that if $\mathbf{C}$ is perpendicular to $\mathbf{t}$, then $\mathbf{t}\cdot\mathbf{C} = 0$ and the expression for F becomes

$$F^2 = \delta^2\Big(1 - \frac{1}{\rho}\mathbf{n}\cdot\mathbf{C}\Big) - \frac{1}{3}\delta^3\Big(-\frac{\rho'}{\rho^2}\mathbf{n} + \frac{1}{\rho\tau}\mathbf{b}\Big)\cdot\mathbf{C},$$

and the degree of contact is now *two*. Moreover, if $\mathbf{n}\cdot\mathbf{C} = \rho$, then

$$F^2 = -\frac{1}{3}\delta^3\Big(-\frac{\rho'}{\rho} + \frac{1}{\rho\tau}\mathbf{b}\cdot\mathbf{C}\Big),$$

and the degree of contact is now *three*.

We could stop here and set $\tilde{\mathbf{C}} = \rho\mathbf{n}$ and the two conditions would be satisfied. The radius of the sphere would equal the radius of curvature at the point $\mathbf{P}$. The center of the sphere would be at the center of curvature and would lie along the normal vector. The intersection of the normal plane with this sphere is called, naturally enough, the circle of curvature.

But this is not the sphere of greatest degree of contact. By setting $\mathbf{b}\cdot\mathbf{C} = \rho'\tau$, the term of degree *three* in the expression for F vanishes and the degree of contact is now at least *four*. The conditions on this sphere are now

$$\mathbf{t}\cdot\mathbf{C} = 0, \qquad \mathbf{n}\cdot\mathbf{C} = \rho, \qquad \mathbf{b}\cdot\mathbf{C} = \rho'\tau,$$

so that

$$\mathbf{C} = \rho\mathbf{n} + \rho'\tau\mathbf{b} \tag{III-7}$$

and

$$r = (\rho^2 + \rho'^2\tau^2)^{1/2}. \tag{III-8}$$

The two equations (III-7) and (III-8) define the center and radius of the *osculating sphere*—that sphere whose degree of contact with the given curve is greatest. Figure III-5 is an attempt to show this.

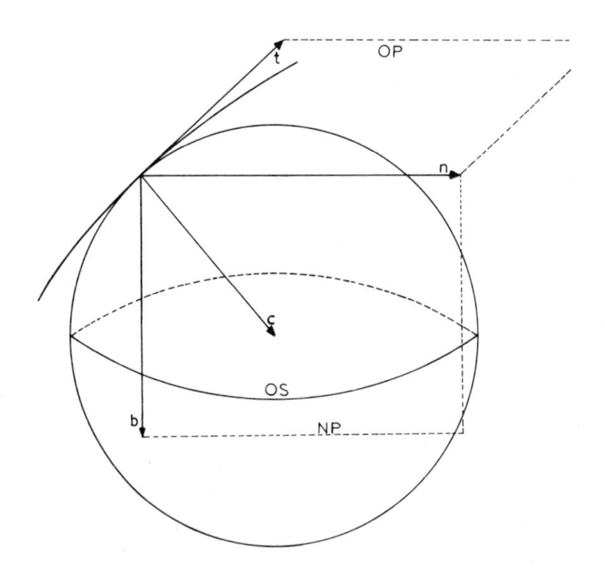

FIG. III-5. The osculating sphere.

C is the vector to the center of the osculating sphere (OS). The *osculating plane* (OP) is the plane determined by the tangent and normal vectors. The *normal plane* (NP) is determined by the normal and binormal vectors. Not identified is the *rectifying plane*, determined by the tangent and binormal vectors.

The third Frenet equation in Eq. (III-5), $\mathbf{n} = -\tau\mathbf{b}'$, when substituted in Eq. (III-7) results in

$$\begin{aligned}\mathbf{C} &= -\rho\tau\mathbf{b}' + \rho'\tau\mathbf{b} \\ &= -\rho^2\tau\left(\frac{1}{\rho}\mathbf{b}\right)'.\end{aligned}$$

Thus

$$\mathbf{C} = -\rho^2\tau\frac{d}{ds}(\mathbf{P}' \times \mathbf{P}''). \tag{III-9}$$

This result will turn out to be relevant to a subsequent chapter. Note that it depends on the geometry of a general space curve and is not dependent in any way on an optical or a physical context.

Now let us have another look at the cylindrical helix. We have calculated the three derivatives of **P** as well as the curvature and torsion, from which it is a simple matter to calculate $\tilde{\mathbf{C}}$, the vector to the center of the circle of curvature, and the vector $\mathbf{P} + \mathbf{C}$. This shows that the center of curvature does not lie on the axis of the cylinder. Moreover, the osculating plane cuts the axis obliquely.

One would like to be able to say that the properties of a general space curve at a point can be described completely in terms of a helix that best fits the given curve in the neighborhood of the point. One could argue that since the curvature and torsion of a helix are constant, a description in these terms would be complete. However, since the curvature of a helix is constant, its derivative is zero and the center of the circle of curvature and that of the osculating sphere would then coincide. This is not a property of a general space curve.

The distinction between the osculating sphere and the circle of curvature is one often missed, particularly by opticists attempting to fit a sphere to a converging wavefront. To illustrate the distinction graphically, inscribe on the surface of a sphere a curve (taking care that it is not also a plane curve). In general, that sphere is the osculating sphere for every point on the inscribed curve, and the normal plane associated with each point on that curve must pass through the sphere's center. On the other hand, the osculating plane and therefore the circle of curvature can lie anywhere.

OPTICAL APPLICATIONS

We have had a rather long look, though by no means a thorough one, at a way of analyzing the geometric properties of a general space curve. At this point we can begin to apply the method to the special class of space curves with which we are primarily concerned, ray paths. Recall that the general equation of a ray in an inhomogeneous medium in which the index of refraction is continuously differentiable, and given by $n(\mathbf{P})$, is, from Eq. (II-19), $d(n\mathbf{P}')/ds = \nabla n$, or, from Eq. (II-20),

$$n\mathbf{P}'' + (\nabla n \cdot \mathbf{P}')\,\mathbf{P}' = \nabla n.$$

Into this we substitute the expressions for the tangent and normal vectors from Eq. (III-2) and get

$$\frac{n}{\rho}\mathbf{n} + (\nabla n \cdot \mathbf{t})\mathbf{t} = \nabla n. \qquad \text{(III-10)}$$

In Eq. (III-10) we begin to realize the fruits of our labors. It shows that the gradient of the refractive index at any point along a ray is a linear combination of the tangent and normal vectors to the ray at that point. Moreover, at that point, the binormal vector must be perpendicular to the gradient of the refractive index. Another way of stating this is to say that the gradient of the index of refraction must always lie in the osculating plane of the ray.

Our next job is to obtain the normal, binormal, curvature, and torsion of a ray in terms of the refractive index and the tangent vector of the ray, the latter of which we will denote in what follows as $\mathbf{P}'$. From Eq. (II-20),

$$\mathbf{P}'' = \frac{1}{n}[\nabla n - (\nabla n \cdot \mathbf{P}')\,\mathbf{P}'] = \frac{1}{n}\mathbf{P}' \times (\nabla n \times \mathbf{P}'),$$

whence

$$\mathbf{n} = \frac{\rho}{n}\mathbf{P}' \times (\nabla n \times \mathbf{P}'). \tag{III-11}$$

Next

$$\mathbf{P}''^2 = \frac{1}{n^2}[(\nabla n)^2 - (\nabla n \cdot \mathbf{P}')^2] = \frac{1}{n^2}(\nabla n \times \mathbf{P}')^2,$$

so

$$\frac{1}{\rho} = \frac{1}{n}[(\nabla n \times \mathbf{P}')^2]^{1/2}. \tag{III-12}$$

Since $\mathbf{b} = \mathbf{t} \times \mathbf{n}$,

$$\mathbf{b} = -\frac{\rho}{n}(\nabla n \times \mathbf{P}'). \tag{III-13}$$

The calculation of torsion is a little more difficult. Differentiating Eq. (II-20), we obtain

$$n\mathbf{P}''' + 2(\nabla n \cdot \mathbf{P}')\,\mathbf{P}'' + (\nabla n \cdot \mathbf{P}'')\,\mathbf{P}' + (\nabla n' \cdot \mathbf{P}')\,\mathbf{P}' = \nabla n'.$$

By taking the scalar product of this with $\mathbf{P}' \times \mathbf{P}''$, we obtain

$$n(\mathbf{P}' \times \mathbf{P}'') \cdot \mathbf{P}''' = (\mathbf{P}' \times \mathbf{P}'') \cdot \nabla n'.$$

But from Eq. (III-13),

$$\mathbf{P}' \times \mathbf{P}'' = -\frac{1}{n}(\nabla n \times \mathbf{P}'),$$

so

$$(\mathbf{P}' \times \mathbf{P}'') \cdot \mathbf{P}''' = -\frac{1}{n^2}(\nabla n \times \mathbf{P}') \cdot \nabla n'.$$

The torsion, by Eq. (III-4), is then

$$\frac{1}{\tau} = \frac{(\mathbf{P}' \times \nabla n) \cdot \nabla n'}{(\nabla n \times \mathbf{P}')^2}. \tag{III-14}$$

It is now a simple matter to find the center of the osculating sphere. From Eq. (III-9),

$$\begin{aligned} \mathbf{C} &= n^2 \frac{(\mathbf{P}' \times \nabla n) \cdot \nabla n'}{(\nabla n \times \mathbf{P}')^4} \frac{d}{ds} \left(\frac{\nabla n \times \mathbf{P}'}{n} \right) \\ &= \frac{(\mathbf{P}' \times \nabla n) \cdot \nabla n'}{(\nabla n \times \mathbf{P}')^4} [n(\nabla n' \times \mathbf{P}') - 2(\nabla n \cdot \mathbf{P}')(\nabla n' \times \mathbf{P}')]. \end{aligned} \tag{III-15}$$

The ray equation (II-20) is used in this last step.

THE DIRECTIONAL DERIVATIVE

Somewhere in this narrative we need to introduce the notation of the directional derivative. Although it looks rather cumbersome, it turns out to be particularly useful in that it can be related, by means of the del operator, to other vector differential operations.

Suppose we have a function $f(\mathbf{P})$ and a vector $\mathbf{V}$. We denote the derivative of f in the direction of $\mathbf{V}$ by

$$(\mathbf{V} \cdot \nabla) f. \tag{III-16}$$

This is exactly equivalent to $\nabla f \cdot \mathbf{V}$, which is the form we have consistently used in writing the derivative of n with respect to s:

$$n' = \nabla n \cdot \mathbf{P}'.$$

Using the notation of the directional derivative, this now can be written as

$$n' = (\mathbf{P}' \cdot \nabla) n.$$

This can be interpreted in the following way. Since n' is the derivative of n with respect to s and since the direction of increasing s is along the ray path whose tangent vector is $\mathbf{P}'$, we can interpret n' as the directional derivative of n in the direction of $\mathbf{P}'$.

The notation is particularly valuable when it is applied to a vector function, say **F**(**P**). The derivative of **F** in the direction **V** is then written as

$$(\mathbf{V} \cdot \nabla)\mathbf{F}.$$

An alternative way of expressing this is

$$\bar{\bar{\nabla}}\mathbf{F} \cdot \mathbf{V},$$

where $\bar{\bar{\nabla}}$ denotes the *vector* gradient, a tensor quantity obtained by taking the gradient of each component of **F**. Notationally and conceptually the vector gradient is awkward. The use of the directional derivative allows us to avoid it completely.

In terms of the directional derivative, the Frenet equations become

$$\begin{aligned}(\mathbf{t} \cdot \nabla)\mathbf{t} &= \frac{1}{\rho}\mathbf{n},\\ (\mathbf{t} \cdot \nabla)\mathbf{n} &= -\frac{1}{\rho}\mathbf{t} + \frac{1}{\tau}\mathbf{b},\\ (\mathbf{t} \cdot \nabla)\mathbf{b} &= -\frac{1}{\tau}\mathbf{n}.\end{aligned} \tag{III-17}$$

Note that in this form there is no longer an explicit dependence on the parameter s. This, for our purposes, is a most useful property of the directional derivative and will play an important role in a subsequent development.

REFERENCES

Blaschke, W. (1930). "Vorlesungen über Differentialgeometrie," Vol. I, 3rd ed. Springer, Berlin. (Reprinted Dover, New York, 1945.)

Hermann, R. (1968). "Differential Geometry and the Calculus of Variations." Academic Press, New York.

Stoker, J. J. (1969). "Differential Geometry." Wiley (Interscience), New York.

Struik, D. J. (1961). "Lectures on Classical Differential Geometry," 2nd ed. Addison-Wesley, Reading, Massachusetts.

Weatherburn, C. E. (1927). "Differential Geometry of Three Dimensions." Cambridge Univ. Press, London and New York.

IV

Applications

Someone, I think it was A. E. Conrady, once made a distinction between real optics and examination optics. Real optics has to do with lenses, mirrors, and general practical problems connected with lens design. Examination optics consists of exactly that—problems that exercise a student's gray cells without having any particular specific application to the methods and techniques of optical design. From this point of view, the problem of the inhomogeneous medium, on which we will spend a great deal of time, belongs entirely in the category of examination optics. However, we are approaching the time when this categorization is no longer valid. The effect of the atmosphere, an inhomogeneous medium of enormous extent and complexity, on astronomical and meteorological observations is certainly approachable from the point of view of geometrical optics. A thermal gradient in a window will affect the accuracy of observations made through that window. These examples show that the study of the inhomogeneous medium is not merely some sort of isometric exercise for students but is a tool capable of providing an approach to the solution of real problems.

A very different kind of an inhomogeneous medium is the electron microscope. Here an electron moves through electrostatic and magnetic fields that affect its velocity and direction at every point. The non-relativistic electron obeys the laws of classical mechanics and therefore satisfies an analog of Fermat's principle. From the fields imposed on the electron a function analogous to refractive index can be deduced. The electron microscope therefore can be treated as an inhomogeneous medium, and the general techniques treated here can be applied to the study of its design and image-forming properties.

The classic example of an inhomogeneous medium is the model of the eye of a fish proposed by Maxwell (1952, Vol. I, pp. 67–69), which first appeared in February 1854 as a solution to a problem posed in the *Cambridge and Dublin Mathematical Journal.* [See also Born and Wolf (1970, pp. 147–149) and Luneburg (1964, pp. 172–182).] The eye of a fish operates in a medium—water—with a relatively high index of refraction. To form images of a size commensurate with its retina, the distance from the lens to the retina would have to be large, yet the eye of a typical fish is flat and dish-shaped. Maxwell suggested that the lens of the eye was of a material in which the rays were arcs of circles. Then rays could be deflected enough by a thin medium to account for the short distance from lens to retina. It is not a particularly realistic model since the vertebrate lens is really a laminate structure. So this portion of our work is really examination optics.

MAXWELL'S FISH EYE

Assume a medium whose refractive index function

$$n(\mathbf{P}) = 1/(1 + \mathbf{P}^2) \tag{IV-1}$$

is spherically symmetric with the center of symmetry located at the coordinate origin. To be realistic this function should involve constants that would ensure that at least part of the medium would have a refractive index greater than unity. By ignoring such constants we make the mathematics a little easier. Besides, fish cannot read; they will never know.

The gradient of the refractive index function is

$$\nabla n = -2\mathbf{P}/(1 + \mathbf{P}^2)^2 \; ; \tag{IV-2}$$

substituting these into the ray equation (II-19) results in

$$(d/ds)[\mathbf{P}'/(1+\mathbf{P}^2)] = -2\mathbf{P}/(1+\mathbf{P}^2)^2,$$

which reduces quickly to

$$(1+\mathbf{P}^2)\,\mathbf{P}'' - 2(\mathbf{P}\cdot\mathbf{P}')\,\mathbf{P}' + 2\mathbf{P} = 0, \tag{IV-3}$$

a second-order ordinary differential equation. Differentiating this results in a third-order equation,

$$(1+\mathbf{P}^2)\,\mathbf{P}''' - 2(\mathbf{P}\cdot\mathbf{P}'')\,\mathbf{P}' = 0. \tag{IV-4}$$

The scalar product of this with $\mathbf{P}''$ gives us

$$\mathbf{P}'''\cdot\mathbf{P}'' = 0 \qquad \text{or} \qquad \mathbf{P}''^2 = \text{const.}$$

Thus the curvature $1/\rho$ of a ray path in this medium is constant. Moreover, the scalar product of Eq. (IV-4) with $\mathbf{P}' \times \mathbf{P}''$ results in

$$\mathbf{P}'''\cdot(\mathbf{P}'\times\mathbf{P}'') = 0,$$

from which we see that the torsion $1/\tau$ is zero at every point along the ray. A ray path in this medium therefore is a plane curve with constant curvature and therefore must be the arc of a circle.

The scalar product of Eq. (IV-3) with $\mathbf{P}''$ gives us

$$(1+\mathbf{P}^2)\,\mathbf{P}''^2 + 2\mathbf{P}\cdot\mathbf{P}'' = 0,$$

from which

$$2\mathbf{P}\cdot\mathbf{P}'' = -(1+\mathbf{P}^2)/\rho^2. \tag{IV-5}$$

When this is substituted into Eq. (IV-4), we get

$$\mathbf{P}''' + \mathbf{P}'/\rho^2 = 0, \tag{IV-6}$$

the general solution of which is

$$\mathbf{P} = \rho[\mathbf{A}\sin(s/\rho) - \mathbf{B}\cos(s/\rho)] + \mathbf{C}, \tag{IV-7}$$

where s is the arc length parameter and **A**, **B**, and **C** are vectors representing the constants of integration. We have already seen that ρ, the radius of curvature, is constant.

Now Eq. (IV-7) is a general solution of Eq. (IV-4), a third-order equation derived from Eq. (IV-3). We are really interested in a general solution of the latter. Although every solution of Eq. (IV-3) satisfies Eq. (IV-4), the converse is not true. We need to extract from Eq. (IV-7) those expressions that also satisfy Eq. (IV-3). We do this by substituting Eq. (IV-7) into Eq. (IV-3) and then finding relationships between the constants of integration, **A**, **B**, and **C**, that cause the resulting expression to vanish identically.

Before doing this we need the derivatives of Eq. (IV-7),

$$\mathbf{P}' = \mathbf{A}\cos(s/\rho) + \mathbf{B}\sin(s/\rho),$$

$$\mathbf{P}'' = (1/\rho)[-\mathbf{A}\sin(s/\rho) + \mathbf{B}\cos(s/\rho)].$$

Note that $\mathbf{P}'^2 = 1$. Therefore $\mathbf{A}^2 = \mathbf{B}^2 = 1$ and $\mathbf{A}\cdot\mathbf{B} = 0$. It follows that **A** and **B** are orthogonal unit vectors. We also need to calculate

$$1 + \mathbf{P}^2 = 1 + \rho^2 + \mathbf{C}^2 + 2\rho[(\mathbf{A}\cdot\mathbf{C})\sin(s/\rho) - (\mathbf{B}\cdot\mathbf{C})\cos(s/\rho)] \quad \text{(IV-8)}$$

as well as

$$\mathbf{P}\cdot\mathbf{P}' = (\mathbf{A}\cdot\mathbf{C})\cos(s/\rho) + (\mathbf{B}\cdot\mathbf{C})\sin(s/\rho).$$

When all of these are substituted into Eq. (IV-3), and after a considerable amount of reduction, we are left with

$$(1/\rho)(1 - \rho^2 + \mathbf{C}^2)[-\mathbf{A}\sin(s/\rho) + \mathbf{B}\cos(s/\rho)]$$
$$-2\mathbf{A}(\mathbf{A}\cdot\mathbf{C}) - 2\mathbf{B}(\mathbf{B}\cdot\mathbf{C}) + 2\mathbf{C} = 0.$$

For this to vanish identically,

$$\mathbf{C} = \mathbf{A}(\mathbf{A}\cdot\mathbf{C}) + \mathbf{B}(\mathbf{B}\cdot\mathbf{C}), \quad \text{(IV-9)}$$

$$\mathbf{C}^2 = \rho^2 - 1. \quad \text{(IV-10)}$$

We already know that **A** and **B** are constant unit vectors perpendicular to one another. The first of these equations (IV-9) tells us that **C** lies in a plane determined by **A** and **B**, and the second (IV-10) shows how its length is related to the constant radius of curvature. The direction of **C** in the **A**, **B** plane remains undetermined. Thus Eq. (IV-7) together with Eqs. (IV-9) and (IV-10) comprises a general solution to Eq. (IV-3) and therefore constitutes an expression for a general ray in the fish eye medium.

We may express this solution and its derivatives in terms of tangent, normal, and binormal vectors,

$$\mathbf{t} = \mathbf{A}\cos(s/\rho) + \mathbf{B}\sin(s/\rho), \quad \text{(IV-11)}$$

$$\mathbf{n} = -\mathbf{A}\sin(s/\rho) + \mathbf{B}\cos(s/\rho), \quad \text{(IV-12)}$$

$$\mathbf{b} = \mathbf{A}\times\mathbf{B}. \quad \text{(IV-13)}$$

The binormal vector is constant, which is to be expected since we are dealing with a plane curve. Note that the normal vector, as well as the tangent vector and $\mathbf{P}$ itself, lies in the plane determined by $\mathbf{A}$ and $\mathbf{B}$. This plane must therefore pass through the origin, which is also the center of symmetry of the medium.

Now let us choose an initial point $\mathbf{P}_0$ and an initial direction $\mathbf{t}_0$. These two initial values must determine uniquely the path of a ray in the medium. Assume that the arc length parameter s is measured along a ray from $\mathbf{P}_0$ so that, at $\mathbf{P}_0$, s equals zero. From Eq. (IV-11) $\mathbf{t}_0 = \mathbf{A}$, and from Eq. (IV-7) $\mathbf{P}_0 = -\rho\mathbf{B} + \mathbf{C}$, so that

$$\mathbf{B} = (1/\rho)(\mathbf{C} - \mathbf{P}_0).$$

Proceeding from this point, note first that

$$\mathbf{A} \cdot \mathbf{C} = \mathbf{P}_0 \cdot \mathbf{t}_0 .$$

Squaring the equation for $\mathbf{P}_0$ above produces

$$\mathbf{P}_0^2 = \rho^2 - 2\rho(\mathbf{B} \cdot \mathbf{C}) + \mathbf{C}^2,$$

which, on applying Eq. (IV-10), yields

$$\mathbf{B} \cdot \mathbf{C} = (1/2\rho)[2\rho^2 - (\mathbf{P}_0^2 + 1)] = \rho - 1/2\rho n_0 ,$$

where n_0 is the value of the refractive index at $\mathbf{P}_0$ and is calculated by using Eq. (IV-1). Finally, using Eq. (IV-9),

$$\mathbf{C} = (\mathbf{P}_0 \cdot \mathbf{t}_0)\, \mathbf{t}_0 + [\rho - 1/2\rho n_0][(1/\rho)(\mathbf{C} - \mathbf{P}_0)],$$

which, in a few short steps, reduces to

$$\begin{aligned} \mathbf{C} &= \mathbf{P}_0 + 2\rho^2 n_0[(\mathbf{P}_0 \cdot \mathbf{t}_0)\, \mathbf{t}_0 - \mathbf{P}_0] \\ &= \mathbf{P}_0 + 2\rho^2 n_0 \mathbf{t}_0 \times (\mathbf{t}_0 \times \mathbf{P}_0). \end{aligned}$$

It follows that

$$\mathbf{B} = 2\rho n_0 \mathbf{t}_0 \times (\mathbf{t}_0 \times \mathbf{P}_0).$$

By squaring this expression we obtain

$$2\rho n_0 = [(\mathbf{t}_0 \times \mathbf{P}_0)^2]^{-1/2}. \tag{IV-14}$$

From this we get an expression relating curvature to the initial values

$$1/\rho = 2n_0[(\mathbf{t}_0 \times \mathbf{P}_0)^2]^{1/2}. \tag{IV-15}$$

In terms of the initial quantities the solution of the ray equation (IV-7) is

$$\mathbf{P} = \mathbf{P}_0 + \rho\mathbf{t}_0 \sin(s/\rho) + 2\rho^2 n_0 \mathbf{t}_0 \times (\mathbf{t}_0 \times \mathbf{P}_0)[1 - \cos(s/\rho)]. \quad \text{(IV-16)}$$

The tangent vector, from Eq. (IV-11), is

$$\mathbf{t} = \mathbf{t}_0 \cos(s/\rho) + 2\rho n_0 \mathbf{t}_0 \times (\mathbf{t}_0 \times \mathbf{P}_0) \sin(s/\rho); \quad \text{(IV-17)}$$

the normal vector, from Eq. (IV-12), is

$$\mathbf{n} = -\mathbf{t}_0 \sin(s/\rho) + 2\rho n_0 \mathbf{t}_0 \times (\mathbf{t}_0 \times \mathbf{P}_0) \cos(s/\rho); \quad \text{(IV-18)}$$

and the binormal vector, from Eq. (IV-13), is

$$\mathbf{b} = -2\rho n_0 \mathbf{t}_0 \times \mathbf{P}_0 . \quad \text{(IV-19)}$$

The curvature $1/\rho$ and n_0 are given by Eqs. (IV-15) and (IV-16), respectively.

Maxwell's fish eye has several fascinating properties. The most obvious of these is the fact that every point within the medium has a perfect image at another point in the medium. To every point $\mathbf{P}_0$ there corresponds a second point $\mathbf{P}^*$ with the property that every ray passing through $\mathbf{P}_0$ also passes through $\mathbf{P}^*$ exactly. This can be seen by expanding the vector product in Eq. (IV-16) and rearranging terms to obtain

$$\begin{aligned} \mathbf{P} &= \mathbf{P}_0\{1 - 2\rho^2 n_0[1 - \cos(s/\rho)]\} \\ &\quad + \rho\mathbf{t}_0\{\sin(s/\rho) + 2\rho n_0(\mathbf{t}_0 \cdot \mathbf{P}_0)[1 - \cos(s/\rho)]\}. \end{aligned} \quad \text{(IV-20)}$$

For $\mathbf{P}^*$ to be a perfect image of $\mathbf{P}_0$, the expression for $\mathbf{P}^*$ must be independent of the initial direction $\mathbf{t}_0$. Therefore s, the arc length parameter, must be such that the coefficient of $\mathbf{t}_0$ in the above expression vanishes:

$$\sin(s/\rho) + 2\rho n_0(\mathbf{t}_0 \cdot \mathbf{P}_0)[1 - \cos(s/\rho)] = 0.$$

We need to solve this for s. Transposing the second term to the right member of the equation and squaring leads to a quadratic in $\cos(s/\rho)$,

$$\begin{aligned} &[1 + 4\rho^2 n_0^2(\mathbf{t}_0 \cdot \mathbf{P}_0)^2] \cos^2(s/\rho) - 8\rho^2 n_0^2(\mathbf{t}_0 \cdot \mathbf{P}_0)^2 \cos(s/\rho) \\ &\quad - [1 - 4\rho^2 n_0^2(\mathbf{t}_0 \cdot \mathbf{P}_0)^2] = 0, \end{aligned}$$

whose solution is

$$\cos(s^*/\rho) = -\frac{1 - 4\rho^2 n_0^2(\mathbf{t}_0 \cdot \mathbf{P}_0)^2}{1 + 4\rho^2 n_0^2(\mathbf{t}_0 \cdot \mathbf{P}_0)^2} .$$

(A second solution of the quadratic is $\cos(s/\rho) = 1$, corresponding to the point $\mathbf{P}_0$ itself.) With the aid of Eq. (IV-14) this expression can be reduced to

$$\cos(s^*/\rho) = -1 + 2(\mathbf{t}_0 \cdot \mathbf{P}_0)^2/\mathbf{P}_0^2. \tag{IV-21}$$

Also

$$\sin(s^*/\rho) = -2(\mathbf{t}_0 \cdot \mathbf{P}_0)[(\mathbf{t}_0 \times \mathbf{P}_0)^2]^{1/2}/\mathbf{P}_0^2. \tag{IV-22}$$

Substituting these expressions back into the formula for $\mathbf{P}$ yields the equation for $\mathbf{P}^*$, the image of $\mathbf{P}_0$,

$$\mathbf{P}^* = -(1/\mathbf{P}_0^2)\,\mathbf{P}_0\,. \tag{IV-23}$$

At first glance it appears a little strange that, although $\mathbf{P}^*$ does not depend on $\mathbf{t}_0$, s^* does. The reason for this becomes clear when we realize that s^* is the length of the circular arc connecting $\mathbf{P}_0$ and $\mathbf{P}^*$. This arc length must necessarily depend on the initial direction. See Fig. IV-1.

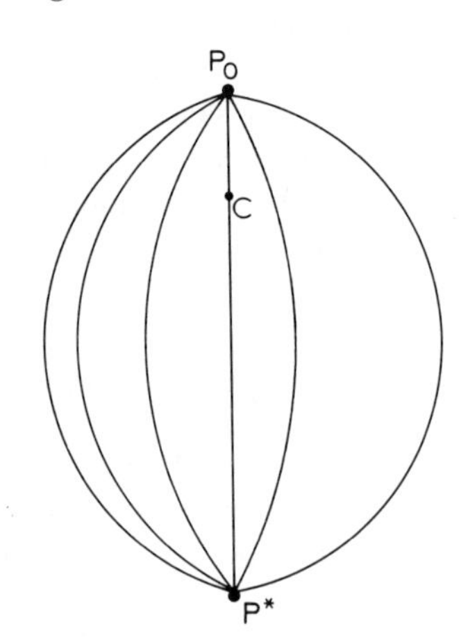

FIG. IV-1. Maxwell's fish eye.

All rays are arcs of circles. Every object point, $\mathbf{P}_0$, is imaged perfectly at some $\mathbf{P}^*$. C represents the center of symmetry of the refractive index.

We can go a little further. Suppose the angle between $\mathbf{P}_0$ and $\mathbf{t}_0$ is θ so that

$$\mathbf{P}_0 \cdot \mathbf{t}_0 = (\mathbf{P}_0^2)^{1/2} \cos\theta \tag{IV-24}$$

and, from Eq. (IV-17),

$$\frac{1}{2\rho} = n_0(\mathbf{P}_0^2)^{1/2} \sin\theta = \frac{(\mathbf{P}_0^2)^{1/2}\sin\theta}{1 + \mathbf{P}_0^2}\,. \tag{IV-25}$$

Then Eqs. (IV-21) and (IV-22) become

$$\cos(s^*/\rho) = \cos 2\theta, \qquad \sin(s^*/\rho) = -\sin 2\theta,$$

so that

$$s^* = -2\rho\theta. \tag{IV-26}$$

To continue even further, from Eq. (IV-25),

$$s^* = -[(1 + \mathbf{P}_0^2)/(\mathbf{P}_0^2)^{1/2}][\theta/\sin\theta]. \tag{IV-27}$$

This provides the arc length of each ray between $\mathbf{P}_0$ and $\mathbf{P}^*$. Its only importance is that it provides a datum necessary for the calculation of the optical path length along a ray between these two points. This is given by

$$I = \int_0^{s^*} n\, ds.$$

To evaluate this integral we need to calculate n, the refractive index function. Using Eq. (IV-16), we find that

$$\begin{aligned} 1 + \mathbf{P}^2 = 1 + \mathbf{P}_0^2 + \rho^2 \sin^2(s/\rho) + 4\rho^4 n_0^2(\mathbf{t}_0 \times \mathbf{P}_0)^2[1 - \cos(s/\rho)]^2 \\ + 2\rho \sin(s/\rho)(\mathbf{t}_0 \cdot \mathbf{P}_0) - 4\rho^2 n_0(\mathbf{t}_0 \times \mathbf{P}_0)^2[1 - \cos(s/\rho)], \end{aligned}$$

which on application of Eq. (IV-14) becomes

$$\begin{aligned} 1 + \mathbf{P}^2 &= 1 + \mathbf{P}_0^2 + \rho^2 \sin^2(s/\rho) + \rho^2[1 - \cos(s/\rho)]^2 \\ &\quad + 2\rho \sin(s/\rho)(\mathbf{t}_0 \cdot \mathbf{P}_0) - (1/n_0)[1 - \cos(s/\rho)] \\ &= 2\rho^2[1 - \cos(s/\rho)] + (1/n_0)[\cos(s/\rho) + 2\rho n_0(\mathbf{t}_0 \cdot \mathbf{P}_0) \sin(s/\rho)]. \end{aligned}$$

The expression for the refractive index now becomes, using Eq. (IV-1),

$$\begin{aligned} n &= 1/(1 + \mathbf{P}^2) \\ &= n_0/[2n_0\rho^2 + (1 - 2n_0\rho^2)\cos(s/\rho) + \cot\theta \sin(s/\rho)]. \end{aligned}$$

Next, using Eq. (IV-25), we find that

$$\begin{aligned} 2n_0\rho^2 &= \frac{1 + \mathbf{P}_0^2}{2\mathbf{P}_0^2 \sin^2\theta}, \\ 1 - 2n_0\rho^2 &= -\frac{1 + \mathbf{P}_0^2 \cos 2\theta}{2\mathbf{P}_0^2 \sin^2\theta}, \end{aligned}$$

and the expression for the index becomes

$$\begin{aligned} n &= \frac{2n_0\mathbf{P}_0^2 \sin^2\theta}{1 + \mathbf{P}_0^2 - (1 + \mathbf{P}_0^2 \cos 2\theta)\cos(s/\rho) + \mathbf{P}_0^2 \sin 2\theta \sin(s/\rho)} \\ &= \frac{2n_0\mathbf{P}_0^2 \sin^2\theta}{1 - \cos(s/\rho) + \mathbf{P}_0^2[1 - \cos(s/\rho + 2\theta)]}. \end{aligned}$$

Through a well-known trigonometric identity, this becomes

$$n = \frac{n_0 \mathbf{P}_0{}^2 \sin^2 \theta}{\sin^2(s/2\rho) + \mathbf{P}_0{}^2 \sin^2(s/2\rho + \theta)},$$

an expression which, with the aid of a table of integrals, gives us the optical path length. It turns out that

$$\int n\, ds = \arc \tan \frac{\sin(s/2\rho)}{(\mathbf{P}_0{}^2)^{1/2} \sin(s/2\rho + \theta)}. \tag{IV-28}$$

When $s = 0$, the argument vanishes; when $s = s^*$, it becomes infinite so that

$$\int_0^{s^*} n\, ds = \pi/2. \tag{IV-29}$$

As one would expect, the optical length between $\mathbf{P}_0$ and $\mathbf{P}^*$ is independent of the initial ray direction and therefore is the same for every pair of points joining these two points. The surprising fact is that the optical path length in the fish eye medium is the same for every pair of conjugate points.

THE HEATED WINDOW

This is a more realistic problem. Suppose a window separates two media of vastly differing temperatures through which accurate optical measurements must be made. For example, we may need to observe the behavior of a model aircraft in a wind tunnel in which the temperature is much greater than the outside where the instruments and the human observers are located. A more exotic example would be the navigational telescope on a spaceship that looks through a window. Here the cabin temperature, which must be such that the crew is reasonably comfortable, is much greater than the outside coldness of deep space.

In both of these cases a thermal gradient exists across the window. Changes in refractive index invariably accompany changes in temperature. It follows that the material of the window is an inhomogeneous optical medium, that its index of refraction at a point depends on the temperature of the medium at that point, which in turn depends on the distance of that point to the heating and cooling surfaces. The problem of tracing a ray through the window is then a problem that can be solved by the methods treated here.

We will make several assumptions. The first is that the refractive index is a linear function of temperature. The second is that the temperature is a linear function of position. The first assumption is a bit drastic; accurate tables giving a relationship between index and temperature very often are quadratic rather than linear. The second assumption is not too unreasonable if we understand the window to have plane parallel surfaces, on each of which the temperature is constant, and if we can neglect edge effects. These two assumptions allow us to write the refractive index function as

$$n(\mathbf{P}) = 1 + \mathbf{A} \cdot \mathbf{P}, \tag{IV-30}$$

where **P** is the usual position vector and where **A** is a vector in the direction of the thermal gradient and is normal to the two parallel surfaces. The vector **A** is also the gradient of the refractive index

$$\nabla n = \mathbf{A}. \tag{IV-31}$$

This is shown in Fig. IV-2.

On substituting the equation for the refractive index (IV-30) and its gradient (IV-22) into the ray equation (II-19), we obtain

$$(d/ds)[(1 + (\mathbf{A} \cdot \mathbf{P})\,\mathbf{P}'] = \mathbf{A},$$

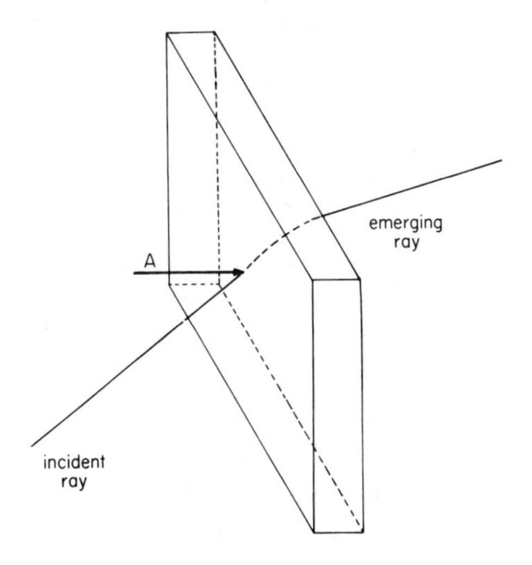

FIG. IV-2. The heated window.
The vector **A** is normal to the front and rear surfaces and is in the direction of ∇n.

which, when integrated, becomes

$$(1 + \mathbf{A} \cdot \mathbf{P})\,\mathbf{P}' = \mathbf{A}s + \mathbf{B}, \tag{IV-32}$$

where $\mathbf{B}$ is a vector representing the constants of integration. By squaring Eq. (IV-23) we obtain

$$(1 + \mathbf{A} \cdot \mathbf{P})^2 = (\mathbf{A}s + \mathbf{B})^2,$$

which, when applied to Eq. (IV-30), gives the index of refraction along a ray as a function of the arc length parameter of the ray:

$$n = [(\mathbf{A}s + \mathbf{B})^2]^{1/2}. \tag{IV-33}$$

On the other hand, by taking the scalar product of Eq. (IV-32) with $\mathbf{P}'$, we obtain

$$n = (1 + \mathbf{A} \cdot \mathbf{P}) = (\mathbf{A}s + \mathbf{B}) \cdot \mathbf{P}'. \tag{IV-34}$$

Now we apply some initial conditions. Suppose we want the path of a ray that starts at the point $\mathbf{P}_0$ with the initial direction $\mathbf{t}_0$. Since the ray path begins at $\mathbf{P}_0$, we may begin the measure of the arc length parameter at that point so that, at $\mathbf{P}_0$, s equals zero. Designating the refractive index of the medium at $\mathbf{P}_0$ by n_0, we obtain from Eqs. (IV-33) and (IV-34),

$$n_0 = (B^2)^{1/2} = \mathbf{B} \cdot \mathbf{t}_0 . \tag{IV-35}$$

We conclude that $\mathbf{B} = n_0\mathbf{t}_0$; we substitute this relation into Eqs. (IV-32) and (IV-33) to obtain

$$(1 + \mathbf{A} \cdot \mathbf{P})\,\mathbf{P}' = \mathbf{A}s + n_0\mathbf{t}_0 \tag{IV-36}$$

and

$$n = [(\mathbf{A}s + n_0\mathbf{t}_0)^2]^{1/2}, \tag{IV-37}$$

whence

$$\mathbf{P}' = \frac{\mathbf{A}s + n_0\mathbf{t}_0}{[(\mathbf{A}s + n_0\mathbf{t}_0)^2]^{1/2}} . \tag{IV-38}$$

The next step is to calculate the geometric properties of the ray by using Eqs. (II-11)–(II-14), into which are substituted Eqs. (IV-31), (IV-37), and (IV-38) to obtain

$$\mathbf{t} = \frac{\mathbf{A}s + n_0\mathbf{t}_0}{[(\mathbf{A}s + n_0\mathbf{t}_0)^2]^{1/2}},$$

$$\mathbf{n} = \frac{(\mathbf{A}s + n_0\mathbf{t}_0) \times (\mathbf{A} \times \mathbf{t}_0)}{[(\mathbf{A}s + n_0\mathbf{t}_0)^2]^{1/2}\,[(\mathbf{A} \times \mathbf{t}_0)^2]^{1/2}},$$

$$\mathbf{b} = -(\mathbf{A} \times \mathbf{t}_0)/[(\mathbf{A} \times \mathbf{t}_0)^2]^{1/2},$$

$$\frac{1}{\rho} = \frac{n_0[(\mathbf{A} \times \mathbf{t}_0)^2]^{1/2}}{(\mathbf{A}s + n_0\mathbf{t}_0)^2}, \qquad \frac{1}{\tau} = 0.$$

Note that the binormal vector is constant and that the torsion is therefore zero by the Frenet equation (III-5), and our ray path is again a plane curve.

At the initial point where $s = 0$, these become

$$\mathbf{n}_0 = \frac{\mathbf{t}_0 \times (\mathbf{A} \times \mathbf{t}_0)}{[(\mathbf{A} \times \mathbf{t}_0)^2]^{1/2}} = \frac{\rho_0}{n_0} \mathbf{t}_0 \times (\mathbf{A} \times \mathbf{t}_0), \tag{IV-39}$$

$$\mathbf{b}_0 = -\frac{\mathbf{A} \times \mathbf{t}_0}{[(\mathbf{A} \times \mathbf{t}_0)^2]^{1/2}} = -\frac{\rho_0}{n_0} (\mathbf{A} \times \mathbf{t}_0), \tag{IV-40}$$

$$1/\rho_0 = (1/n_0)[(\mathbf{A} \times \mathbf{t}_0)^2]^{1/2}. \tag{IV-41}$$

These are then applied to the preceding more general expressions to obtain

$$\mathbf{t} = (n_0/n)\, \mathbf{t}_0 + (s/n)\mathbf{A}, \tag{IV-42}$$

$$\mathbf{n} = (n_0/n)\, \mathbf{n}_0 - (s/n)\, \mathbf{A} \times \mathbf{b}_0\,, \tag{IV-43}$$

$$\mathbf{b} = \mathbf{b}_0\,, \tag{IV-44}$$

$$1/\rho = (n_0^2/n^2)(1/\rho_0). \tag{IV-45}$$

The last of these leads to a nice invariant relation between the curvature and the refractive index at any point along the ray

$$n^2/\rho = n_0^2/\rho_0 = \text{const.} \tag{IV-46}$$

Now to complete the problem. There remains outstanding the first-order differential equation (IV-38),

$$\mathbf{P}' = \frac{\mathbf{A}s + n_0\mathbf{t}_0}{[(\mathbf{A}s + n_0\mathbf{t}_0)^2]^{1/2}}\,.$$

The variables separate, and we may write it in the form

$$d\mathbf{P} = \mathbf{A}\frac{s\,ds}{[(\mathbf{A}s + n_0\mathbf{t}_0)^2]^{1/2}} + \mathbf{t}_0\frac{n_0\,ds}{[(\mathbf{A}s + n_0\mathbf{t}_0)^2]^{1/2}}\,,$$

which, with the aid of a table of integrals, results in

$$\mathbf{P}(s) = \frac{1}{\mathbf{A}^2}\Big\{[(\mathbf{A}s + n_0\mathbf{t}_0)^2]^{1/2}\,\mathbf{A} + \frac{n_0}{(\mathbf{A}^2)^{1/2}}\,\mathbf{A} \times (\mathbf{t}_0 \times \mathbf{A}) \log\{\mathbf{A} \cdot (\mathbf{A}s + n_0\mathbf{t}_0)$$

$$+ (\mathbf{A}^2)^{1/2}[(\mathbf{A}s + n_0\mathbf{t}_0)^2]^{1/2}\}\Big\} + \mathbf{C}, \tag{IV-47}$$

where $\mathbf{C}$ represents a constant of integration. This solution can be simplified considerably by applying Eqs. (IV-39)–(IV-46) to obtain

$$\mathbf{P}(s) = \frac{1}{\mathbf{A}^2}\left\{n\mathbf{A} + \frac{{n_0}^2}{\rho_0(\mathbf{A}^2)^{1/2}}(\mathbf{A} \times \mathbf{b}_0)\log[n_0(\mathbf{A}\cdot\mathbf{t}_0) + n(\mathbf{A}^2)^{1/2} + s\mathbf{A}^2]\right\} + \mathbf{C}. \tag{IV-48}$$

The value of $\mathbf{C}$ is obtained at the initial point $\mathbf{P}_0$ by setting $s = 0$,

$$\mathbf{C} = \mathbf{P}_0 - \frac{1}{\mathbf{A}^2}\left\{n_0\mathbf{A} + \frac{{n_0}^2}{\rho_0(\mathbf{A}^2)^{1/2}}(\mathbf{A} \times \mathbf{b}_0)\log[n_0(\mathbf{A}\cdot\mathbf{t}_0) + n_0(\mathbf{A}^2)^{1/2}]\right\},$$

so that the solution, in its final form, is

$$\mathbf{P}(s) = \mathbf{P}_0 + \frac{1}{\mathbf{A}^2}\left\{(n - n_0)\mathbf{A} + \frac{{n_0}^2}{\rho_0(\mathbf{A}^2)^{1/2}}(\mathbf{A} \times \mathbf{b}_0)\log\frac{n_0(\mathbf{A}\cdot\mathbf{t}_0) + n(\mathbf{A}^2)^{1/2} + s\mathbf{A}^2}{n_0(\mathbf{A}\cdot\mathbf{t}_0) + n_0(\mathbf{A}^2)^{1/2}}\right\}. \tag{IV-49}$$

Now suppose that our inhomogeneous medium is a slab of thickness d with plane parallel surfaces. Because the thermal gradient, and therefore the index gradient, is perpendicular to the surfaces, the vector $\mathbf{A}/(\mathbf{A}^2)^{1/2}$ is a unit normal vector to the two surfaces. Let $\mathbf{P}_0$ lie on the first surface and let $\mathbf{P}^*$ lie on the second. As before, let n_0 be the refractive index on the first surface and let n^* be that on the second. Then $(\mathbf{P}^* - \mathbf{P}_0)\cdot\mathbf{A}/(\mathbf{A}^2)^{1/2}$ is a projection of the ray onto a line normal to the two surfaces. The length of this line is exactly equal to the thickness d, so

$$(\mathbf{P}^* - \mathbf{P}_0)\cdot\mathbf{A} = d(\mathbf{A}^2)^{1/2}.$$

Applying this to our final solution, Eq. (IV-49), results in

$$n^* - n_0 = d(\mathbf{A}^2)^{1/2}. \tag{IV-50}$$

Using this and Eq. (IV-29), we get

$$[(\mathbf{A}s + n_0\mathbf{t}_0)^2]^{1/2} = n_0 + d(\mathbf{A}^2)^{1/2},$$

which leads to a quadratic in s,

$$\mathbf{A}^2s^2 + 2n_0(\mathbf{A}\cdot\mathbf{t}_0)s - [\mathbf{A}^2d^2 + 2n_0(\mathbf{A}^2d)^{1/2}] = 0. \tag{IV-51}$$

Now let the angle between $\mathbf{A}$ and $\mathbf{t}_0$ be θ. Note that θ is exactly the angle of refraction of the entering ray. Then $\mathbf{A} \cdot \mathbf{t}_0 = (\mathbf{A}^2)^{1/2} \cos \theta$, and the quadratic becomes

$$(\mathbf{A}^2)^{1/2} s^2 + 2n_0 s \cos \theta - d[(\mathbf{A}^2)^{1/2} d + 2n_0] = 0,$$

whose discriminant is

$$D^2 = \mathbf{A}^2 d^2 + 2n_0(\mathbf{A}^2)^{1/2} d + n_0^2 \cos^2 \theta,$$

which factors into

$$D^2 = [(\mathbf{A}^2)^{1/2} d + n_0(1 + \sin \theta)][(\mathbf{A}^2)^{1/2} d + n_0(1 - \sin \theta)].$$

The solution of the quadratic, Eq. (IV-43), then becomes

$$s^* = \frac{1}{(\mathbf{A}^2)^{1/2}} \{-n_0 \cos \theta \pm [(\mathbf{A}^2)^{1/2} d + n_0(1 + \sin \theta)]^{1/2} \times [(\mathbf{A}^2)^{1/2} d + n_0(1 - \sin \theta)]^{1/2}\},$$

which can be written as

$$s^* = \frac{1}{(\mathbf{A}^2)^{1/2}} \{-[(\mathbf{A}^2)^{1/2} d + n_0(1 + \cos \theta)] + \tfrac{1}{2}([(\mathbf{A}^2)^{1/2} d + n_0(1 + \sin \theta)]^{1/2} \pm [(\mathbf{A}^2)^{1/2} d + n_0(1 + \sin \theta)]^{1/2})^2\}.$$

To remove the ambiguity in sign, note that, when $d = 0$, s^* is zero only if the positive branch of the solution is chosen. We therefore replace $\pm$ by $+$.

We now use this calculation of s^* in the argument of the logarithm in Eq. (IV-49). We make use of Eq. (IV-50):

$$\frac{n_0(\mathbf{A} \cdot \mathbf{t}_0) + n^*(\mathbf{A}^2)^{1/2} + s^*\mathbf{A}^2}{n_0(\mathbf{A} \cdot \mathbf{t}_0) + n_0(\mathbf{A}^2)^{1/2}} = \frac{n_0 \cos \theta + n^* + (\mathbf{A}^2)^{1/2} s^*}{n_0(1 + \cos \theta)} = \frac{\{[(\mathbf{A}^2)^{1/2} d + n_0(1 + \sin \theta)]^{1/2} + [(\mathbf{A}^2)^{1/2} d + n_0(1 - \sin \theta)]^{1/2}\}^2}{2n_0(1 + \cos \theta)}.$$

Thus the expression for $\mathbf{P}^*$ becomes

$$\mathbf{P}^* = \mathbf{P}_0 + \frac{1}{\mathbf{A}^2} \left\{ d(\mathbf{A}^2)^{1/2}\mathbf{A} + \frac{n_0^2}{\rho_0(\mathbf{A}^2)^{1/2}} (\mathbf{A} \times \mathbf{b}_0) \times \log \frac{\{[(\mathbf{A}^2)^{1/2} d + n_0(1 + \sin \theta)]^{1/2} + [(\mathbf{A}^2)^{1/2} dn_0(1 - \sin \theta)]^{1/2}\}^2}{2n_0(1 + \cos \theta)} \right\}.$$

However, the location of $\mathbf{P}^*$ is only part of the story. What is needed in addition is $\mathbf{t}^*$, the direction of the ray at the second surface. This we get from Eq. (IV-42):

$$\mathbf{t}^* = \frac{1}{n^*}(n_0\mathbf{t}_0 + s^*\mathbf{A}).$$

If θ' is the angle of incidence of the ray at the second surface, then

$$\cos\theta' = \frac{\mathbf{t}^* \cdot \mathbf{A}}{(\mathbf{A}^2)^{1/2}} = \frac{n_0(\mathbf{t}_0 \cdot \mathbf{A}) + s^*\mathbf{A}^2}{n^*(\mathbf{A}^2)^{1/2}}.$$

From this we calculate the sine

$$\sin^2\theta' = \frac{1}{n^{*2}\mathbf{A}^2}[n^{*2}\mathbf{A}^2 - n_0^2(\mathbf{t}_0 \cdot \mathbf{A})^2 - 2n_0(\mathbf{t}_0 \cdot \mathbf{A})\,\mathbf{A}^2 s^* - \mathbf{A}^4 s^{*2}].$$

Applying to this Eq. (IV-50) and rearranging terms results in

$$\sin^2\theta' = \frac{1}{n^{*2}\mathbf{A}^2}\{n_0^2[\mathbf{A}^2 - (\mathbf{t}_0 \cdot \mathbf{A})^2] - \mathbf{A}^2[\mathbf{A}^2 s^{*2} + 2n_0(\mathbf{t}_0 \cdot \mathbf{A})\,s^* - \mathbf{A}^2 d^2 - 2n_0 d(\mathbf{A}^2)^{1/2}]\}.$$

Since s^* is a solution of Eq. (IV-51), the quantity within the second set of brackets vanishes and we are left with

$$\sin\theta' = (n_0/n^*)\sin\theta.$$

Now suppose a ray enters the first surface of the slab at $\mathbf{P}_0$ at an angle of incidence θ_a from a homogeneous medium of refractive index n_a. Then, by Snell's law, which we have not covered yet but which can be found in Eq. (V-19) and Eq. (V-20),

$$n_a \sin\theta_a = n_0 \sin\theta.$$

Suppose this ray, after passing through the inhomogeneous medium, exits through the second surface at $\mathbf{P}^*$ into a homogeneous medium of refractive index n_b. Then

$$n^* \sin\theta' = n_b \sin\theta_b.$$

Combining the last three equations, we obtain

$$n_a \sin\theta_a = n_b \sin\theta_b.$$

If n_a and n_b are equal, the ray is undeflected.

But this we knew already! A plane homogeneous slab will not change the direction of a ray passing through it. We could have approximated the inhomogeneous medium by a series of homogeneous plane laminae. Since each lamina would leave a ray undeflected, the net effect of all the laminae would be that, although there would be a lateral displacement of a ray, its direction would be unchanged.

Our astrogator is then in good shape. All the light rays entering the window on his spaceship from a star will be parallel and will be undeviated by the inhomogeneity of the window. He may have to take into account the difference between the refractive index of the air in his cabin and that of deep space but he need not worry about his window as long as its surfaces are plane and parallel and as long as he does not sight through a region too near its edge.

The window in the wind tunnel is a different matter. Here the light reaching the window comes from points relatively near by. Although the rays' directions are undeviated, the rays themselves are displaced laterally by an amount given by $|\mathbf{P}^* - \mathbf{P}_0|$. In this case the magnitude of the displacement will determine whether or not the quality of the images seen through the window and the accuracy of the measurements made therethrough will be affected.

THE BRACHISTOCHRONE

In Chapter I we began the problem of the brachistochrone but never completed it. Lest we be accused of failing to drop the other shoe (or, in this case, the bead), we will complete it here. The starting point is the variational integral in Eq. (I-5),

$$I = \int_{a_1}^{a_2} \{(1 + y'^2)/[2g(y - b_2) + v_A^2]\}^{1/2}\, dx.$$

This we transform into parametric form and, at the same time, pass to the three-dimensional case. Let a curve be described in terms of the vector function $\mathbf{P}(t)$ and let the fixed initial and final points be $\mathbf{P}(t_1)$ and $\mathbf{P}(t_2)$, respectively. Also let $\mathbf{G}$ be a vector in the direction of the force of gravity so that $\mathbf{G}^2 = g^2$. Then

$$I = \int_{t_2}^{t_1} \{\mathbf{P}_t^2/[2\mathbf{G} \cdot (\mathbf{P} - \mathbf{P}_1) + v_A^2]\}^{1/2}\, dt. \qquad \text{(IV-52)}$$

The vector $\mathbf{P}_1$ is of course $\mathbf{P}(t_1)$. With no difficulty we can recognize that this is in the form of Eq. (I-8),

$$I = \int n\, ds,$$

where $n = [2\mathbf{G} \cdot (\mathbf{P} - \mathbf{P}_1) + v_A{}^2]^{-1/2}$ and where $ds = (\mathbf{P}^2)^{1/2}\, dt$. The ray equation provides the differential equation for the extremals. The gradient is

$$\nabla n = -\mathbf{G}[2\mathbf{G} \cdot (\mathbf{P} - \mathbf{P}_1) + v_A{}^2]^{-3/2},$$

and we use Eq. (II-19) to obtain

$$(d/ds)\{\mathbf{P}_1'[2\mathbf{G} \cdot (\mathbf{P} - \mathbf{P}_1) + v_A{}^2]^{-1/2}\} = -\mathbf{G}[2\mathbf{G} \cdot (\mathbf{P} - \mathbf{P}_1) + v_A{}^2]^{-3/2}.$$

Multiplying by $\mathbf{G}$ and noting that

$$(d/ds)[2\mathbf{G} \cdot (\mathbf{P} - \mathbf{P}_1) + v_A{}^2]^{1/2} = (\mathbf{G} \cdot \mathbf{P}')[2\mathbf{G} \cdot (\mathbf{P} - \mathbf{P}_1) + v_A{}^2]^{-1/2},$$

we obtain

$$(d^2/ds^2)\{[2\mathbf{G} \cdot (\mathbf{P} - \mathbf{P}_1) + v_A{}^2]^{1/2}\} = -g^2[2\mathbf{G} \cdot (\mathbf{P} - \mathbf{P}_1) + v_A{}^2]^{-3/2}.$$

Now let

$$u = [2\mathbf{G} \cdot (\mathbf{P} - \mathbf{P}_1) + v_A{}^2]^{1/2}, \tag{IV-53}$$

and the above becomes $u^3(d^2u/ds^2) + g^2 = 0$. In differential equations of this type, in which the independent variable does not appear explicitly, the use of a transformation is frequently the most convenient approach to take. Define a new dependent variable, $p = du/ds$, and consider it a function of u rather than of s. Then

$$d^2u/ds^2 = dp/ds = (dp/du)(du/ds) = p(dp/du).$$

Substitution yields

$$u^3p\, dp/du + g^2 = 0.$$

This separates nicely, leaving us with

$$p\, dp + g^2\, du/u^3 = 0.$$

A quadrature results in

$$p^2 - g^2/u^2 = a,$$

where a is a constant of integration. But $p = du/ds$. This leads to

$$u\,du/(g^2 + au^2)^{1/2} = ds,$$

which, when integrated, yields

$$(g^2 + au^2)^{1/2} = a(s + b),$$

where b is a constant of integration. Solving for u, we obtain

$$u = \{[a^2(s + b)^2 - g^2]/a\}^{1/2}. \tag{IV-54}$$

Recalling that $u = [2\mathbf{G} \cdot (\mathbf{P} - \mathbf{P}_1) + v_A{}^2]^{1/2}$, we substitute for the latter expression into the original vector equation and obtain

$$(d/ds)\{a^{1/2}\mathbf{P}'[a^2(s + b)^2 - g^2]^{-1/2}\} = -a^{3/2}\mathbf{G}[a^2(s + b)^2 - g^2]^{-3/2}.$$

Note that

$$-\frac{1}{g^2}\frac{d}{ds}\left\{\frac{(s + b)}{[a^2(s + b)^2 - g^2]^{1/2}}\right\} = [a^2(s + b)^2 - g^2]^{-3/2},$$

whence

$$(d/ds)\{\mathbf{P}'[a^2(s + b)^2 - g^2]^{-1/2}\} = (a/g^2)(d/ds)\{(s + b)[a^2(s + b)^2 - g^2]^{-1/2}\}\mathbf{G}$$

so that

$$\mathbf{P}' = (a/g^2)(s + b)\mathbf{G} + [a^2(s + b)^2 - g^2]^{1/2}\mathbf{B},$$

where $\mathbf{B}$ is a constant vector. However, $\mathbf{P}'^2$ must still equal unity. Squaring the above expression, collecting terms, and reducing, we get

$$(1 + g^2\mathbf{B}^2)[a^2(s + b)^2 - g^2]^{1/2} + 2a(s + b)\mathbf{G} \cdot \mathbf{B} = 0,$$

an awkward situation in which $\mathbf{B}$ needs to be complex. We escape by changing the sign of the radicand in the expression for $\mathbf{P}'$, and we find that $\mathbf{B}$ can now be a real vector. Then $\mathbf{B}^2 = 1/g^2$ and $\mathbf{G} \cdot \mathbf{B} = 0$. Thus

$$\mathbf{P}' = (a/g^2)(s + b)\mathbf{G} + [g^2 - a^2(s + b)^2]^{1/2}\mathbf{B}. \tag{IV-55}$$

A final integration produces

$$\mathbf{P} = (a/2g^2)(s + b)^2\mathbf{G} + (1/2a)\{g^2 \arcsin[a(s + b)/g] + a(s + b)[g^2 - a^2(s + b)^2]^{1/2}\}\,\mathbf{B} + \mathbf{C}, \tag{IV-56}$$

where $\mathbf{C}$ is a constant of integration.

At the starting point $\mathbf{P}_1$, the arc length parameter, is zero. By setting $s = 0$, we obtain from Eqs. (IV-56)

$$\mathbf{P}_1 = (ab^2/2g^2)\mathbf{G} + (1/2a)[g^2 \arcsin(ab/g) + ab(g^2 - a^2b^2)^{1/2}]\,\mathbf{B} + \mathbf{C}.$$

The initial direction, which we will denote by $\mathbf{P}_1'$, comes from Eq.(IV-55):

$$\mathbf{P}_1' = (ab/g^2)\mathbf{G} + (g^2 - a^2b^2)^{1/2}\mathbf{B}.$$

Also from Eqs. (IV-53) and (IV-54), recalling that the sign of the radicand in the latter has been changed, and setting $s = 0$, we obtain

$$v_A = (g^2 - a^2b^2)^{1/2}.$$

Now let $ab/g = \sin\theta$. Then $v_A = g\cos\theta$,

$$\mathbf{P}_1 = (1/2a)[\sin^2\theta\,\mathbf{G} + g(g\theta + v_A\sin\theta)\mathbf{B}] + \mathbf{C},$$

and

$$\mathbf{P}_1' = (1/g)(\sin\theta\,\mathbf{G} + v_A\mathbf{B}).$$

Thus, the constants of integration are determined by the initial position, direction, and velocity of the bead, or the ray, or the electron, depending on what the context is. Solving these expressions for **B** and **C** in terms of $\mathbf{P}_1$ and $\mathbf{P}_1'$ and substituting into Eqs. (IV-55) and (IV-56), we get the solution of the brachistochrone with the appropriate initial conditions.

It is not difficult to see that this is a plane curve and that it is a cycloid. Differentiating Eq. (IV-55) gives us the normal vector. A vector product yields the binormal vector, which turns out to be constant. The curve is therefore a plane curve. One needs only to write Eqs. (IV-56) as scalar equations and compare with a standard form as, for example, in Drábek (1969, pp. 165–168) to see that the curve is indeed a cycloid. An illustration appears in Fig. IV-3.

The reader may now wish to calculate the value of the variational integral. If we take $\mathbf{P}_1$ as corresponding to the value of **P** when $s = 0$, we can get

$$I = \log \frac{[v_A^2 + as(s + 2b)]^{1/2} - a(s + b)}{v_A - ab}$$

as the length along a brachistochrone curve from $\mathbf{P}_1$ to $\mathbf{P}(s)$.

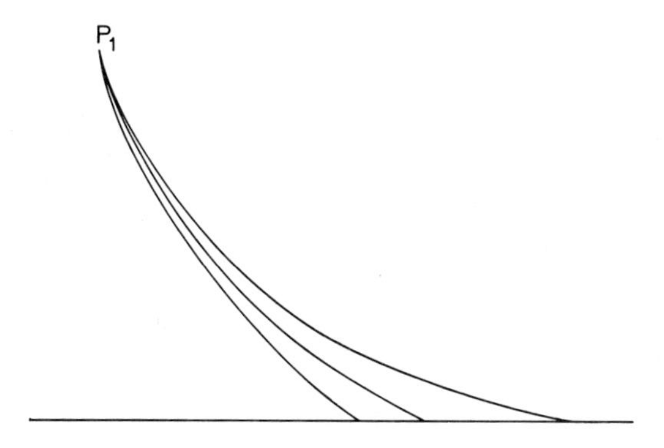

FIG. IV-3. The brachistochrone solution.

The solution consists of several members of a class of cycloids joining an initial point $\mathbf{P}_1$ with a straight line, or rays from a point $\mathbf{P}_1$ in a medium in which the refractive index is a function of position:

$$n(\mathbf{P}) = \mathbf{P}^2[2\mathbf{G} \cdot (\mathbf{P} - \mathbf{P}_1) + v_A{}^2]^{-1/2}.$$

REFERENCES

Born, M., and Wolf, E. (1970). "Principles of Optics," 4th ed. Pergamon, Oxford.

Drábek, K. (1969). Plane curves and constructions, *in* "Survey of Applicable Mathematics" (K. Rektorys, ed.). MIT Press, Cambridge, Massachusetts.

Luneburg, R. K. (1964). "Mathematical Theory of Optics." Univ. of California Press, Berkeley.

Maxwell, J. C. (1952). "Scientific Papers" (W. D. Niven, ed.). Dover, New York.

V

The Hilbert Integral

Up to this point we have been examining the properties of ray paths in inhomogeneous, isotropic media in which the refractive index is a continuous function of position. We have disregarded completely those media in which the refractive index is discontinuous. Yet such media comprise almost the entire range of problems of practical importance in optics. Consider two pieces of glass cemented together. The cemented surface separates two regions in which the refractive index is constant (and therefore continuous), but at that surface it is discontinuous. From this point of view a conventional lens system can be considered as an inhomogeneous medium in which the refractive index function is sectionally continuous (in fact, sectionally constant) and suffers jump discontinuities at the various air–glass and glass–glass surfaces. We are now in a position to consider media that are partitioned into two regions in which the refractive index is continuous but suffers a jump discontinuity on the surface separating the two regions. Our primary goal is to derive Snell's law. In doing so we will be beneficiaries to a host of spinoffs that will lead to some important information concerning wavefronts applicable to both continuous and discontinuous refractive indices.

There is one blatant misconception that enjoys a certain amount of currency in the optical community concerning the derivation of Snell's law. The argument is that, since a ray at a refracting surface undergoes a sudden change of direction, it is proper to invoke the corner condition to arrive at Snell's law. However, Zatzkis (1965) has shown that the correct application of the corner condition leads to an absurdity. This is due to the fact that the derivation of the corner condition depends on the continuity of the function f appearing in the variational integral. In the case we are considering, since f depends on the refractive index that is discontinuous, f itself is discontinuous and the corner condition cannot properly be applied.

What then can be said about ray paths in a medium in which discontinuities in refractive index occur? By Fermat's principle the optical path length along a ray from some initial point to some terminal point must be an extremum. Moreover, for such a ray path to be an extremum, its segments between discontinuities must be extremals and satisfy the Euler equations within the regions where the refractive index is continuous. What remains to be shown is what happens at these discontinuities on the refracting surfaces. The instrument we will use to settle this question is the Hilbert integral, the main topic of this chapter.

THE HILBERT INTEGRAL DEFINED

Suppose a variational integral is given,

$$I = \int_a^b f(x, y, z, y', z')\,dx,$$

in which the limits of integration must be fixed and the integrand function $f(x, y, z, y', z')$ must satisfy all the required continuity conditions. In what follows, let

$$y(x),\ z(x)$$

represent an extremal, which therefore must satisfy the Euler equations (II-9),

$$\frac{d}{dx}\frac{\partial f}{\partial y'} = \frac{\partial f}{\partial y}, \qquad \frac{d}{dx}\frac{\partial f}{\partial z'} = \frac{\partial f}{\partial z}.$$

Let some curve C be defined. Suppose the coordinates of the points of C are given by three single-valued functions of some parameter u:

$$C: \quad \bar{x}(u), \bar{y}(u), \bar{z}(u).$$

Choose two points on C, say P_1 and P_2, which are associated with two parameter values, u_1 and u_2. The Hilbert integral is defined as

$$J^*(C) = \int_{u_1}^{u_2} \left[\left(f - y' \frac{\partial f}{\partial y'} - z' \frac{\partial f}{\partial z'}\right) \frac{d\bar{x}}{du} + \frac{\partial f}{\partial y'} \frac{d\bar{y}}{du} + \frac{\partial f}{\partial z'} \frac{d\bar{z}}{du}\right] du, \quad \text{(V-1)}$$

where y' and z' are the derivatives of the two extremal functions and where f and its partial derivatives have as arguments these extremal functions (Bliss, 1946, pp. 18–20; Gelfand and Fomin, 1963, pp. 145–146; Clegg, 1968, Chapter 4).

The Hilbert integral depends on two things: the solution to a variational problem expressed in terms of the two functions $y(x)$ and $z(x)$, and the arbitrary curve C, the nature of which is left entirely unspecified.

The Hilbert integral, when written as a line integral along the curve C, takes the form

$$J^*(C) = \int_{P_1}^{P_2} \left[\left(f - y' \frac{\partial f}{\partial y'} - z' \frac{\partial f}{\partial z'}\right) d\bar{x} + \frac{\partial f}{\partial y'} d\bar{y} + \frac{\partial f}{\partial z'} d\bar{z}\right], \quad \text{(V-2)}$$

where $d\bar{x}$, $d\bar{y}$, and $d\bar{z}$ represent the three components of a differential vector along C.

One quite obvious property of the Hilbert integral is that, along an extremal, it degenerates into the variational integral. If C is an extremal, then

$$d\bar{x} = dx, \qquad d\bar{y} = dy, \qquad d\bar{z} = dz,$$

and $J^*(C) = I$.

A final definition before proceeding to the main theorem: If C intersects an extremal and if at the point of intersection

$$\left(f - y' \frac{\partial f}{\partial y'} - z' \frac{\partial f}{\partial z'}\right) d\bar{x} + \frac{\partial f}{\partial y'} d\bar{y} + \frac{\partial f}{\partial z'} d\bar{z} = 0, \quad \text{(V-3)}$$

then the direction $(d\bar{x}, d\bar{y}, d\bar{z})$ is said to be *transversal* to the extremal (Bliss, 1946, p. 25). If each member of a family of extremals is transversal to some curve C, then that curve is said to be transversal to the family.

This definition can be turned around. Equation (V-3) can be regarded as the differential equation for a curve transversal to the given one-parameter family of extremals. In fact, as we shall see later, the existence of a solution to Eq. (V-3) ensures the existence of the transversal curve.

What has been said here can be generalized to two-parameter families of extremals. We can define a transversal surface as a surface on which the Hilbert integral vanishes independently of the path of integration on that surface. In this case Eq. (V-3) can be regarded as a total differential equation for a transversal surface, and, moreover, the existence of a solution to Eq. (V-3) in this case assures the existence of the transversal surface.

A LEMMA

This rather simple statement is one that students are often unaware of. Let $f(x, t)$ be a function differentiable in t. The integral

$$I(t) = \int_a^b f(x, t)\, dx$$

is obviously a function of t. It can also be considered a function of a and b, the limits of integration, and we may write

$$I(t, a, b) = \int_a^b f(x, t)\, dx,$$

so that

$$\partial I/\partial a = -f(a, t), \qquad \partial I/\partial b = f(b, t).$$

If, in addition, a and b are themselves functions of t,

$$I(t, a(t), b(t)) = \int_{a(t)}^{b(t)} f(x, t)\, dx,$$

then

$$\begin{aligned}\frac{dI}{dt} &= \frac{\partial I}{\partial t} + \frac{\partial I}{\partial a}\frac{da}{dt} + \frac{\partial I}{\partial b}\frac{db}{dt} \\ &= \int_{a(t)}^{b(t)} \frac{\partial f}{\partial t}\, dx - f(a(t), t)\frac{da}{dt} + f(b(t), t)\frac{db}{dt}\,.\end{aligned}$$

THE MAIN THEOREM

Suppose f is a function that is continuous and possesses continuous partial derivatives of at least the second order. Then the Euler equations,

$$\frac{d}{dx}\frac{\partial f}{\partial y'} = \frac{\partial f}{\partial y}, \qquad \frac{d}{dx}\frac{\partial f}{\partial z'} = \frac{\partial f}{\partial z},$$

are defined and, in general, possess a four-parameter family of solutions,

$$y(x; \alpha, \beta, \gamma, \delta), \qquad z(x; \alpha, \beta, \gamma, \delta),$$

which are extremals for the variational integral

$$I = \int f(x, y, z, y', z')\, dx.$$

Specifying an initial point and a terminal point not only assigns limits to the variational integral but provides boundary conditions for the Euler equations as well. This is tantamount to extracting out of the four-parameter family of extremals one that passes through the two points.

Next we take two arbitrary space curves, C_1 and C_2, lying within the four-parameter field of extremals and some means of correlating the points of one with those of the other. For example, we could describe the two curves parametrically,

$$C_1\colon\ x_1(u), y_1(u), z_1(u), \qquad C_2\colon\ x_2(u), y_2(u), z_2(u).$$

Each value of the parameter u corresponds to a point on each curve. Each time a value of u is specified, a different pair of points, one on each curve, is determined, and since C_1 and C_2 both lie in the field of extremals an extremal connecting these two points is always specified. Thus, we generate a one-parameter family of extremals connecting corresponding points on C_1 and C_2. We write this one-parameter family of extremals as

$$y(x, u), \qquad z(x, u)$$

with the understanding that the endpoints of each extremal lie on the curves C_1 and C_2. Thus

$$\begin{aligned} y(x_1(u), u) &= y_1(u), & z(x_1(u), u) &= z_1(u), \\ y(x_2(u), u) &= y_2(u), & z(x_2(u), u) &= z_2(u). \end{aligned} \tag{V-4}$$

All this is illustrated in Fig. V-1.

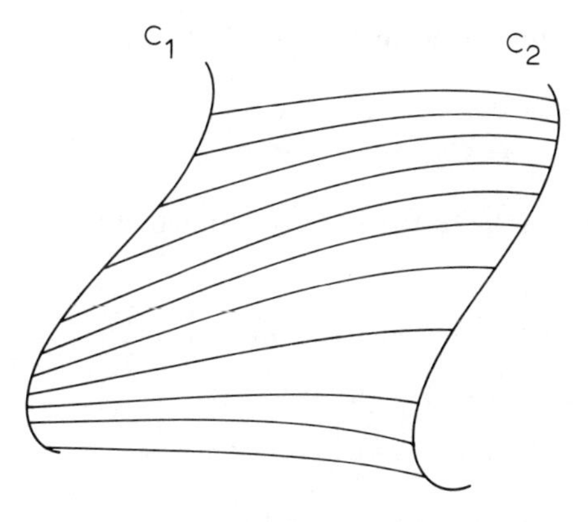

FIG. V-1.
A one-parameter family of extremals connecting the curves C_1 and C_2.

The variational integral itself is now a function of the parameter u,

$$I(u) = \int_{x_1(u)}^{x_2(u)} f(x, y(x, u), z(x, u), y'(x, u), z'(x, u))\, dx,$$

where y' and z' now denote the partial derivatives of y and z with respect to x. By differentiating $I(u)$, making use of the lemma, we obtain

$$\begin{aligned} dI/du &= f(x_2(u), y_2(u), z_2(u), y_2'(u), z_2'(u))\, dx_2/du \\ &\quad - f(x_1(u), y_1(u), z_1(u), y_1'(u), z_1'(u))\, dx_1/du \\ &\quad + \int_{x_1(u)}^{x_2(u)} \left[\frac{\partial f}{\partial y}\frac{\partial y}{\partial u} + \frac{\partial f}{\partial z}\frac{\partial z}{\partial u} + \frac{\partial f}{\partial y'}\frac{\partial y'}{\partial u} + \frac{\partial f}{\partial z'}\frac{\partial z'}{\partial u}\right] dx. \quad \text{(V-5)} \end{aligned}$$

In this expression

$$y_2'(u) = y'(x_2(u), u), \qquad \text{etc.}$$

Next note that

$$\frac{d}{dx}\left(\frac{\partial f}{\partial y'}\frac{\partial y}{\partial u}\right) = \frac{d}{dx}\left(\frac{\partial f}{\partial y'}\right)\frac{\partial y}{\partial u} + \frac{\partial f}{\partial y'}\frac{d}{dx}\left(\frac{\partial y}{\partial u}\right).$$

By applying the first Euler equation to the first term and interchanging the order of differentiation in the second, we get

$$\frac{d}{dx}\left(\frac{\partial f}{\partial y'}\frac{\partial y}{\partial u}\right) = \frac{\partial f}{\partial y}\frac{\partial y}{\partial u} + \frac{\partial f}{\partial y'}\frac{\partial y'}{\partial u}.$$

In like manner, using the second of the Euler equations, we get

$$\frac{d}{dx}\left(\frac{\partial f}{\partial z'}\frac{\partial z}{\partial u}\right) = \frac{\partial f}{\partial z}\frac{\partial z}{\partial u} + \frac{\partial f}{\partial z'}\frac{\partial z'}{\partial u}.$$

Applying these relations to the integral appearing in Eq. (V-5), we obtain

$$\int_{x_1(u)}^{x_2(u)} \left[\frac{\partial f}{\partial y}\frac{\partial y}{\partial u} + \frac{\partial f}{\partial z}\frac{\partial z}{\partial u} + \frac{\partial f}{\partial y'}\frac{\partial y'}{\partial u} + \frac{\partial f}{\partial z'}\frac{\partial z'}{\partial u}\right] dx$$

$$= \int_{x_1(u)}^{x_2(u)} \frac{d}{dx}\left[\frac{\partial f}{\partial y'}\frac{\partial y}{\partial u} + \frac{\partial f}{\partial z'}\frac{\partial z}{\partial u}\right] dx$$

$$= \left[\frac{\partial f}{\partial y'}\frac{\partial y}{\partial u} + \frac{\partial f}{\partial z'}\frac{\partial z}{\partial u}\right]_{x_1(u)}^{x_2(u)}.$$

The derivative of I is therefore, from the above and Eq. (V-5),

$$\frac{dI}{du} = \left[f\frac{dx}{du} + \frac{\partial f}{\partial y'}\frac{\partial y}{\partial u} + \frac{\partial f}{\partial z'}\frac{\partial z}{\partial u}\right]_{x_1(u)}^{x_2(u)}.$$

However, when the identities

$$\frac{dy}{du} = y'\frac{dx}{du} + \frac{\partial y}{\partial u}, \qquad \frac{dz}{du} = z'\frac{dx}{du} + \frac{\partial z}{\partial u}$$

are solved for $\partial y/\partial u$ and $\partial z/\partial u$ and substituted into the above expression, we get

$$\frac{dI}{du} = \left[\left(f - y'\frac{\partial f}{\partial y'} - z'\frac{\partial f}{\partial z'}\right)\frac{dx}{du} + \frac{\partial f}{\partial y'}\frac{dy}{du} + \frac{\partial f}{\partial z'}\frac{dz}{du}\right]_{x_1(u)}^{x_2(u)},$$

which may also be written in the form of a differential,

$$dI = \left[\left(f - y'\frac{\partial f}{\partial y'} - z'\frac{\partial f}{\partial z'}\right) dx + \frac{\partial f}{\partial y'} dy + \frac{\partial f}{\partial z'} dz\right]_{x_1(u)}^{x_2(u)}$$

$$= \left[\left(f_2 - y_2'\frac{\partial f_2}{\partial y'} - z_2'\frac{\partial f_2}{\partial z'}\right) dx_2 + \frac{\partial f_2}{\partial y'} dy_2 + \frac{\partial f_2}{\partial z'} dz_2\right]$$

$$- \left[\left(f_1 - y_1'\frac{\partial f_1}{\partial y'} - z_1'\frac{\partial f_1}{\partial z'}\right) dx_1 + \frac{\partial f_1}{\partial y'} dy_1 + \frac{\partial f_1}{\partial z'} dz_1\right],$$

where $f_1 = f(x_1, y_1, z_1, y_1', z_1')$, and where dx_1, dy_1, dz_1 are differentials along the curve C_1, etc. By comparing these with Eq. (V-2) we find that

$$dI = dJ^*(C_2) - dJ^*(C_1). \tag{V-6}$$

Next we choose two extremals from the one-parameter family. This is tantamount to selecting two values of the parameter u, say u_1 and u_2. By integrating Eq. (V-6) with respect to u between the limits u_1 and u_2, we get

$$I(u_2) - I(u_1) = J^*(C_2) - J^*(C_1). \qquad \text{(V-7)}$$

This relationship between the variational integral and the Hilbert integral is our main result. In fact, this line of reasoning is the only motivation for the definition of the Hilbert integral. As we shall shortly see, Eqs. (V-6) and (V-7) are of great importance in geometrical optics.

In summary, let two curves, C_1 and C_2, be connected by a one-parameter family of extremals. Select a pair of these extremals, E_1 and E_2, joining C_1 and C_2 as shown in Fig. V-2. These are determined by the

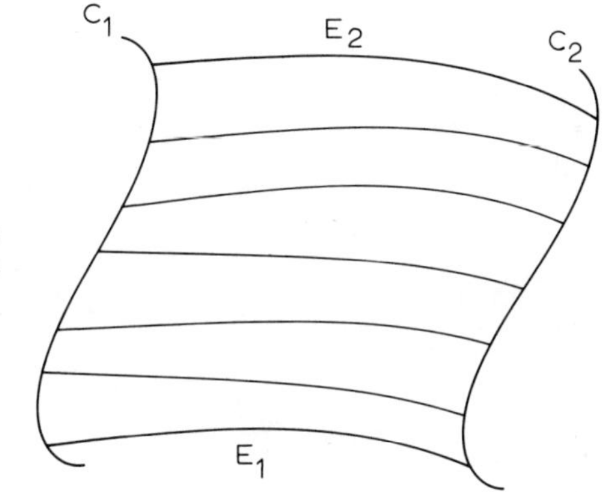

FIG. V-2.
The Hilbert integral is defined on the two curves C_1 and C_2 and on the two extremals E_1 and E_2.

two parameter values u_1 and u_2. Calculate the values of the two variational integrals along E_1 and E_2. These are designated as $I(u_1)$ and $I(u_2)$. Finally, calculate the two Hilbert integrals along C_1 and C_2 between E_1 and E_2, obtaining $J^*(C_1)$ and $J^*(C_2)$. These four quantities will then satisfy Eq. (V-7).

In deriving Eq. (V-7) we could have used a fixed point instead of one of the curves. If, instead of the curve C_1, we assumed that a one-parameter family of extremals connected a point with the curve C_2, we would have obtained, in place of Eq. (V-7),

$$I(u_2) - I(u_1) = J^*(C_2). \qquad \text{(V-8)}$$

On the other hand, if C_2 were replaced by a point, we would obtain

$$I(u_2) - I(u_1) = -J^*(C_1). \qquad \text{(V-9)}$$

If both C_1 and C_2 are replaced by points and if there are still two distinct extremals connecting them, then

$$I(u_2) - I(u_1) = 0. \tag{V-10}$$

In other words, if there is more than one extremal connecting two points, then the value of the variational integral calculated between the two points is the same for each extremal. The variational integral therefore is independent of the path. These cases are shown in Fig. V-3.

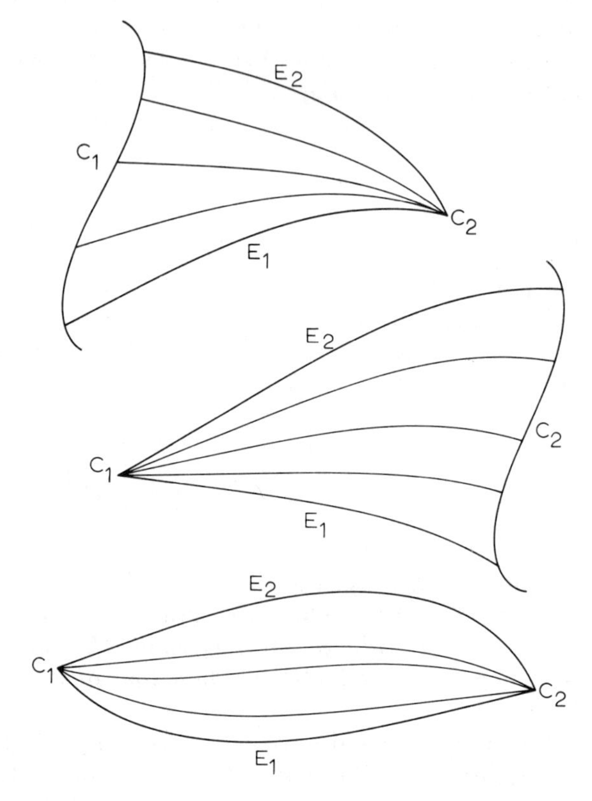

FIG. V-3.

These figures are obtained by allowing either or both of the curves C_1 and C_2 to degenerate to a point.

Refer once again to Fig. V-2 and calculate the Hilbert integral around the closed path C_2, E_2, C_1, E_1. We obtain

$$J^*(C_2) + J^*(E_2) + J^*(C_1) + J^*(E_1).$$

However, along an extremal the Hilbert integral degenerates into the variational integral so that the above expression becomes

$$J^*(C_2) + I(u_2) - J^*(C_1) - I(u_1),$$

which is zero by Eq. (V-7). Therefore, under the conditions for which the conclusions of the main theorem are valid, the Hilbert integral, calculated around a closed curve, vanishes. The Hilbert integral is then independent of the path of integration.

Finally, if both C_1 and C_2 are curves transversal to the one-parameter family of extremals, then the Hilbert integral vanishes on the two curves and Eq. (V-7) reduces to $I(u_2) = I(u_1)$. The variational integral is constant along the extremals connecting two transversal curves.

AN ILLUSTRATION

Earlier we used the calculus of variations to show that the shortest distance between two points is a straight line. We were led to the variational integral

$$I = \int (1 + y'^2 + z'^2)^{1/2}\, dx,$$

which had associated with it the Euler equations

$$\frac{d}{dx}\frac{y'}{(1 + y'^2 + z'^2)^{1/2}} = 0, \qquad \frac{d}{dx}\frac{z'}{(1 + y'^2 + z'^2)^{1/2}} = 0.$$

These led to the four-parameter family of straight lines, which are the extremals for this problem,

$$y = \frac{ax + c}{(1 - a^2 - b^2)^{1/2}}, \qquad z = \frac{bx + d}{(1 - a^2 - b^2)^{1/2}},$$

where the four parameters a, b, c, and d arise as constants of integration.

We also used the helix,

$$\bar{x} = r\cos u, \qquad \bar{y} = r\sin u, \qquad \bar{z} = ku,$$

to illustrate some properties of space curves. Suppose we select a point on the x axis,

$$\mathbf{P}\colon\quad (-h, 0, 0),$$

and ask at which points on the helix the distance from **P** to the helix assumes stationary values (see Fig. V-4). First we must find the one-

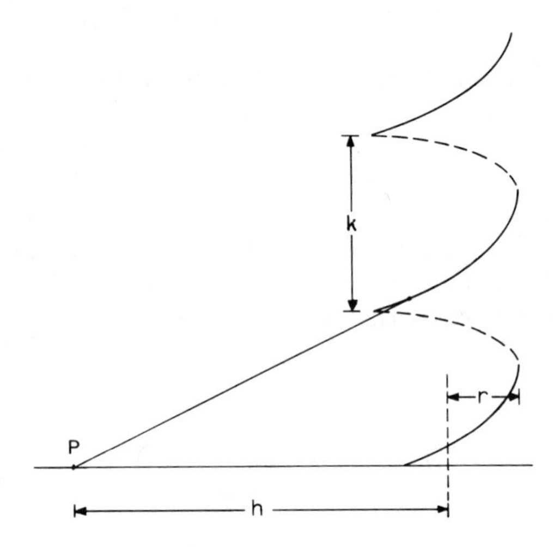

FIG. V-4. The helix problem.

The problem is to find the points on the helix at which the distance from the helix to the fixed point **P** is an extremum.

parameter family of extremals that passes through both **P** and the helix. It is a simple matter to show that they are given by

$$y = \frac{r(x+h)\sin u}{h + r\cos u}, \qquad z = \frac{k(x+h)u}{h + r\cos u}.$$

Using Eq. (V-6), the differential form of the Hilbert integral calculated along the helix, we obtain

$$\begin{aligned} dI &= \left(f - y'\frac{\partial f}{\partial y'} - z'\frac{\partial f}{\partial z'}\right) d\bar{x} + \frac{\partial f}{\partial y'}\, d\bar{y} + \frac{\partial f}{\partial z'}\, d\bar{z} \\ &= \frac{(r\cos u + h)(-r\sin u)\, du + (r\sin u)(r\cos u)\, du + (ku)k\, du}{[(r\cos u + h)^2 + r^2\sin^2 u + k^2u^2]^{1/2}}. \end{aligned}$$

Here, f is the integrand of the variational integral and is equal to $(1 + y'^2 + z'^2)^{1/2}$.

For the distance between the point **P** and a point on the helix to be stationary, $dI = 0$, which leads to the equation

$$hr\sin u = k^2 u.$$

Those values of u for which $(\sin u)/u = k^2/hr$ provide the points on the helix that have the desired properties. See Fig. V-5. Note that, if

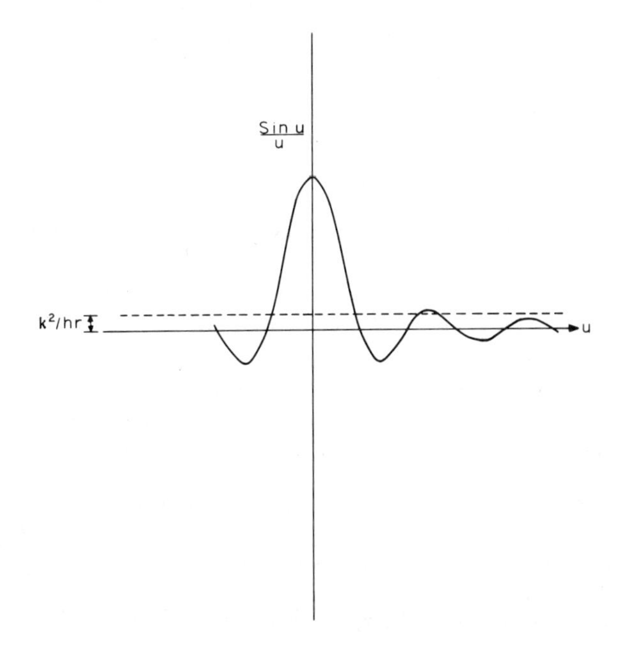

FIG. V-5. Graphical solution to the helix problem.
In the case shown there are exactly six solutions.

$h < k^2/r$, there are no solutions and the distance from **P** to the helix increases monotonically as u increases. On the other hand, if $h > k^2/r$, there will be a finite number of points on the helix at which the distance from **P** has a stationary value.

THE PARAMETRIC CASE

The representation of the Hilbert integral where the extremals are given in parametric form is particularly useful in the application of the calculus of variations to optics. The methods and concepts are identical to those already encountered in the nonparametric case. From Eq. (II-18) the Euler equation is

$$d(\nabla_t f)/dt = \nabla f,$$

a second-order differential equation in the vector function $\mathbf{P}$ of the parameter t. As always, $f = f(\mathbf{P}, \mathbf{P}_t)$, the integrand of the variational integral. In general, this differential equation possesses a four-parameter family of solutions that can be represented perhaps as $\mathbf{P}(t; \alpha, \beta, \gamma, \delta)$, which defines a field of extremals of the variational integral

$$I = \int f(\mathbf{P}, \mathbf{P}_t)\, dt.$$

As before, we select two curves, defined by two vector functions of the parameter u, $\mathbf{V}(u)$, $\mathbf{W}(u)$, so that each value of u determines one point on each curve. We must add the proviso that the two curves lie in a region covered by the field of extremals. Each value of u therefore determines an extremal curve $\mathbf{P}(t, u)$ connecting the two points. We have thus extracted out of the four-parameter family of extremals a one-parameter subfamily connecting corresponding points on the two curves $\mathbf{V}$ and $\mathbf{W}$. Finally, we choose the parameter t in such a way that $\mathbf{P}(t_1, u)$ always lies on the curve $\mathbf{V}$ and that $\mathbf{P}(t_2, u)$ always lies on $\mathbf{W}$, so that

$$\mathbf{P}(t_1, u) = \mathbf{V}(u), \qquad \mathbf{P}(t_2, u) = \mathbf{W}(u) \tag{V-11}$$

for all values of u. For convenience we define

$$\mathbf{P}_t(t_1, u) = \mathbf{V}_t(u), \qquad \mathbf{P}_t(t_2, u) = \mathbf{W}_t(u). \tag{V-12}$$

With these ground rules established we now note that the variational integral, as before, is a function of the parameter u,

$$I(u) = \int_{t_1}^{t_2} f(\mathbf{P}(t, u), \mathbf{P}_t(t, u))\, dt.$$

Moreover, the limits of integration are fixed so that its derivative is simply

$$\begin{aligned}
\frac{dI}{du} &= \int_{t_1}^{t_2} \left[\nabla f \cdot \frac{\partial \mathbf{P}}{\partial u} + \nabla_t f \cdot \frac{\partial \mathbf{P}_t}{\partial u}\right] dt \\
&= \int_{t_1}^{t_2} \left[\left(\frac{d}{dt}(\nabla_t f)\right) \cdot \frac{\partial \mathbf{P}}{\partial u} + \nabla_t f \cdot \left(\frac{d}{dt}\frac{\partial \mathbf{P}}{\partial u}\right)\right] dt \\
&= \int_{t_1}^{t_2} \frac{d}{dt}\left[\nabla_t f \cdot \frac{\partial \mathbf{P}}{\partial u}\right] dt \\
&= \left[\nabla_t f \cdot \frac{\partial \mathbf{P}}{\partial u}\right]_{t_1}^{t_2}.
\end{aligned}$$

Note the use of the Euler equation.

We may write this expression out in more detail:

$$\frac{dI}{dt} = \nabla_t f(\mathbf{P}(t_2, u), \mathbf{P}_t(t_2, u)) \cdot \frac{\partial \mathbf{P}(t_2, u)}{\partial u}$$

$$-\nabla_t f(\mathbf{P}(t_1, u), \mathbf{P}_t(t_1, u)) \cdot \frac{\partial \mathbf{P}(t_1, u)}{\partial u}$$

$$= \nabla_t f(\mathbf{W}, \mathbf{W}_t) \cdot \frac{d\mathbf{W}}{du} - \nabla_t f(\mathbf{V}, \mathbf{V}_t) \cdot \frac{d\mathbf{V}}{du}. \tag{V-13}$$

Here the derivative of the variational integral is expressed entirely in terms of the two curves **V** and **W**.

Now, choosing two values of the parameter u, say u_1 and u_2, and integrating the above expression, we obtain

$$I(u_2) - I(u_1) = \int_{u_1}^{u_2} \nabla_t f(\mathbf{W}, \mathbf{W}_t) \cdot \frac{d\mathbf{W}}{du}\, du - \int_{u_1}^{u_2} \nabla_t f(\mathbf{V}, \mathbf{V}_t) \cdot \frac{d\mathbf{V}}{du}\, du, \tag{V-14}$$

which puts us in position to define the Hilbert integral for the parametric case:

$$J^*(\mathbf{V}) = \int_{u_1}^{u_2} f(\mathbf{V}, \mathbf{V}_t) \frac{d\mathbf{V}}{du}\, du, \tag{V-15}$$

where **V** is any curve. Alternatively, we may write the Hilbert integral as a line integral

$$J^*(\mathbf{V}) = \int_{\mathbf{V}_1}^{\mathbf{V}_2} f(\mathbf{V}, \mathbf{V}_t) \cdot d\mathbf{V}. \tag{V-16}$$

In either case, Eq. (V-11) may be written in the same form as Eq. (V-7),

$$I(u_2) - I(u_1) = J^*(\mathbf{W}) - J^*(\mathbf{V}). \tag{V-17}$$

Of course the properties of the Hilbert integral shown to hold for the nonparametric case carry over to the parametric case.

We have already encountered the notion of *transversality*. If the curve **V** intersects an extremal and if at the point of intersection

$$\nabla_t f(\mathbf{V}, \mathbf{V}_t) \cdot d\mathbf{V} = 0,$$

then the direction of the curve at that point is said to be *transversal* to the extremal. If **V** is transversal to every member of a one-parameter family of extremals, then **V** is called a *transversal curve* to the family and the Hilbert integral vanishes when calculated between every pair of points on **V**. The idea of transversality can be extended to two-parameter families of extremals and a surface intercepting each member of the family. If the Hilbert integral vanishes over the surface, then that surface is called a *transversal* surface for the two-parameter family.

Suppose a one-parameter family of extremals possesses an *envelope*. By an *envelope* we mean a curve $\mathbf{E}$ tangent at each point to an extremal belonging to the family such that each member of the family is tangent to $\mathbf{E}$ at some point. The Hilbert integral along $\mathbf{E}$ is

$$J^*(\mathbf{E}) = \int \nabla_t f(\mathbf{E}, \mathbf{E}_t) \cdot d\mathbf{E}.$$

Through each point of $\mathbf{E}$ there passes, tangent to $\mathbf{E}$, an extremal $\mathbf{P}$. At some such point we may write $\mathbf{E} = \mathbf{P}$. From the definition of $\mathbf{P}_t$ given in Eq. (V-12), $\mathbf{E}_t = \mathbf{P}_t$. Moreover, $d\mathbf{E} = d\mathbf{P}$ so that at every point of $\mathbf{E}$,

$$\nabla_t f(\mathbf{E}, \mathbf{E}_t) \cdot d\mathbf{E} = \nabla_t f(\mathbf{P}, \mathbf{P}_t) \cdot d\mathbf{P} = f(\mathbf{P}, \mathbf{P}_t)\, dt = f(\mathbf{E}, \mathbf{E}_t)\, dt.$$

Here we make use of the fact that $f(\mathbf{P}, \mathbf{P}_t)$ is homogeneous in $\mathbf{P}_t$. What happens at each point happens at every point on $\mathbf{E}$, so

$$J^*(\mathbf{E}) = \int f(\mathbf{E}, \mathbf{E}_t)\, dt = I,$$

and we have proved that the Hilbert integral along an envelope degenerates into a variational integral.

The idea of the envelope can be extended. A two-parameter family of extremals may have a surface as an envelope. As we will see shortly, in geometrical optics this leads to the caustic surface (Bliss, 1946, pp. 24–36).

SNELL'S LAW

We are now prepared to turn our hand to the application of the Hilbert integral to the optics of inhomogeneous media. Our first task is to obtain Snell's law. We have seen that, if the refractive index is given by $n = n(\mathbf{P})$, the variational integral is

$$I = \int n(\mathbf{P}_t^2)^{1/2}\, dt,$$

so that

$$f = n(\mathbf{P})(\mathbf{P}_t^2)^{1/2} \qquad \text{and} \qquad \nabla_t f = n(\mathbf{P})\,\mathbf{P}_t/(\mathbf{P}_t^2)^{1/2} = n\mathbf{P}.$$

The Hilbert integral is therefore

$$J^*(\mathbf{V}) = \int n\mathbf{P}' \cdot d\mathbf{V},$$

where $\mathbf{V}$ represents any curve in the medium.

Suppose, however, that the medium has a discontinuous index of refraction, that there is a surface, $\beta(\mathbf{P}) = 0$, such that the refractive index of the medium on one side is given by $n_1(\mathbf{P})$ and on the other side is given by $n_2(\mathbf{P})$. If $\bar{\mathbf{P}}$ is a point on the surface $\beta = 0$, then

$$n_2(\bar{\mathbf{P}}) - n_1(\bar{\mathbf{P}}) \neq 0.$$

In words, the refractive index suffers a jump discontinuity on the surface $\beta = 0$.

Take a pair of points, $\mathbf{A}_1$ and $\mathbf{A}_2$, on opposite sides of the discontinuity and a point $\bar{\mathbf{P}}$ on the discontinuity. Let $\mathbf{P}_1$ be a ray path connecting $\mathbf{A}_1$ with $\bar{\mathbf{P}}$ and let $\mathbf{P}_2$ be a ray going from $\bar{\mathbf{P}}$ to $\mathbf{A}_2$. The two rays $\mathbf{P}_1$ and $\mathbf{P}_2$ are of course extremals in the medium in which they lie. Recalling that $\mathbf{P}'$ denotes the tangent vector to a ray $\mathbf{P}(s)$, we let $\bar{\mathbf{P}}_1'$ denote $\mathbf{P}_1'$ at $\bar{\mathbf{P}}$, and we define $\bar{\mathbf{P}}_2'$ in the same manner.

Let I_1 be the optical path length, identical to the value of the variational integral, along $\mathbf{P}_1$ from $\mathbf{A}_1$ to $\bar{\mathbf{P}}$, and let I_2 be the path length for $\mathbf{P}_2$ from $\bar{\mathbf{P}}$ to $\mathbf{A}_2$. See Fig. V-6. Then the total optical path length from $\mathbf{A}_1$ to $\mathbf{A}_2$ via $\bar{\mathbf{P}}$ is given by

$$I = I_1 + I_2.$$

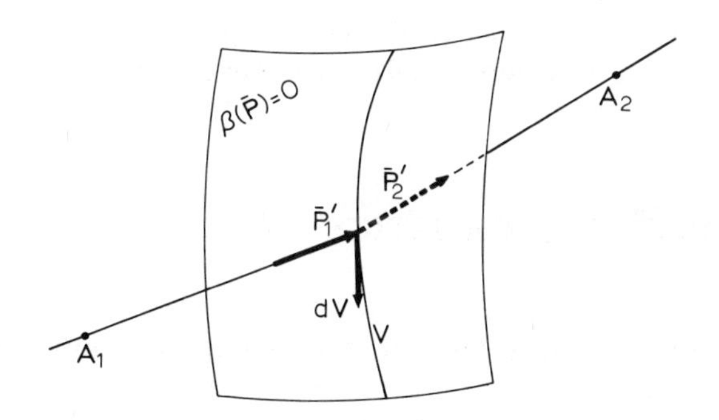

FIG. V-6. Derivation of Snell's law using the Hilbert integral.

If $\mathbf{V}$ is any curve on $\beta = 0$, then

$$dI_1 = n_1\bar{\mathbf{P}}_1' \cdot d\mathbf{V},$$

from Eqs. (V-6) and (V-8). From Eqs. (V-6) and (V-9) we get

$$dI_2 = -n_2\bar{\mathbf{P}}_2' \cdot d\mathbf{V}.$$

Fermat's principle states that the optical path length along a ray path must be stationary. Using that here we get

$$dI = 0 = dI_1 + dI_2 = n_1\bar{\mathbf{P}}_1' \cdot d\mathbf{V} - n_2\bar{\mathbf{P}}_2' \cdot d\mathbf{V}.$$

If we write this as

$$(n_1\bar{\mathbf{P}}_1' - n_2\bar{\mathbf{P}}_2') \cdot d\mathbf{V} = 0,$$

we can see that the vector quantity $(n_1\bar{\mathbf{P}}_1' - n_2\bar{\mathbf{P}}_2')$ is perpendicular to the curve $\mathbf{V}$ at $\bar{\mathbf{P}}$. Since $\mathbf{V}$ can be any curve on $\beta = 0$ through $\bar{\mathbf{P}}$, the above vector quantity is perpendicular to every such curve and must therefore be perpendicular to the surface itself. Let $\mathbf{N}$ be a unit normal vector to the surface $\beta = 0$ at the point $\bar{\mathbf{P}}$. Then

$$n_1\bar{\mathbf{P}}_1' - n_2\bar{\mathbf{P}}_2' = k\mathbf{N} \tag{V-18}$$

since the vector quantity on the left must be parallel to $\mathbf{N}$. Here k is a suitable constant of proportionality. Taking the vector product of the above expression with $\mathbf{N}$, we obtain the vector form of Snell's law,

$$n_1(\bar{\mathbf{P}}_1' \times \mathbf{N}) = n_2(\bar{\mathbf{P}}_2' \times \mathbf{N}). \tag{V-19}$$

Let i_1 and i_2 be the angles between $\mathbf{N}$ and $\bar{\mathbf{P}}_1'$ and $\bar{\mathbf{P}}_2'$, respectively. Then, from Eq. (V-19), since $\bar{\mathbf{P}}_1'$, $\bar{\mathbf{P}}_2'$, and $\mathbf{N}$ are all unit vectors,

$$n_1 \sin i_1 = n_2 \sin i_2 , \tag{V-20}$$

the familiar scalar form of Snell's law. It is quite clear from their definitions that i_1 and i_2 are the angles of incidence and refraction, respectively.

The treatment given here is essentially that of Zatzkis (1965).

WAVEFRONTS AND CAUSTICS

The Hilbert integral applied to the geometrical optics of inhomogeneous media is

$$J^*(\mathbf{V}) = \int n\mathbf{P}' \cdot d\mathbf{V},$$

defined for any curve **V**. If **V** lies on a surface transversal to a two-parameter family of rays, then $n\mathbf{P}' \cdot d\mathbf{V} = 0$. But this is exactly a statement that the curve **V** is *orthogonal* to the family of rays. Therefore, in the context of geometrical optics, *transversal* is synonymous with *orthogonal* and we can say that the Hilbert integral vanishes on a surface orthogonal to a family of rays.

Now suppose a two-parameter family of rays possesses two orthogonal surfaces. Select two curves, **V** and **W**, each lying on one of the surfaces and connected by a one-parameter family of rays. Next, pick two rays out of the family. We are now ready to apply Eq. (V-14),

$$I(u_1) - I(u_2) = J^*(\mathbf{W}) - J^*(\mathbf{V}),$$

where $I(u_1)$ and $I(u_2)$ are the values of the variational integral along each of the two rays connecting **V** and **W** and where $J^*(\mathbf{W})$ and $J^*(\mathbf{V})$ are the Hilbert integrals calculated along **W** and **V** between the two extremals. Since **V** and **W** lie on orthogonal surfaces, $J^*(\mathbf{V})$ and $J^*(\mathbf{W})$ vanish, and we are left with

$$I(u_1) = I(u_2).$$

Since I is defined as the optical path length along a ray, we have proved that the optical path length between any pair of orthogonal surfaces is the same for all rays.

We stretch our intuition a little bit and use this fact to justify our calling orthogonal surfaces *wavefronts*. The equation

$$n\mathbf{P}' \cdot d\mathbf{V} = 0 \tag{V-21}$$

can now be considered a differential equation for a wavefront should one exist.

If the curve **E** is an envelope to a family of rays, we have seen that the Hilbert integral along **E** degenerates into a variational integral

$$\int n\mathbf{P}' \cdot d\mathbf{E} = \int n\, ds.$$

At every point of **E**, $n\mathbf{P}' \cdot d\mathbf{E} = n\, ds$. This can be considered as a differential equation of the envelope.

In geometrical optics the envelope to a family of rays transmitted by a lens or reflected off a mirror is called the *caustic*, from the diminutive form of the Greek word for "burning iron." The caustic is one of the few things in geometrical optics that has any physical reality. Wavefronts and rays are not realizable; they are just convenient symbols on which we can hang our ideas. The caustic on the other hand is real and becomes visible by blowing a cloud of smoke in the region of the focus of a lens.

REFERENCES

Bliss, G. A. (1946). "Lectures on the Calculus of Variations." Univ. of Chicago Press, Chicago.

Clegg, J. C. (1968). "Calculus of Variations." Wiley, New York.

Gelfand, I. M., and Fomin, S. V. (1963). "Calculus of Variations" (R. A. Silverman, transl.). Prentice-Hall, Englewood Cliffs, New Jersey.

Zatzkis, H. (1965). Snell's law and the calculus of variations. *J. Opt. Soc. Amer.* **55**, 59–61.

VI

Ray Tracing

It has been said that Snell's law is all anyone needs to know about geometrical optics in order to do optical design. This canard has been uttered so many times by so many influential people that it has been accepted as true. And, like some cleverer lies, the statement contains just enough truth to make rebuttal difficult and unconvincing.

Ray tracing is the single most important tool used in optical design. Snell's law is the main ingredient in formulating procedures for tracing rays. Indeed, lens design as we know it is inconceivable without Snell's law. A knowledge of Snell's law is therefore a necessary condition for designing lenses. But as any mathematician knows, necessity does not imply sufficiency. It is simply not enough for a lens designer to know where his rays are going. He also needs to know where they ought to go and how to get them there.

However, our concern in this chapter is the mechanics of ray tracing. To trace a ray one begins at an object point and selects a ray direction. Next, the point where the ray intersects the first surface of the optical system is found. At that point the normal to the surface is calculated.

This normal and the incident ray determine the *plane of incidence*. The *angle of incidence* i is the angle between the incident ray and the surface normal. Knowing the refractive indices and using Snell's law, one determines the direction of the *refracted ray*. The angle between the refracted ray and the surface normal is the *angle of refraction* i'. Starting with the point of intersection of the ray with the surface and knowing the direction of the refracted ray from that point, one finds the point of intersection of that ray with the next surface. One repeats the process until one runs out of refracting surfaces.

The ray is now traced through the optical system. What is now done with it will not concern us here. In this chapter our primary concern is with the mechanics of the process.

It is clear that in ray tracing two operations are involved. One, *refraction*, determines the direction of the refracted ray. The inputs to this operation are: (1) the direction of the surface normal; (2) two refractive indices, one on each side of the refracting surface; and, of course, (3) the direction of the incident ray. Snell's law is applied.

The second process, which we will call *transfer*, involves finding the point of intersection of a ray with the refracting surface. Its inputs are: (1) the shape of the refracting surface, and (2) the position of the starting point of the ray relative to this surface, as well as (3) the ray's direction. The output of this process is not only the location of the point of intersection but the direction of the surface normal at that point as well. In addition, one may require the distance along the ray from its starting point to its intersection with the refracting surface. We will analyze these two processes separately.

REFRACTION

Before beginning, we need to make a change of notation. Up to now we have used $\mathbf{P}'$ or $\mathbf{t}$ to denote the tangent vector to a ray path. Also, rather than use $\mathbf{P}(s)$ to denote the vector function describing a ray path, we will use the vector function $\mathbf{W}$. This notation will avoid confusion later on when we will need to designate the normal to a wavefront and a vector function on that wavefront.

So let $\mathbf{N}$ be the direction vector for the incident ray and let $\mathbf{N}^*$ be the direction vector for the refracted ray. Moreover, let $\bar{\mathbf{N}}$ be the normal to the refracting surface. Finally let μ be the ratio of the refractive indices on opposite sides of the refracting surface. (That is, if n is the index of refraction on the incident-ray side of the refracting surface and n^* is that on the refracted-ray side, then $\mu = n/n^*$.) The vector form of Snell's law, from Eq. (V-19), is

$$\mathbf{N}^* \times \bar{\mathbf{N}} = \mu(\mathbf{N} \times \bar{\mathbf{N}}),$$

and the scalar version, from Eq. (V-20), is

$$\sin i^* = \mu \sin i,$$

where i and i^* are the angles of incidence and refraction, respectively. From the vector form we have

$$(\mathbf{N}^* - \mu\mathbf{N}) \times \bar{\mathbf{N}} = 0,$$

so that the vectors $(\mathbf{N}^* - \mu\mathbf{N})$ and $\bar{\mathbf{N}}$ are parallel. Therefore there exists a scalar quantity γ such that $\mathbf{N}^* - \mu\mathbf{N} = \gamma\bar{\mathbf{N}}$, which leads at once to the ray tracing formula for refraction,

$$\mathbf{N}^* = \mu\mathbf{N} + \gamma\bar{\mathbf{N}}. \tag{VI-1}$$

Here γ is yet to be determined. Since $\mathbf{N}^*$, $\mathbf{N}$, and $\bar{\mathbf{N}}$ are all unit vectors, the square of the above equation yields

$$1 = \mu^2 + \gamma^2 + 2\mu\gamma(\mathbf{N} \cdot \bar{\mathbf{N}}),$$

a quadratic in γ whose solution is

$$\gamma = -\mu(\mathbf{N} \cdot \bar{\mathbf{N}}) \pm \{1 - \mu^2[1 - (\mathbf{N} \cdot \bar{\mathbf{N}})^2]\}^{1/2}.$$

The ambiguity in sign is removed by noting that at normal incidence the incident ray is undeviated by refraction and that in this case $\mathbf{N}^* = \mathbf{N} = \bar{\mathbf{N}}$. Then $\mathbf{N} \cdot \bar{\mathbf{N}} = 1$ and therefore $\gamma = -\mu \pm 1$. Substituting this into Eq. (VI-1) shows that the positive branch must be used. Thus we have following equivalent expressions for γ:

$$\begin{aligned} \gamma &= -\mu(\mathbf{N} \cdot \bar{\mathbf{N}}) + \{1 - \mu^2[1 - (\mathbf{N} \cdot \bar{\mathbf{N}})^2]\}^{1/2} \\ &= -\mu(\mathbf{N} \cdot \bar{\mathbf{N}}) + \{1 - \mu^2(\mathbf{N} \times \bar{\mathbf{N}})^2\}^{1/2} \\ &= -\mu(\mathbf{N} \cdot \bar{\mathbf{N}}) + \{1 - (\mathbf{N}^* \times \bar{\mathbf{N}})^2\}^{1/2} \\ &= -\mu \cos i + \{1 - \mu^2 \sin^2 i\}^{1/2} \\ &= -\mu \cos i + \cos i^*. \end{aligned} \tag{VI-2}$$

So, for refracting a ray we need only apply Eq. (VI-1) and one of the forms of Eq. (VI-2).

REFLECTION

Reflection is best treated as a special case of refraction. One way to do this is to use Eqs. (VI-1) and (VI-2) and take $n^* = -n$, so that

$$\mu = -1. \tag{VI-3}$$

Thus, from Eqs. (VI-2) we obtain $\gamma = 2 \cos i$ and from Eq. (VI-1)

$$\mathbf{N}^* = -\mathbf{N} + 2 \cos i \bar{\mathbf{N}}. \tag{VI-4}$$

In this calculation the direction of $\mathbf{N}^*$ is through the reflecting surface. The transfer operation to the next surface requires a negative value for the distance to that surface. This makes sense since, after a reflection, the direction of general propagation of the ray has been reversed.

TRANSFER

Most optical systems are rotationally symmetric. This is not to say that *all* optical systems are rotationally symmetric or even that all important or interesting optical systems are rotationally symmetric. Anamorphotic lenses, which include lenses for Cinemascope cameras and projectors, as well as ophthalmic lenses for correcting astigmatism, are optical systems that do not have rotational symmetry. Rotationally symmetric systems are the simplest to deal with, and therefore we will consider them first.

In rotationally symmetric systems, the refracting and reflecting surfaces are surfaces of revolution. These are arranged so that the axes of revolution all lie on a common straight line called, simply, the *axis*. The plane determined by the axis and an object point is called a *meridian plane*; a ray lying on this plane is called a *meridian ray*. Any other ray not confined to the meridian plane is called a *skew ray*. A property of rotationally symmetric systems is that a meridian ray remains a meridian ray throughout the entire lens. As a consequence, a skew ray can never become a meridian ray but must always intersect the meridian plane at some nonzero angle.

For convenience, the equation for a refracting surface is given with reference to a coordinate system in which the z axis coincides with the axis of the optical system. The x and y axes lie in a plane perpendicular to the axis and tangent to the refracting surface at the point where the axis and the surface intersect. This point, for some strange reason, is called the *vertex* of the surface.

If $\mathbf{W} = (x, y, z)$ is the starting point of a ray and if $\mathbf{N} = (X, Y, Z)$ represents the direction of that ray, then the equation of any point $\mathbf{W}_\lambda$ on that ray is given by $\mathbf{W}_\lambda = \mathbf{W} + \lambda\mathbf{N}$, where λ is the distance from the starting point to $\mathbf{W}_\lambda$. In scalar form the equations are

$$x_\lambda = x + \lambda X, \qquad y_\lambda = y + \lambda Y, \qquad z_\lambda = z + \lambda Z.$$

We need to calculate the intersection of this ray with the next surface. Since the next surface is described in terms of its own coordinate system, we need to make a change of coordinate axes. Let t be the distance between the coordinate origin used in specifying the point $\mathbf{W}$ and the vertex of the next surface. Then a translation of the coordinate axes results in the following equations for a point on the ray:

$$x_\lambda = x + \lambda X, \qquad y_\lambda = y + \lambda Y, \qquad z_\lambda = z - t + \lambda Z.$$

To put this in vector form, we need to define a constant unit vector in the direction of the axis of the optical system: $\mathbf{A} = (0, 0, 1)$. Then in vector form the above equations become

$$\mathbf{W}_\lambda = \mathbf{W} - t\mathbf{A} + \lambda\mathbf{N}. \tag{VI-5}$$

Let the equation of the refracting surface be given by

$$\phi(\bar{x}, \bar{y}, \bar{z}) = \phi(\bar{\mathbf{R}}) = 0, \tag{VI-6}$$

where $\bar{\mathbf{R}}$ is a vector to any point on that surface. Note that the unit normal vector to the surface at $\bar{\mathbf{R}}$ is

$$\bar{\mathbf{N}} = \nabla\phi/[(\nabla\phi)^2]^{1/2}. \tag{VI-7}$$

Substituting the equation for $\mathbf{W}_\lambda$, Eq. (VI-5), into the expression for the refracting surface, Eq. (VI-6), results in an equation for λ,

$$\phi(\mathbf{W} - t\mathbf{A} + \lambda\mathbf{N}) = 0, \tag{VI-8}$$

whose solution, let us call it $\bar{\lambda}$, represents the distance from the ray's starting point to where it intersects the refracting surface. This point is then given by

$$\bar{\mathbf{R}} = \mathbf{W} - t\mathbf{A} + \bar{\lambda}\mathbf{N}; \tag{VI-9}$$

substituting this into Eq. (VI-7) gives us the normal to the refracting surface at the point of incidence.

It is interesting to note that the derivative of Eq. (VI-8) with respect to λ, calculated at $\bar{\lambda}$, gives

$$\left.\frac{d\phi}{d\lambda}\right|_{\lambda=\bar{\lambda}} = \nabla\phi \cdot \mathbf{N} = [(\nabla\phi)^2]^{1/2} \cos i, \tag{VI-10}$$

where i is the angle of incidence.

It is unfortunate that the equation for $\bar{\lambda}$, Eq. (VI-8), cannot be solved explicitly. We will have to consider a battery of special cases. The easiest of these is the spherical surface.

THE SPHERE

The equation of a sphere passing through the origin with its center on the z axis is

$$x^2 + y^2 + (z - r)^2 = r^2,$$

where r is its radius. This simplifies quickly to

$$x^2 + y^2 + z^2 - 2rz = 0.$$

For our purposes, it is more convenient to use the curvature of the sphere, $c = 1/r$, rather than the radius. In this case, the plane refracting surface can be treated as a special case of the sphere by setting $c = 0$, which is considerably less awkward than setting $r = \infty$. We will also adopt the convention that, if $c > 0$, the center of curvature is to the right of the surface in the general direction of ray propagation.

The sphere's equation is then

$$\phi = c(x^2 + y^2 + z^2) - 2z = 0$$

or, using the more compact vector notation,

$$\phi = c\bar{\mathbf{R}}^2 - 2(\mathbf{A} \cdot \bar{\mathbf{R}}) = 0, \tag{VI-11}$$

where $\mathbf{A}$ is the unit vector in the direction of the z axis. From Eq. (VI-7), the unit normal to the sphere at the point of intersection $\bar{\mathbf{R}}$ is given by

$$\bar{\mathbf{N}} = -c\bar{\mathbf{R}} + \mathbf{A}. \tag{VI-12}$$

Note that we have changed the sign here. This is so the direction of the surface normal will be in the general direction of ray propagation and at the same time conform to the convention concerning the location of the center of curvature.

Substituting Eq. (VI-9) into Eq. (VI-11) provides the equation for $\bar{\lambda}$,

$$c(\mathbf{W} - t\mathbf{A} + \lambda\mathbf{N})^2 - 2\mathbf{A} \cdot (\mathbf{W} - t\mathbf{A} + \lambda\mathbf{N}) = 0,$$

which reduces to

$$c\lambda^2 - 2[\mathbf{A} \cdot \mathbf{N}(1 + ct) - c\mathbf{N} \cdot \mathbf{W}]\lambda + c(\mathbf{W} - t\mathbf{A})^2 - 2\mathbf{A} \cdot (\mathbf{W} - t\mathbf{A}) = 0, \qquad \text{(VI-13)}$$

a quadratic. The solution is

$$\bar{\lambda} = (1/c)[\mathbf{A} \cdot \mathbf{N}(1 + ct) - c\mathbf{N} \cdot \mathbf{W} \pm \mathbf{R}], \qquad \text{(VI-14)}$$

where the discriminant is

$$\begin{aligned} R^2 &= [\mathbf{A} \cdot \mathbf{N}(1 + ct) - c\mathbf{N} \cdot \mathbf{W}]^2 - c^2(\mathbf{W} - t\mathbf{A})^2 + 2c\mathbf{A} \cdot (\mathbf{W} - t\mathbf{A}) \\ &= 1 - [\mathbf{A} \times \mathbf{N}(1 + ct) + c\mathbf{N} \times \mathbf{W}]^2. \qquad \text{(VI-15)} \end{aligned}$$

Substituting Eq. (VI-14) into Eq. (VI-9), we get the equation for the points of intersection

$$\begin{aligned} \bar{\mathbf{R}} &= \mathbf{W} - t\mathbf{A} + (1/c)[\mathbf{A} \cdot \mathbf{N}(1 + ct) - c\mathbf{N} \cdot \mathbf{W} \pm R]\mathbf{N} \\ &= \mathbf{N} \times [(\mathbf{W} - t\mathbf{A}) \times \mathbf{N}] + (1/c)(\mathbf{A} \cdot \mathbf{N})\mathbf{N} \pm (1/c)\, R\mathbf{N}. \end{aligned}$$

Thus, the two branches of the solution of Eq. (VI-13) correspond to the two points of intersection of a straight line with a sphere. Of these we are interested in only one, that nearest the sphere's vertex at the coordinate origin. If we choose the ray's starting point at its origin, $\mathbf{W} = (0, 0, 0)$, and its direction along the z axis, so that $\mathbf{N} = \mathbf{A}$, then Eqs. (VI-14) and (VI-15) become

$$\bar{\lambda} = (1/c)[(1 + ct) \pm R], \qquad \text{where} \quad R^2 = 1,$$

and the expression for the point of intersection becomes

$$\bar{\mathbf{R}} = (1/c)(1 \pm 1)\mathbf{A}.$$

Clearly, for this point to represent the vertex of the refracting surface we must choose the negative branch of the solution. This then gives us the points of intersection on the side of the sphere nearest the vertex. The equation for $\bar{\lambda}$ then becomes

$$\bar{\lambda} = (1/c)[(\mathbf{A} \cdot \mathbf{N})(1 + ct) - c(\mathbf{N} \cdot \mathbf{W}) - R]. \qquad \text{(VI-16)}$$

Finally, note that the angle of incidence can be calculated directly from R:

$$\cos i = \mathbf{N} \cdot \bar{\mathbf{N}} = R. \tag{VI-17}$$

These equations are algebraically correct; they have been derived with the usual mathematical rigor and vigor. However, when $c = 0$, the case when the refracting surface is a plane, the expression for $\bar{\lambda}$, Eq. (VI-16), becomes indeterminate, causing no end of anguish and embarrassment to the poor soul who must program these equations for a computer. The solution is extremely simple; we need only multiply the numerator and denominator of Eq. (VI-16) by

$$U = (\mathbf{A} \cdot \mathbf{N})(1 + ct) - c(\mathbf{N} \cdot \mathbf{W}) + R \tag{VI-18}$$

to obtain

$$\bar{\lambda} = (1/U)[c(\mathbf{W} - t\mathbf{A})^2 - 2\mathbf{A} \cdot (\mathbf{W} - t\mathbf{A})]. \tag{VI-19}$$

Collecting together Eqs. (VI-15), (VI-18), (VI-19), and (VI-12), we have the transfer equations for spherical surfaces:

$$R^2 = 1 - [(\mathbf{A} \times \mathbf{N})(1 + ct) + c(\mathbf{N} \times \mathbf{W})]^2, \tag{VI-20a}$$

$$U = (\mathbf{A} \cdot \mathbf{N})(1 + ct) - c(\mathbf{N} \cdot \mathbf{W}) + R, \tag{VI-20b}$$

$$\bar{\lambda} = (1/U)[c(\mathbf{W} - t\mathbf{A})^2 - 2\mathbf{A} \cdot (\mathbf{W} - t\mathbf{A})], \tag{VI-20c}$$

$$\bar{\mathbf{R}} = \mathbf{W} - t\mathbf{A} + \bar{\lambda}\mathbf{N}, \tag{VI-20d}$$

$$\bar{\mathbf{N}} = -c\bar{\mathbf{R}} + \mathbf{A}. \tag{VI-20e}$$

CONIC SURFACES

The term *conic surface* as used here, in spite of what the words imply, does not mean the surface of a cone. A better term might be *conic section of revolution,* meaning surfaces generated by rotating a conic section about an axis. Thus a parabola, a hyperbola, or an ellipse would generate a paraboloid, a hyperboloid, or an ellipsoid of revolution. It is such surfaces that we shall call *conic surfaces*. The choice of words is a little unfortunate, but the term is short and moreover has become accepted in the optical community so we shall stick with it. Brueggeman (1968) is an excellent source of practical information on the application of these surfaces to the design of reflecting systems.

It is a simple matter to consider the paraboloid, the hyperboloid, and the ellipsoid as a single class of quadratic surfaces just as all conic sections can be treated as a single class of curves described by one equation,

$$x^2 = 4pz - (1 - \epsilon^2)\, z^2. \qquad \text{(VI-21)}$$

(See Korn and Korn, 1968, pp. 41–52.) Such a curve is symmetric with respect to the z axis and passes through the coordinate origin, where it is tangent to the x axis. Here ϵ, which is always taken to be a non-negative quantity, stands for the *eccentricity* of the curve and $|4p|$ is the length of the *latus rectum*. For a general conic section, *eccentricity* is defined as the constant ratio of the distance from a point on the curve to a fixed line (the *directrix*) and the distance from that point to a fixed point (the *focus*). The *latus rectum* is a chord through the focus and perpendicular to the curve's axis of symmetry, in this case the z axis. These terms are not particularly useful to us in this context.

What follows *is* useful. The general conic section can be thought of as having two foci, the z coordinates of which are given by

$$z_1 = 2p/(1 + \epsilon), \qquad z_2 = 2p/(1 - \epsilon). \qquad \text{(VI-22)}$$

The center of symmetry is given by

$$z_c = 2p/(1 - \epsilon^2),$$

the mean of the coordinates of the two foci. The curve crosses the z axis at

$$z_0 = 4p/(1 - \epsilon^2)$$

as well as at $z = 0$. If $p > 0$, the curve, at least that portion of it in the neighborhood of the coordinate origin, lies on the positive side of the z axis.

When $\epsilon = 0$, the curve becomes a circle; the two foci and the center of symmetry all coincide at the circle's center, and the radius equals $|2p|$.

When $0 < \epsilon < 1$, the curve is an ellipse. Both foci lie on the positive side of the z axis if $p > 0$.

As ϵ approaches unity, the center of symmetry and one of the foci move farther and farther out on the positive side of the z axis until, in the limit, they become infinite. The curve then becomes a parabola.

Finally, when $\epsilon > 1$, the focus and center of symmetry that became infinite in the case of the parabola now reappear on the negative side of the z axis. A second sheet of the curve appears intersecting the negative side of the z axis at $z_0 = 4p/(1 - \epsilon^2)$. The curve now has the form of the hyperbola. These curves are shown in Fig. VI-1.

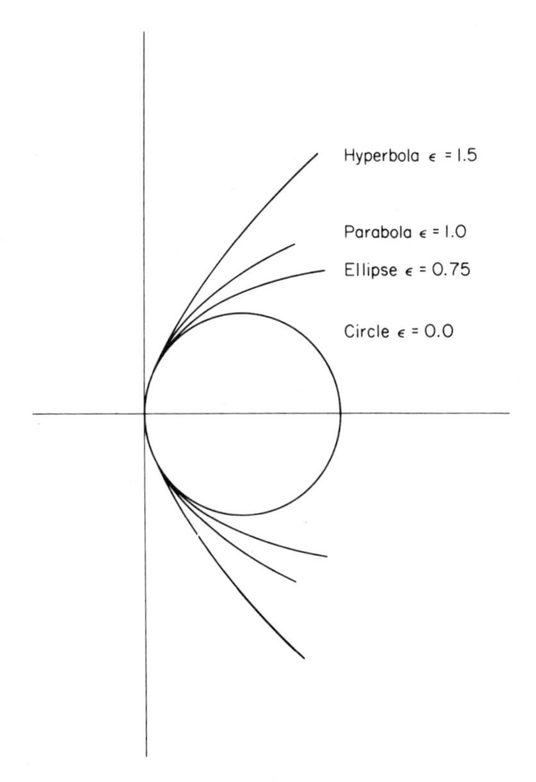

FIG. VI-1. The family of conic sections.
The eccentricity is ϵ.

Throughout this discussion we have tacitly assumed that $p > 0$. All of these situations are reversed when p is negative.

In dealing with the sphere, we saw that it was advantageous to specify curvature rather than radius. The same is true for the general conic. Using a well-known formula for the curvature of a plane curve, we find that

$$c = -4p^2/(4p^2 + \epsilon^2 y^2)^{3/2}.$$

At the origin the vertex curvature is therefore $c = -1/2p$. At this point we adopt the convention that the curvature will be positive if the center of curvature is on the positive side of the z axis. Then $c = +1/2p$. Introducing this into the equation for the general conic section, Eq. (VI-21), results in

$$cx^2 = 2z - c(1 + \epsilon^2)\, z^2. \tag{VI-23}$$

It is sometimes convenient to specify a conic section in terms of its foci rather than in terms of eccentricity and the length of the latus rectum. It is trivial task to obtain ϵ and p in terms of z_1 and z_2 from Eq. (VI-22),

$$\epsilon = \left|\frac{z_1 - z_2}{z_1 + z_2}\right|, \qquad \frac{1}{c} = \frac{2z_1 z_2}{z_1 + z_2}. \tag{VI-24}$$

When these are substituted into the equation for the general conic, Eq. (VI-21), we obtain

$$(z_1 + z_2)^2 x^2 = 4z_1 z_2[(z_1 + z_2)z - z^2]. \tag{VI-25}$$

To convert these equations for conic sections to equations for conic surfaces, we need only replace x^2 by $(x^2 + y^2)$. This results in

$$\phi = c(1 - \epsilon^2)\, z^2 + c(x^2 + y^2) - 2z = 0 \tag{VI-26}$$

or

$$\phi = 4z_1 z_2 z^2 + (z_1 + z_2)^2(x^2 + y^2) - 4z_1 z_2(z_1 + z_2)z = 0. \tag{VI-27}$$

It is convenient to write these in vector form:

$$\phi = c\bar{\mathbf{R}}^2 - c\epsilon^2(\bar{\mathbf{R}} \cdot \mathbf{A})^2 - 2(\bar{\mathbf{R}} \cdot \mathbf{A}) = 0, \tag{VI-28}$$

$$\phi = (z_1 + z_2)^2\bar{\mathbf{R}}^2 - (z_1 - z_2)^2(\bar{\mathbf{R}} \cdot \mathbf{A})^2 - 4z_1 z_2(z_1 + z_2)(\bar{\mathbf{R}} \cdot \mathbf{A}) = 0, \tag{VI-29}$$

where $\mathbf{A}$, as before, is the unit vector in the direction of the z axis and $\bar{\mathbf{R}}$ is the vector designating a point on the conic surface.

Another way to express these surfaces is as follows. Considering Eq. (VI-26) as a quadratic in z, its solution is

$$z = \frac{1}{c(1 - \epsilon^2)}\{1 - [1 - c^2(1 - \epsilon^2)(x^2 + y^2)]^{1/2}\},$$

where, as in the case of the sphere, we use only the negative branch of the solution of the quadratic. Also, as in the case of the sphere, we arrange this formula so that when $c = 0$ (when the surface degenerates into a plane), the expression does not become indeterminate:

$$z = \frac{c(x^2 + y^2)}{1 + [1 - c^2(1 - \epsilon^2)(x^2 + y^2)]^{1/2}}. \tag{VI-30}$$

Finally, we need to calculate the unit surface normal. From Eq. (VI-28),

$$\nabla\phi = 2[c\bar{\mathbf{R}} - c\epsilon^2(\mathbf{A} \cdot \bar{\mathbf{R}})\mathbf{A} - \mathbf{A}],$$

$$(\nabla\phi)^2 = 4[1 + c^2\epsilon^2(\bar{\mathbf{R}} \cdot \mathbf{A})^2].$$

Again we choose the direction of the normal to be in the general direction of ray propagation so that

$$\bar{\mathbf{N}} = \frac{-c\bar{\mathbf{R}} + c\epsilon^2(\mathbf{A} \cdot \bar{\mathbf{R}})\mathbf{A} + \mathbf{A}}{[1 + c^2\epsilon^2(\bar{\mathbf{R}} \cdot \mathbf{A})^2]^{1/2}}. \qquad \text{(VI-31)}$$

Note that, when $\epsilon = 0$, the conic surface becomes a sphere and Eq. (VI-31) reduces to Eq. (VI-12).

TRANSFER FOR CONIC SURFACES

To obtain the equations for transfer for conic surfaces we use exactly the same procedures that applied to the spherical refracting surface. Substituting the expression for $\bar{\mathbf{R}}$ in Eq. (VI-9) into the equation for the conic surface yields

$$\begin{aligned} &c[1 - \epsilon^2(\mathbf{N} \cdot \mathbf{A})^2]\,\lambda^2 \\ &\quad - 2[(\mathbf{A} \cdot \mathbf{N})(1 + ct) - c(\mathbf{N} \cdot \mathbf{W}) + c\epsilon^2\mathbf{A} \cdot (\mathbf{W} - t\mathbf{A})(\mathbf{A} \cdot \mathbf{N})]\lambda \\ &\quad + c(\mathbf{W} - t\mathbf{A})^2 - 2\mathbf{A} \cdot (\mathbf{W} - t\mathbf{A}) - c\epsilon^2[\mathbf{A} \cdot (\mathbf{W} - t\mathbf{A})]^2 = 0, \qquad \text{(VI-32)} \end{aligned}$$

another quadratic in λ. Again note that if $\epsilon = 0$ this reduces to Eq. (VI-13), the corresponding equation for the sphere. The solution is

$$\bar{\lambda} = \frac{\mathbf{A} \cdot \mathbf{N}(1 + ct) - c\mathbf{N} \cdot \mathbf{W} + c\epsilon^2\mathbf{A} \cdot (\mathbf{W} - t\mathbf{A})(\mathbf{A} \cdot \mathbf{N}) - R}{c[1 - \epsilon^2(\mathbf{N} \cdot \mathbf{A})^2]}, \qquad \text{(VI-33)}$$

where

$$\begin{aligned} R^2 = {} & [1 - \epsilon^2(\mathbf{N} \cdot \mathbf{A})^2]\{1 - [(\mathbf{A} \times \mathbf{N})(1 + ct) + c(\mathbf{N} \times \mathbf{W})]^2\} \\ & + \epsilon^2\{1 - (\mathbf{A} \times \mathbf{N}) \cdot [(\mathbf{A} \times \mathbf{N})(1 + ct) + c(\mathbf{N} \times \mathbf{W})]\}^2. \qquad \text{(VI-34)} \end{aligned}$$

Once again note that Eqs. (VI-34) and (VI-35) reduce to Eqs. (VI-14) and (VI-15), respectively, when $\epsilon = 0$. As in the case of the sphere we have taken the negative branch of the solution of the quadratic, Eq. (VI-32). We also rearrange Eq. (VI-34) so that the expression is not indeterminate when $c = 0$. In this way we can, as in the case of the sphere, include the plane refracting surface as a special case of the conic surface. This leads to

$$\begin{aligned} V &= (\mathbf{A} \cdot \mathbf{N})(1 + ct) - c(\mathbf{N} \cdot \mathbf{W}) + R + c\epsilon^2\mathbf{A} \cdot (\mathbf{W} - t\mathbf{A})(\mathbf{A} \cdot \mathbf{N}), \\ \bar{\lambda} &= (1/V)\{c(\mathbf{W} - t\mathbf{A})^2 - 2\mathbf{A} \cdot (\mathbf{W} - t\mathbf{A}) - c\epsilon^2[\mathbf{A} \cdot (\mathbf{W} - t\mathbf{A})]^2\}. \end{aligned}$$

Summing up, we repeat the equations for transfer for a conic refracting surface:

$$R^2 = [1 - \epsilon^2(\mathbf{N}\cdot\mathbf{A})^2]\{1 - [(\mathbf{A}\times\mathbf{N})(1+ct) + c(\mathbf{N}\times\mathbf{W})]^2\} + \epsilon^2\{1 - (\mathbf{A}\times\mathbf{N})\cdot[(\mathbf{A}\times\mathbf{N})(1+ct) + c(\mathbf{N}\times\mathbf{W})]\}^2, \tag{VI-35a}$$

$$V = (\mathbf{A}\cdot\mathbf{N})(1+ct) - c(\mathbf{N}\cdot\mathbf{W}) + R + c\epsilon^2\mathbf{A}\cdot(\mathbf{W} - t\mathbf{A})(\mathbf{A}\cdot\mathbf{N}), \tag{VI-35b}$$

$$\bar{\lambda} = (1/V)\{c(\mathbf{W} - t\mathbf{A})^2 - 2\mathbf{A}\cdot(\mathbf{W} - t\mathbf{A}) - c\epsilon^2[\mathbf{A}\cdot(\mathbf{W} - t\mathbf{A})]^2\}, \tag{VI-35c}$$

$$\bar{\mathbf{R}} = \mathbf{W} - t\mathbf{A} + \bar{\lambda}\mathbf{N}, \tag{VI-35d}$$

$$\bar{\mathbf{N}} = \frac{-c\bar{\mathbf{R}} + \mathbf{A} + c\epsilon^2(\mathbf{A}\cdot\bar{\mathbf{R}})\mathbf{A}}{[1 - c^2\epsilon^2(\bar{\mathbf{R}}\cdot\mathbf{A})^2]^{1/2}}. \tag{VI-35e}$$

Conic surfaces are particularly important in mirror optics. As we shall see later, the salient property of two conjugate points—two points that are perfect images of one another—is that the optical path length of all rays connecting them is the same. Now recall the geometrical definition of the ellipse: the locus of all points such that the sum of the distances from each point to two fixed points (the foci) is constant. An optical interpretation of this definition is that the optical path length between the two foci along any ray path that includes a point on the ellipse is constant. The two foci must therefore be optically conjugate points. A point source located at one focus must be imaged perfectly at the other.

In the case of the parabola, one focus has become infinite. We can interpret this in terms of geometrical optics by saying that an aggregate of rays, parallel to each other and to the axis of a paraboloid (or emanating from an infinite object point), after reflection by the paraboloid, will pass through its focus. This fact is the basis of the design of the Newtonian telescope, in which a paraboloid is used to collect and focus light from distant objects. The only extraneous feature is a small plane mirror that reflects the focus to one side so that the observer's head will not obstruct the incoming rays.

The operation of the hyperboloid is essentially the same except that the second focus is on the wrong side of the mirror. An aggregate of rays converging to the second focus will be interpreted by the hyperbolic mirror before reaching the second focus and will be reflected to the first focus. The hyperboloid and the paraboloid are used together in the design of the Cassegrainian telescope, in which the parallel rays from an infinite object point are collected by the paraboloid and brought to a focus (see Fig. VI-2). This focus is made to coincide with the second focus of a hyperboloid, which in turn reflects the rays to the hyperboloid's first focus. A small hole is cut through the paraboloid since the hyperboloid's first focus is a considerable distance from its surface.

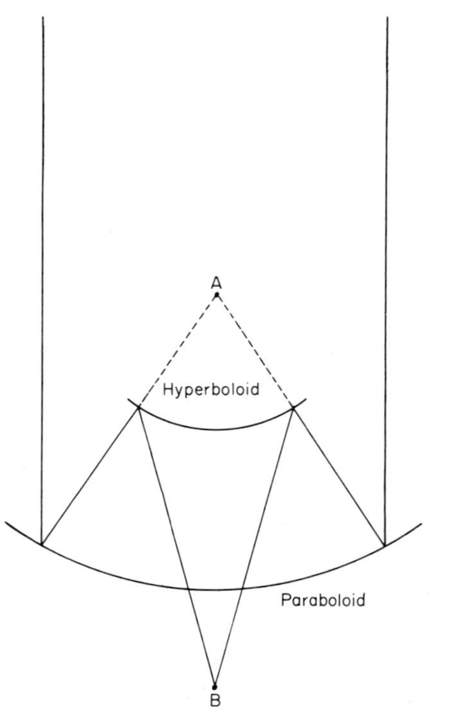

FIG. VI-2. The Cassegrainian configuration. Point A is where the focus of the paraboloid and the virtual focus of the hyperboloid coincide. Point B is the real focus of the hyperboloid.

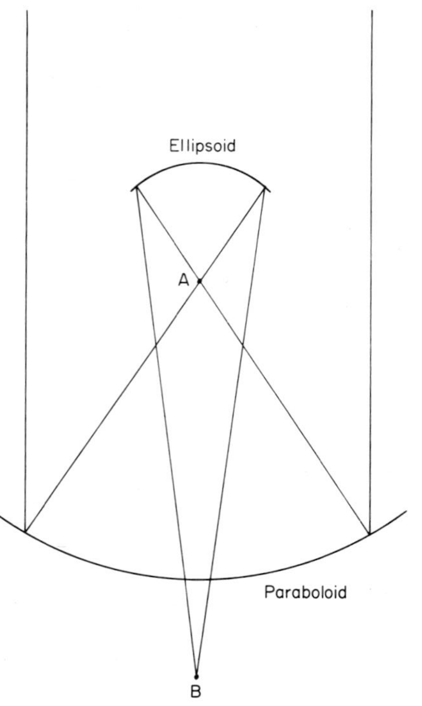

FIG. VI-3. The Gregorian configuration. Point A is where the focus of the paraboloid and one of the foci of the ellipsoid coincide. Point B is the other focus of the ellipsoid.

If the hyperboloid is replaced by an ellipsoid, we obtain the configuration of the Gregorian telescope (see Fig. VI-3). In this case the focus of the paraboloid is made to coincide with one of the foci of the ellipsoid. The rays from the paraboloid then pass through that focus before reaching the ellipsoid, where they are then reflected through the ellipsoid's other focus.

Thus, conic surfaces have remarkably useful optical properties when reflecting systems are being considered. They have refracting counterparts that are far less useful owing to the difficulty of their manufacture.

THE CYLINDRICAL LENS

The commonest anamorphotic lens element is the cylinder, which we will analyze next. As before, the axis of the optical system is taken to be the z axis of the coordinate system, and the unit vector in this direction is **A**. Let the direction of the axis of the cylinder by the unit vector **B**. The origin of the coordinate axes is at the intersection of the z axis with the surface of the cylinder. The x axis is parallel to the axis of the cylinder **B**.

In cylindrical coordinates, the equation of the cylinder relative to the described coordinate system is

$$\bar{\mathbf{R}} = (x, r \sin t, r(1 - \cos t))$$

and $\mathbf{A} = (0, 0, 1)$ and $\mathbf{B} = (1, 0, 0)$. It is a simple matter to obtain the vector equation

$$\phi(\bar{\mathbf{R}}) = c\bar{\mathbf{R}}^2 - c(\bar{\mathbf{R}} \cdot \mathbf{B})^2 - 2(\bar{\mathbf{R}} \cdot \mathbf{A}) = 0 \tag{VI-36}$$

in the form of Eq. (VI-6) and analogous to Eqs. (VI-11) and (VI-21) for the sphere and the conic surface, respectively. From this we obtain the gradient

$$\nabla\phi = 2[c\bar{\mathbf{R}} - c(\bar{\mathbf{R}} \cdot \mathbf{B})\mathbf{B} - \mathbf{A}],$$

which, with Eq. (VI-36), leads to the unit normal vector to the surface

$$\bar{\mathbf{N}} = -c\bar{\mathbf{R}} + c(\bar{\mathbf{R}} \cdot \mathbf{B})\mathbf{B} + \mathbf{A}. \tag{VI-37}$$

Once again we have changed signs to conform to the optical convention.

To obtain the transfer equations, we proceed in the same way as in the preceding sections. By substituting Eq. (VI-9) into Eq. (VI-36), we obtain the quadratic in $\bar{\lambda}$,

$$c\bar{\lambda}^2[1-(\mathbf{N}\cdot\mathbf{B})^2]+2\bar{\lambda}[c(\mathbf{W}-t\mathbf{A})\cdot\mathbf{N}-c(\mathbf{W}\cdot\mathbf{B})(\mathbf{N}\cdot\mathbf{B})-(\mathbf{N}\cdot\mathbf{A})] \\ +[c(\mathbf{W}-t\mathbf{A})^2-c(\mathbf{W}\cdot\mathbf{B})^2-2(\mathbf{W}\cdot\mathbf{A})+2t]=0$$

or

$$c\bar{\lambda}^2(\mathbf{N}\times\mathbf{B})^2+2\bar{\lambda}[c(\mathbf{W}\times\mathbf{B})\cdot(\mathbf{N}\times\mathbf{B})-(1+ct)(\mathbf{N}\cdot\mathbf{A})] \\ +[c(\mathbf{W}\times\mathbf{B})^2-2(1+ct)(\mathbf{W}\cdot\mathbf{A})+t(ct+2)]=0,$$

the discriminant of which is

$$R^2=-c^2[(\mathbf{W}\times\mathbf{B})\times(\mathbf{N}\times\mathbf{B})]^2 \\ +2c(1+ct)[\mathbf{A}\times(\mathbf{N}\times\mathbf{W})]\cdot[\mathbf{B}\times(\mathbf{N}\times\mathbf{B})] \\ +(1+ct)^2(\mathbf{N}\cdot\mathbf{A})^2-ct(ct+2)(\mathbf{N}\times\mathbf{B})^2$$

so that

$$\bar{\lambda}=\frac{1}{c(\mathbf{N}\times\mathbf{B})^2}[-c(\mathbf{W}\times\mathbf{B})\cdot(\mathbf{N}\times\mathbf{B})+(1+ct)(\mathbf{N}\cdot\mathbf{A})-R]$$

or

$$\bar{\lambda}=\frac{c(\mathbf{W}\times\mathbf{B})^2-2(1+ct)(\mathbf{W}\cdot\mathbf{A})+c(ct+2)}{(1+ct)(\mathbf{N}\cdot\mathbf{A})-c(\mathbf{W}\times\mathbf{B})\cdot(\mathbf{N}\times\mathbf{B})+R}.$$

Thus the equations for transfer for a cylindrical surface are

$$R^2=-c^2[(\mathbf{W}\times\mathbf{B})\times(\mathbf{N}\times\mathbf{B})]^2+2c(1+ct)[\mathbf{A}\times(\mathbf{N}\times\mathbf{W})] \\ \cdot[\mathbf{B}\times(\mathbf{N}\times\mathbf{B})]+(1+ct)^2(\mathbf{N}\cdot\mathbf{A})^2-ct(ct+2)(\mathbf{N}\times\mathbf{B})^2, \quad \text{(VI-38a)}$$

$$V=(1+ct)(\mathbf{N}\cdot\mathbf{A})-c(\mathbf{W}\times\mathbf{B})\cdot(\mathbf{N}\times\mathbf{B})+R, \quad \text{(VI-38b)}$$

$$\bar{\lambda}=(1/V)[c(\mathbf{W}\times\mathbf{B})^2-2(1+ct)(\mathbf{W}\cdot\mathbf{A})+c(ct+2)], \quad \text{(VI-38c)}$$

$$\bar{\mathbf{R}}=\mathbf{W}-t\mathbf{A}+\bar{\lambda}\mathbf{N}, \quad \text{(VI-38d)}$$

$$\bar{\mathbf{N}}=-c\bar{\mathbf{R}}+c(\bar{\mathbf{R}}\cdot\mathbf{B})\mathbf{B}+\mathbf{A}. \quad \text{(VI-38e)}$$

A more general and perhaps more useful anamorphotic surface is the torus, of which the cylinder is a special case. The toric surface is found wherever anamorphotic optical systems abound, from spectacle lenses to wide-screen cameras and projectors. Their treatment goes along the same lines as the cases already studied. They are quartic surfaces, and therefore to find the transfer equation a quartic polynomial must be solved. An algebraic solution seems hardly worth the effort especially since numerical solutions are so easily available from high-speed computers.

CARTESIAN OVALS

We have seen that, when reflection is considered, conic surfaces are capable of forming perfect images of certain isolated points. We have also seen that reflection can be considered as a special case of refraction. Now we complete the syllogism. There exists a class of refracting surfaces that are capable of forming perfect images of isolated points that reduce to conic surfaces in the special case of reflection. (See Luneburg, 1964, Chapter 3; Herzberger, 1958, Chapter 17.)

These surfaces are called Cartesian ovals after their discoverer René Descartes. Because they are virtually impossible to manufacture, they have not found any useful application and therefore are not well known. About every ten years or so someone rediscovers them and then goes to the trouble of writing a journal article or reading a paper at a meeting or even preparing a patent application, thus affording those of us who are a little more knowledgeable an opportunity to show off our cleverness and erudition.

We seek a surface separating two media of refractive index n and n^* such that every ray from a fixed object point $\mathbf{P}$ refracted by the surface passes through a single fixed image point $\mathbf{P}^*$. We represent the refracting surface by some vector function $\bar{\mathbf{P}}$. It is obvious that the refracting surface $\bar{\mathbf{P}}$ is rotationally symmetric, the axis of symmetry, being the straight line joining $\mathbf{P}$ and $\mathbf{P}^*$. Let $\mathbf{A}$ be a unit vector along this axis. Choose the origin of the coordinate system at the point where the axis of symmetry intersects the refracting surface $\bar{\mathbf{P}}$. If z and z^* are the distances of the points $\mathbf{P}$ and $\mathbf{P}^*$ from the origin, then

$$\mathbf{P} = -z\mathbf{A}, \qquad \mathbf{P}^* = z^*\mathbf{A}.$$

This is shown in Fig. (VI-4).

If $\bar{\mathbf{P}}$ is any point on the refracting surface, then $(\bar{\mathbf{P}} - \mathbf{P})$ represents a vector from $\mathbf{P}$ to $\bar{\mathbf{P}}$ whose length is equal to

$$[(\bar{\mathbf{P}} - \mathbf{P})^2]^{1/2} = [(\bar{\mathbf{P}} + z\mathbf{A})^2]^{1/2}.$$

In like manner, $[(\bar{\mathbf{P}} - z\mathbf{A})^2]^{1/2}$ is the length of the vector from $\bar{\mathbf{P}}$ to the image point $\mathbf{P}^*$. It follows that the optical path length of a ray from $\mathbf{P}$ to $\mathbf{P}^*$ via $\bar{\mathbf{P}}$ is exactly

$$n[(\bar{\mathbf{P}} + z\mathbf{A})^2]^{1/2} + n^*[(\bar{\mathbf{P}} - z^*\mathbf{A})^2]^{1/2}.$$

If $\mathbf{P}$ and $\mathbf{P}^*$ are to be perfect images of one another, then the optical path length must be the same for all rays. The above quantity is therefore

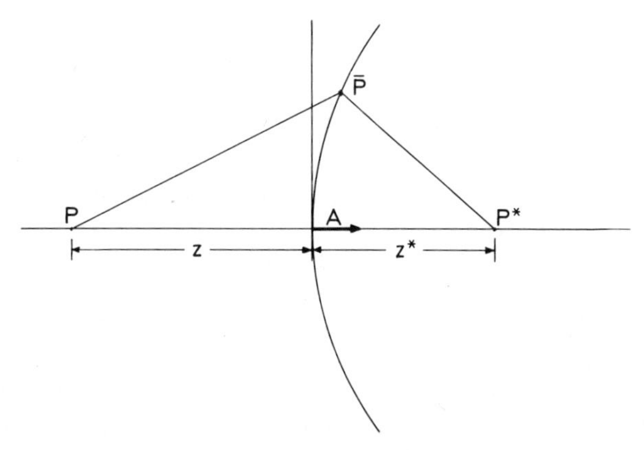

FIG. VI-4. The Cartesian oval construction.

Points **P** and **P*** represent the object and image points, respectively. **A** represents a unit vector in the direction of **P***–**P**. $\bar{\mathbf{P}}$ is an arbitrary point on the refracting surface.

$$\mathbf{P} = -z\mathbf{A} \qquad \text{and} \qquad \mathbf{P}^* = z^*\mathbf{A}.$$

a constant. Of all possible rays from **P** to **P*** through $\bar{\mathbf{P}}$, let us consider the one that lies along the axis **A**. For this ray, $\bar{\mathbf{P}} = \mathbf{0}$ and the above expression reduced to $nz + n^*z^*$. These two expressions must be equal. Thus

$$n[(\bar{\mathbf{P}} + z\mathbf{A})^2]^{1/2} + n^*[(\bar{\mathbf{P}} - z^*\mathbf{A})^2]^{1/2} = nz + n^*z^* \tag{VI-39}$$

becomes the equation for the Cartesian oval.

To obtain a more useful expression we square both sides. This then reduces to

$$\begin{aligned}\bar{\mathbf{P}}^2(n^2 + n^{*2}) - 2(zn^2 - z^*n^{*2})(\bar{\mathbf{P}} \cdot \mathbf{A}) - 2nn^*zz^*\\ = -2nn^*[(\bar{\mathbf{P}} - z\mathbf{A})^2]^{1/2}[(\bar{\mathbf{P}} + z^*\mathbf{A})^2]^{1/2}.\end{aligned}$$

Again squaring and collecting terms, we obtain

$$\begin{aligned}&\bar{\mathbf{P}}^4(n^2 - n^{*2})^2 + 4\bar{\mathbf{P}}^2(\bar{\mathbf{P}} \cdot \mathbf{A})(n^2 - n^{*2})(n^2z + n^{*2}z^*)\\ &\quad + 4(\bar{\mathbf{P}} \cdot \mathbf{A})^2(n^2z + n^{*2}z^*)^2 - 4\bar{\mathbf{P}}^2nn^*(nz + n^*z^*)(n^*z + nz^*)\\ &\quad - 8(\bar{\mathbf{P}} \cdot \mathbf{A})\, nn^*zz^*(n - n^*)(nz + n^*z^*) = 0.\end{aligned}$$

This expression very considerately allows itself to be written in the more compact form,

$$\begin{aligned}&[\bar{\mathbf{P}}^2(n^2 - n^{*2}) + 2(\bar{\mathbf{P}} \cdot \mathbf{A})(n^2z + n^{*2}z^*)]^2\\ &\quad -4nn^*(nz + n^*z^*)[\bar{\mathbf{P}}^2(n^*z + nz^*) + 2(\bar{\mathbf{P}} \cdot \mathbf{A})\, zz^*(n - n^*)] = 0. \qquad \text{(VI-40)}\end{aligned}$$

Such a surface is shown in Fig. VI-5.

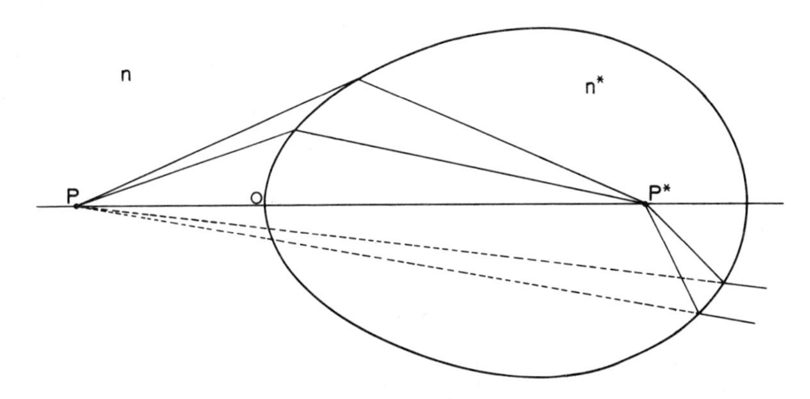

FIG. VI-5. The general Cartesian oval.

n is the refractive index outside the oval and n^*, in its interior. **P** and **P*** are the two conjugate points. Note that **P** is a real image of **P*** relative to that portion of the oval lying between **P** and **P***. With respect to the portion of the oval lying to the right of **P***, **P** is a virtual image of **P***.

First, we look at reflection as a special case of refraction. In this equation set $n^* = -n$. Further, let $\bar{\mathbf{P}} = (\bar{x}, \bar{y}, \bar{z})$ and let $\mathbf{A} = (0, 0, 1)$. Comparing the result with Eq. (VI-25), we see that, apart from the sign of z^*, the two expressions are identical. This proves that in the special case of reflection Cartesian ovals degenerate into conic sections.

Luneburg (1964, pp. 132–134) shows that, when the object point is at infinity, the Cartesian ovals degenerate in a decidedly different way to conic surfaces. In particular, when $n^* > n$, the surface is an ellipsoid and when $n^* < n$, it is a hyperboloid.

This can be seen by letting the object point approach infinity. Divide Eq. (VI-40) by z^2 and get

$$[(1/z)(n^2 - n^{*2})\,\bar{\mathbf{P}} + 2(\bar{\mathbf{P}}\cdot\mathbf{A})(n^2 + n^{*2}z^*/z)]^2$$
$$-4nn^*(n + n^*z^*/z)[\bar{\mathbf{P}}^2(n^* + nz^*/z) + 2(\bar{\mathbf{P}}\cdot\mathbf{A})\,z^*(n - n^*)] = 0.$$

When $z \to \infty$, terms of degree greater than *two* drop out leaving

$$n^2(\bar{\mathbf{P}}\cdot\mathbf{A})^2 - n^{*2}\bar{\mathbf{P}}^2 - 2n^*z^*(n - n^*)(\bar{\mathbf{P}}\cdot\mathbf{A}) = 0.$$

This is a quadratic and therefore is the equation for a conic surface. Translating the expression into a scalar notation, we have

$$n^2\bar{z}^2 - n^{*2}(\bar{x}^2 + \bar{y}^2 + \bar{z}^2) - 2n^*z^*(n - n^*)\bar{z} = 0,$$

which can be rewritten as

$$\left(\frac{n+n^*}{n^*z^*}\right)^2\left(\bar{z}-\frac{n^*z^*}{n+n^*}\right)^2+\frac{n+n^*}{(n^*-n)\,z^{*2}}(\bar{x}^2+\bar{y}^2)=1.$$

This is an ellipsoid when $n^* > n$ and a hyperboloid when $n^* < n$. The center is at

$$z_c = n^*z^*/(n+n^*)$$

and the curve crosses the z axis when $z = 0$ and $z = 2n^*z^*/(n+n^*)$. These cases are illustrated in Fig. VI-6.

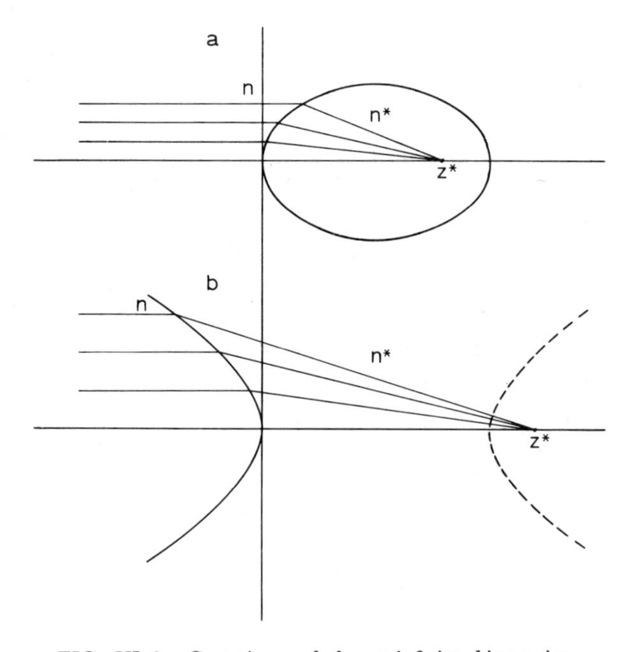

FIG. VI-6. Cartesian ovals for an infinite object point.

In case (a), $n^* > n$ and the oval is an ellipsoid. When $n^* < n$, shown in (b), the oval is a hyperboloid.

The general Cartesian surface degenerates in yet another way when the object and image distances are finite and are related to the refractive indices in a certain way. Refer to Eq. (VI-40). If the quantity in the first bracket is proportional to the quantity in the second bracket, then the equation can be made to factor. Let k be some arbitrary constant of proportionality and set

$$n^2 - n^{*2} = k(n^*z + nz^*),$$

$$n^2z + n^{*2}z^* = kzz^*(n - n^*).$$

Satisfying this pair of equations ensures that Eq. (VI-40) will factor. Solving it for z and z^*, two sets of solutions are obtained,

$$z = -n^*/k, \qquad z^* = n/k, \tag{VI-41a}$$

$$z = -(n + n^*)/k, \qquad z^* = (n + n^*)/k. \tag{VI-41b}$$

Substituting Eqs. (VI-41a) into Eq. (VI-40), we find that the second term vanishes and it therefore is a perfect square. The equation reduces to

$$k(n + n^*)\,\bar{\mathbf{P}}^2 - 2nn^*(\bar{\mathbf{P}} \cdot \mathbf{A}) = 0,$$

which in scalar form is the sphere,

$$\left(\bar{z} - \frac{nn^*}{k(n + n^*)}\right)^2 + \bar{x}^2 + \bar{y}^2 = \left(\frac{nn^*}{k(n + n^*)}\right)^2,$$

with center at $z_c = nn^*/k(n + n^*)$ and with radius $r = nn^*/k(n + n^*)$. The sphere, of course, passes through the origin from which the object point and image point are measured. Usually these are measured from the sphere's center. Let z_a and $z_a{}^*$ be these distances. Then

$$z_a = z + z_c = -n^{*2}/k(n + n^*),$$

$$z_a{}^* = z^* - z_c = n^2/k(n + n^*).$$

In terms of the sphere's radius, these quantities become the more familiar

$$z_a = -r(n^*/n), \qquad z_b = r(n/n^*). \tag{VI-42}$$

These points are called the *aplanatic points* of a sphere. The general cartesian surface degenerates into a sphere, and its object and image become these aplanatic points. However, the sphere is a surface of radial symmetry. The line on which lie the aplanatic points also passes through the sphere's center. It follows that on any line through the center there will lie a pair of aplanatic points. All such aplanatic points will constitute two conjugate *aplanatic surfaces* that actually are a pair of concentric spheres. Both of these surfaces are on the same side of the spherical refracting surface, so one must be regarded as virtual. This is illustrated in Fig. VI-7.

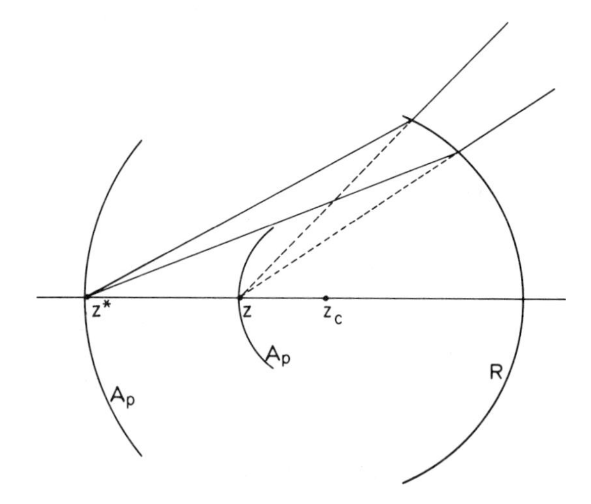

FIG. VI-7. The aplanatic surfaces of a sphere.

When the Cartesian oval degenerates into a sphere, the two conjugates, z^* and z, are called its *aplanatic points*. Point z_c is the sphere's center. The spheres marked A_p, also centered at z_c, are the *aplanatic surfaces*.

An interesting property of the aplanatic surfaces is that the optical path length between them is always zero. This can be seen from the fact that $nz + n^*z^*$ vanishes when Eqs. (VI-41a) are used.

The second case, corresponding to Eqs. (VI-41b), is disappointingly trivial. Note first that z and z^* coincide. When these values from Eqs. (VI-41b) are substituted into Eq. (VI-40), we find that the two factors are

$$k\bar{\mathbf{P}}^2 - 2(n + n^*)(\bar{\mathbf{P}} \cdot \mathbf{A}) \qquad \text{and} \qquad k^2\bar{\mathbf{P}}^2 - 2k(n + n^*)(\bar{\mathbf{P}} \cdot \mathbf{A}) + 4nn^*.$$

These give rise to two spheres

$$\left(\bar{z} - \frac{n + n^*}{k}\right)^2 + (\bar{x}^2 + \bar{y}^2) = \left(\frac{n + n^*}{k}\right)^2,$$

$$\left(\bar{z} - \frac{n + n^*}{k}\right)^2 + (\bar{x}^2 + \bar{y}^2) = \left(\frac{n - n^*}{k}\right)^2.$$

Although the radii are different, their centers coincide with the object and image point. This result is of course tautologous; the center of a refracting sphere is always imaged perfectly on itself.

REFERENCES

Brueggemann, H. F. (1968). "Conic Mirrors." Focal Press, London and New York.

Herzberger, M. (1958). "Modern Geometrical Optics." Wiley (Interscience), New York.

Korn, G. A., and Korn, T. M. (1968). "Mathematical Handbook for Scientists and Engineers," 2nd ed. McGraw-Hill, New York.

Luneburg, R. K. (1964). "Mathematical Theory of Optics." Univ. of California Press, Berkeley.

VII

Orthotomic Systems of Rays

In Chapter V we found that, if a two-parameter family of rays was orthogonal to a wavefront **W**, then from Eq. (V-21)

$$n\mathbf{P}' \cdot d\mathbf{W} = 0$$

and this expression could be considered as a differential equation for that wavefront. The left-hand member of this equation is the integrand of the Hilbert integral, which vanishes on a path connecting any pair of points on the wavefront. The vanishing of this line integral ensures that the above differential equation is exact, i.e., that there exists a function of position $\phi(\mathbf{P})$ such that

$$d\phi = n\mathbf{P}' \cdot d\mathbf{W}.$$

The equation for the wavefront then becomes $d\phi = 0$, the integral of which is simply $\phi = \text{const}$.

Such a collection of rays possessing an orthogonal wavefront is termed an *orthotomic system*. An alternative term for the same idea is *normal congruence*. We prefer the former.

The problem that this chapter is primarily concerned with is finding the conditions that ensure that rays from a given family form an orthotomic system. When does there exist a surface intersecting each member of the family orthogonally? Stated a little differently, exactly when does an arbitrary collection of rays possess a wavefront? (See Born and Wolf, 1970, pp. 126–127.)

Suppose we have an aggregate of rays, each represented by the vector function $\mathbf{P}(s)$, which in turn must satisfy the ray equation, Eq. (II-19):

$$(d/ds)(n\mathbf{P}') = \nabla n.$$

Under what circumstances does there exist a function ϕ such that

$$d\phi = n\mathbf{P}' \cdot d\mathbf{W}?$$

The answer is that $n\mathbf{P}' \cdot d\mathbf{W}$ must be an exact differential. Those conditions on the aggregate of rays that ensure that this expression is an exact differential are exactly the conditions that ensure that the rays possess an orthogonal wavefront. We are led to some rather simple existence theorems for total differential equations, which provide us with our requirements.

THE TOTAL DIFFERENTIAL EQUATION

A total differential equation in three independent variables is an expression in the form

$$P\,dx + Q\,dy + R\,dz = 0, \tag{VII-1}$$

where P, Q, and R are functions of x, y, and z. For a general discussion, see Bateman, 1966, Chapter VII; Forsyth, 1929, pp. 309–320; Rektorys, 1969, pp. 748–752. In general, a total differential equation need not depend on exactly three variables, but three variables are exactly right for our purposes. The key to differential equations of this type is the following: Does there exist a continuous function ϕ such that

$$d\phi = P\,dx + Q\,dy + R\,dz? \tag{VII-2}$$

If such a function exists, then the solution to the total differential equation is simply $\phi =$ const. If ϕ is any function whatsoever, its total differential is

$$d\phi = (\partial\phi/\partial x)\, dx + (\partial\phi/\partial y)\, dy + (\partial\phi/\partial z)\, dz;$$

moreover, if ϕ and its derivatives are continuous, then the second derivatives are independent of the order of differentiation. In other words,

$$\partial^2\phi/\partial x\, \partial y = \partial^2\phi/\partial y\, \partial x, \qquad \text{etc.}$$

Now if

$$d\phi = P\, dx + Q\, dy + R\, dz,$$

then the coefficients of corresponding differentials must be equal, and we obtain

$$\partial\phi/\partial x = P, \qquad \partial\phi/\partial y = Q, \qquad \partial\phi/\partial z = R.$$

Differentiating and equating equal second partial derivatives, we get

$$\partial P/\partial y = \partial Q/\partial x, \qquad \partial P/\partial z = \partial R/\partial x, \qquad \partial Q/\partial z = \partial R/\partial y. \tag{VII-3}$$

Therefore, if a function ϕ exists that satisfies Eq. (VII-2), then P, Q, and R must satisfy Eq. (VII-3). It turns out that the converse is also true: if P, Q, and R satisfy Eq. (VII-3), then there exists a ϕ such that Eq. (VII-2) is satisfied.

Let us say that any total differential equation (VII-1) whose coefficients satisfy Eq. (VII-2), is *exact*.

Now suppose that Eq. (VII-1) is not exact, that the coefficients P, Q, and R do not satisfy Eq. (VII-3). It may happen that there exists a function μ of x, y, and z that has the property that, when Eq. (VII-1) is multiplied by μ, it becomes exact. That is, Eq. (VII-1),

$$P\, dx + Q\, dy + R\, dz = 0,$$

is not exact but

$$(\mu P)\, dx + (\mu Q)\, dy + (\mu R)\, dz = 0 \tag{VII-4}$$

is, and therefore μP, μQ, and μR satisfy Eq. (VII-3). Substituting these into Eq. (VII-3) yields, after some rearranging,

$$\begin{aligned}
&\qquad\qquad\quad - R(\partial\mu/\partial y) + Q(\partial\mu/\partial z) &&= \mu[(\partial R/\partial y) - (\partial Q/\partial z)],\\
&R(\partial\mu/\partial x) \qquad\qquad\qquad - P(\partial\mu/\partial z) &&= -\mu[(\partial R/\partial x) - (\partial P/\partial z)],\\
&-Q(\partial\mu/\partial x) + P(\partial\mu/\partial y) &&= \mu[(\partial Q/\partial x) - (\partial P/\partial y)],
\end{aligned} \tag{VII-5}$$

a system of linear partial differential equations in μ. One would expect that one could solve Eq. (VII-5) algebraically for the three partial derivatives of μ. However, it is a simple matter to show that the determinant of coefficients of the left-hand members is zero. The three left-hand members are therefore linearly dependent. A solution does not exist unless the three right-hand members are also linearly dependent.

The linear dependence relationship of the three left-hand members is simple to demonstrate. Multiply the first of the three equations by P, the second by Q, and the third by R, then add; the sum is zero. The linear dependence relationship that must be satisfied by the three right-hand members must be the same and results in

$$P(\partial R/\partial y - \partial Q/\partial z) - Q(\partial R/\partial x - \partial P/\partial z) + R(\partial Q/\partial x - \partial P/\partial y) = 0. \quad \text{(VII-6)}$$

Should P, Q, and R satisfy Eq. (VII-6), then Eq. (VII-5), considered as an algebraic equation in the partial derivatives of μ, possesses a solution from which, in principle, μ can be determined. It turns out that Eq. (VII-6) is both a necessary and a sufficient condition for the existence of the integrating factor μ. When Eq. (VII-1) is multiplied by μ, it becomes exact and the existence of the solution ϕ is ensured.

The function μ, if it should exist, is called an *integrating factor*. We will say that a total differential equation whose coefficients satisfy Eq. (VII-6) is *integrable*. Note that these definitions of *exactness* and *integrability* do not conform to the usual terminology.

What we have shown here are *conditions* for the existence of solutions; the satisfaction of Eqs. (VII-3) guarantees that a ϕ exists that satisfies Eq. (VII-2); if Eq. (VII-6) is satisfied, then there is an integrating factor μ such that a solution ϕ exists that satisfies Eq. (VII-4). These conditions tell us that there is something there. Very little help is provided to assist us in finding it.

EXACTNESS AND INTEGRABILITY

In terms of the vector operator ∇, defined previously, the conditions for exactness and integrability can be expressed in a compact, easy-to-use notation. Let $\mathbf{V} = (\alpha, \beta, \gamma)$ be some vector function of x, y, and z. We define the *curl* of $\mathbf{V}$,

$$\nabla \times \mathbf{V} = \left(\frac{\partial}{\partial x}, \frac{\partial}{\partial y}, \frac{\partial}{\partial z}\right) \times (\alpha, \beta, \gamma) = \left(\frac{\partial \gamma}{\partial y} - \frac{\partial \beta}{\partial z}, -\frac{\partial \gamma}{\partial x} + \frac{\partial \alpha}{\partial z}, \frac{\partial \beta}{\partial x} - \frac{\partial \alpha}{\partial y}\right).$$

At the same time, although we will not be using it immediately, we define the *divergence* of $\mathbf{V}$,

$$\nabla \cdot \mathbf{V} = \left(\frac{\partial}{\partial x}, \frac{\partial}{\partial y}, \frac{\partial}{\partial z}\right) \cdot (\alpha, \beta, \gamma) = \frac{\partial \alpha}{\partial x} + \frac{\partial \beta}{\partial y} + \frac{\partial \gamma}{\partial z}.$$

These definitions of curl and divergence, which make use of the del operator, as well as the definition given for the gradient, can best be described as operational definitions. They may also be defined in terms of integrals from which the above can be derived. The integral definitions have the advantage of being stated in a form that is independent of the choice of coordinate system and therefore are sometimes called invariant definitions. The operational definitions are most suited for our purposes.

Now let the vector $\mathbf{V} = (P, Q, R)$, where P, Q, and R are the three functions appearing in Eq. (VII-1), and let $\mathbf{E} = (x, y, z)$ so that $d\mathbf{E} = (dx, dy, dz)$. Then the total differential equation, Eq. (VII-1), can be written in the form

$$\mathbf{V} \cdot d\mathbf{E} = 0. \tag{VII-7}$$

If a function ϕ exists such that

$$d\phi = \nabla\phi \cdot d\mathbf{E} = \mathbf{V} \cdot d\mathbf{E},$$

then, by equating corresponding coefficients of differentials, $\nabla\phi = \mathbf{V}$. Since the curl of a gradient is zero, the curl of the above exprcssion yields

$$\nabla \times \mathbf{V} = 0, \tag{VII-8}$$

the condition for exactness. Its expansion exactly the three equations of Eqs. (VII-3).

If μ is an integrating factor for Eq. (VII-7), then $\nabla \times (\mu\mathbf{V}) = \mathbf{0}$, which when expanded yields $\nabla\mu \times \mathbf{V} + \mu\nabla \times \mathbf{V} = \mathbf{0}$. Taking the scalar product of this with $\mathbf{V}$ yields

$$\mathbf{V} \cdot (\nabla \times \mathbf{V}) = 0, \tag{VII-9}$$

the condition for integrability, which, when expanded, is identical to Eq. (VII-6). It is in this vector form that we will apply these conditions to the problems of optics.

ORTHOTOMIC SYSTEMS

We are now in a position to consider the problem of the orthotomic system of rays. The differential equation from Eq. (V-21),

$$n\mathbf{P}' \cdot d\mathbf{W} = 0,$$

is a total differential equation, and the condition that it be exact is that, from Eq. (VII-8),

$$\nabla \times (n\mathbf{P}') = \mathbf{0}. \tag{VII-10}$$

If this is satisfied, then a solution exists and that solution put equal to a constant is an equation for the wavefront. Equation (VII-10) is therefore a condition that a system of rays, given by a family of solutions of the ray equation, Eq. (II-19),

$$(d/ds)(n\mathbf{P}') = \nabla n,$$

forms an orthotomic system.

The del operator is, of course, a field operator; the derivatives contained in it can be thought of as limits of difference quotients of adjacent rays. The operator d/ds, on the other hand, is derived from a difference quotient obtained from adjacent points on the same ray. It is clear that the two operators commute.

There is yet another route to Eq. (VII-10). Since, in the original total differential equation, n appears as a nonzero factor, we can drop it, leaving

$$\mathbf{P}' \cdot d\mathbf{W} = 0. \tag{VII-11}$$

If we apply the condition of integrability, Eq. (VII-9), to this form of the equation, we get

$$\mathbf{P}' \cdot (\nabla \times \mathbf{P}') = 0. \tag{VII-12}$$

Now $\nabla \times n\mathbf{P}' = \nabla n \times \mathbf{P}' + n\nabla \times \mathbf{P}'$, so multiplying the above equation by n yields

$$\begin{aligned} n\mathbf{P}' \cdot (\nabla \times \mathbf{P}') &= \mathbf{P}' \cdot (n\nabla \times \mathbf{P}') \\ &= \mathbf{P}' \cdot (\nabla \times n\mathbf{P}' - \nabla n \times \mathbf{P}') \\ &= \mathbf{P}' \cdot \nabla \times n\mathbf{P}' = 0. \end{aligned}$$

Differentiating this with respect to s gives

$$\mathbf{P}'' \cdot (\nabla \times n\mathbf{P}') + \mathbf{P}' \cdot [\nabla \times (d/ds)(n\mathbf{P}')] = 0.$$

But by the ray equation,

$$\nabla \times (d/ds)(n\mathbf{P}') = \nabla \times \nabla n = \mathbf{0}$$

since the curl of a gradient is always zero. Thus,

$$\mathbf{P}'' \cdot [\nabla \times (n\mathbf{P}')] = 0.$$

Differentiating once again and going through exactly the same steps result in

$$\mathbf{P}''' \cdot (\nabla \times n\mathbf{P}') = 0.$$

In Chapter III it was shown that $\mathbf{P}'''$ could be written as a linear combination of the normal vector, the tangent vector, and the binormal vector (see Eq. (III-6)), and therefore, for some α, β, and γ,

$$\mathbf{P}''' = \alpha\mathbf{P}' + \beta\mathbf{P}'' + \gamma(\mathbf{P}' \times \mathbf{P}'').$$

Substituting this in the last equation and making use of the preceding two, result in

$$(\mathbf{P}' \times \mathbf{P}'') \cdot (\nabla \times n\mathbf{P}') = 0.$$

Thus the scalar product of $\nabla \times n\mathbf{P}'$ with each of three orthogonal vectors vanishes. It follows that

$$\nabla \times n\mathbf{P}' = \mathbf{0}. \qquad \text{(VII-13)}$$

What we have shown is that the refractive index function n is an integrating factor for the total differential equation (VII-11).

SOME PROPERTIES

By expanding Eq. (VII-10), we get

$$\nabla n \times \mathbf{P}' + n\nabla \times \mathbf{P}' = \mathbf{0}.$$

By taking the vector product of the ray equation and $\mathbf{P}'$, we obtain

$$\nabla n \times \mathbf{P}' = n(\mathbf{P}'' \times \mathbf{P}'),$$

so that, for an orthotomic system of rays,

$$\nabla \times \mathbf{P}' = \mathbf{P}' \times \mathbf{P}''. \qquad \text{(VII-14)}$$

Since the divergence of a curl is zero, it follows from Eq. (VII-14) that

$$\nabla \cdot (\mathbf{P}' \times \mathbf{P}'') = 0. \qquad \text{(VII-15)}$$

Finally, differentiating Eq. (VII-14), we get

$$\nabla \times \mathbf{P}'' = (d/ds)(\mathbf{P}' \times \mathbf{P}''). \qquad \text{(VII-16)}$$

The expansion of Eq. (VII-16) leads to

$$\nabla \times \mathbf{P}'' = \mathbf{P}' \times \mathbf{P}'''.$$

Applying Eq. (III-6) (the formula for $\mathbf{P}'''$) to this results in

$$\begin{aligned}\nabla \times \mathbf{P}'' &= -(\rho'/\rho)\,\mathbf{P}' \times \mathbf{P}'' + (1/\tau)\,\mathbf{P}' \times (\mathbf{P}' \times \mathbf{P}'')\\ &= -(1/\tau)\,\mathbf{P}'' - (\rho'/\rho)\,\mathbf{P}' \times \mathbf{P}''.\end{aligned}$$

Comparing this with Eq. (III-11), we obtain

$$\nabla \times \mathbf{P}'' = -(1/\rho^2\tau)\mathbf{C}, \qquad \text{(VII-17)}$$

where $\mathbf{C}$ is the vector from a point on a ray to the center of its osculating sphere at that point. Now $\mathbf{P}''$ is the vector to the center of the circle of curvature. Thus Eq. (VII-17) states that the curl of the vector to the center of curvature is proportional to the vector to the center of the osculating sphere.

Now the derivative in the right-hand member of Eq. (VII-16) is with respect to the arc length parameter s. By introducing a time parammeter t we may write

$$d/ds = (dt/ds)(d/dt) = (1/v)(d/dt),$$

where v represents velocity. But velocity here refers to the velocity of light in the medium in which the light is traveling and can be expressed in terms of the refractive index and c, the velocity of light *in vacuo*. Thus, $v = c/n$. Equation (VII-16) thus becomes

$$\nabla \times \mathbf{P}'' = (n/c)(d/dt)(\mathbf{P}' \times \mathbf{P}''). \qquad \text{(VII-18)}$$

Now $\mathbf{P}''$ and $\mathbf{P}' \times \mathbf{P}''$ are vectors perpendicular to $\mathbf{P}'$, the direction of propagation of light. In the Maxwell equations, the vectors $\mathbf{E}$ and $\mathbf{H}$ are also perpendicular to the direction of propagation. Equations (VII-15) and (VII-18) therefore, in a certain sense, mimic two of the four Maxwell equations. We will defer further discussion of this situation to a later chapter.

THE ORTHOGONAL SURFACE

From Eq. (VII-10), which states that the curl of $n\mathbf{P}'$ is zero, there must exist a function ϕ such that

$$n\mathbf{P}' = \nabla\phi. \qquad \text{(VII-19)}$$

This function ϕ provides the solution to the original total differential equation of the wavefront

$$n\mathbf{P}' \cdot d\mathbf{W} = \nabla\phi \cdot d\mathbf{W} = 0,$$

so that $\phi =$ const. is our solution. We still have no clue as to what ϕ is. We have merely replaced one problem with another. By squaring Eq. (VII-19), we obtain the very important equation, which Luneburg calls the eikonal equation,

$$(\nabla\phi)^2 = n^2, \tag{VII-20}$$

which relates the orthogonal surface (the wavefront, if you will) to only the refractive index of the medium. The orthotomic system of rays does not appear in this equation, and therefore the solutions of Eq. (VII-20) are descriptive of all wavefronts in the medium in which the refractive index is given by n. Of course, boundary conditions that enable one to extract particular solutions are obtained from the rays themselves.

Taking the scalar product of Eq. (VII-19) with $\mathbf{P}'$ yields $\nabla\phi \cdot \mathbf{P}' = n$, which can be interpreted, by means of the directional derivative indicated in Eq. (III-16), as

$$d\phi/ds = n. \tag{VII-21}$$

The divergence of Eq. (VII-19) is

$$\nabla^2\phi = \nabla \cdot (n\mathbf{P}') = \nabla n \cdot \mathbf{P}' + n\nabla \cdot \mathbf{P}' = dn/ds + n\nabla \cdot \mathbf{P}'.$$

From Eq. (VII-21), this can be written as

$$\nabla^2\phi = d^2\phi/ds^2 + n\nabla \cdot \mathbf{P}', \tag{VII-22}$$

which bears an obvious resemblance to the wave equation. The divergence of $\mathbf{P}'$ is, in general, not equal to zero, a fact that will subsequently be shown. At any rate, Eq. (VII-22) provides us with a little further justification for calling the orthogonal surfaces wavefronts.

CANONICAL VARIABLES AND THE HAMILTON–JACOBI EQUATIONS

The best-known route to the notion of orthotomic systems of rays and to the eikonal equation, distinctly different from the approach taken so far, is along the path first blazed by Hamilton more than a century ago. The innovation he used involved the introduction of a pair of new variables, the so-called canonical variables, and the Hamiltonian function H. (For a full treatment, see Bliss, 1946, Chapter III; or Rund, 1966.) Our destination is the Hamilton–Jacobi differential equations. To reach this point we begin with first principles, the general variational problem

$$I = \int f(x, y, z, y', z')\, dx$$

and the associated Euler equations

$$(d/dx)(\partial f/\partial y') = \partial f/\partial y, \qquad (d/dx)(\partial f/\partial z') = \partial f/\partial z.$$

We define the canonical variables

$$u = \partial f/\partial y', \qquad v = \partial f/\partial z'. \tag{VII-23}$$

The partial derivatives are function of x, y, z, y', and z', and under certain conditions these two equations can be solved for y' and z'. Thus

$$y' = P(x, y, z, u, v), \qquad z' = Q(x, y, z, u, v). \tag{VII-24}$$

The condition that applies here comes from the theory of implicit functions and is

$$(\partial^2 f/\partial y'^2)(\partial^2 f/\partial z'^2) - (\partial^2 f/\partial y' \partial z')^2 \neq 0.$$

The Hamiltonian is next defined:

$$\begin{aligned} H(x, y, z, u, v) &= y'(\partial f/\partial y') + z'(\partial f/\partial z') - f \\ &= Pu + Qv - f. \end{aligned} \tag{VII-25}$$

Its total differential is

$$\begin{aligned} dH &= (\partial H/\partial x)\, dx + (\partial H/\partial y)\, dy + (\partial H/\partial z)\, dz + (\partial H/\partial u)\, du + (\partial H/\partial v)\, dv \\ &= P\, du + u\, dP + Q\, dv + v\, dQ \\ &\quad -(\partial f/\partial x)\, dx - (\partial f/\partial y)\, dy - (\partial f/\partial z)\, dz - (\partial f/\partial y')\, dP - (\partial f/\partial z')\, dQ, \end{aligned}$$

which, on application of Eq. (VII-23), reduces to

$$dH = P\, du + Q\, dv - (\partial f/\partial x)\, dx - (\partial f/\partial y)\, dy - (\partial f/\partial z)\, dz.$$

On comparing coefficients of corresponding differentials, we get

$$\partial H/\partial x = -\partial f/\partial x, \qquad \partial H/\partial y = -\partial f/\partial y, \qquad \partial H/\partial z = -\partial f/\partial z,$$

$$\partial H/\partial u = P, \qquad \partial H/\partial v = Q.$$

The first of these is identical to Eq. (II-10). Applying the Euler equations to the second, we obtain

$$\partial H/\partial y = -\partial f/\partial y = -(d/dx)(\partial f/\partial y') = -du/dx. \tag{VII-26a}$$

In exactly the same way we obtain from the third equation

$$\partial H/\partial z = -dv/dx. \tag{VII-26b}$$

Finally, from Eq. (VII-24), we may write the fourth and fifth equations as

$$\partial H/\partial u = dy/dx, \qquad \partial H/\partial v = dz/dx. \tag{VII-26c, d}$$

Equations (VII-26) are known as the canonical equations. Recall that the equation for a transversal surface from Eq. (VII-3) makes use of the definition of the Hamiltonian, Eq. (VII-25), and the definitions of the canonical variables u and v, Eqs. (VII-23),

$$-Hd\bar{x} + ud\bar{y} + vd\bar{z} = 0. \tag{VII-27}$$

Now note that the canonical equations, Eqs. (VII-26), form the condition that Eq. (VII-27) be exact. This can be seen from Eq. (VII-3).

Since the equation for the transversal surface, Eq. (VII-27), is exact, there must exist a function ϕ, depending on $\bar{x}$, $\bar{y}$, and $\bar{z}$, such that

$$\partial\phi/\partial\bar{x} = -H, \qquad \partial\phi/\partial\bar{y} = u, \qquad \partial\phi/\partial\bar{z} = v,$$

where $H = H(x, y, z, u, v)$. Eliminating u and v results in the Hamilton–Jacobi partial differential equation for the transversal surface

$$\partial\phi/\partial\bar{x} + H(x, y, z, \partial\phi/\partial\bar{y}, \partial\phi/\partial\bar{z}) = 0. \tag{VII-28}$$

The equation for the transversal surface is therefore $\phi = \text{const}$.

The canonical equations can be regarded as a system of ordinary differential equations whose solutions lead to extremals that have the property that they are associated with a transversal surface. Solutions to the Euler equations, on the other hand, are by definition also extremals. But the latter are not associated with a transversal surface unless they are coupled with a condition, in the form of Eq. (VII-3) or Eq. (VII-6), assuring its existence.

THE APPLICATION TO GEOMETRICAL OPTICS

The application of our preceding results, the Hamiltonian function, the canonical variables, and the Hamilton–Jacobi differential equation, to the geometrical optics of inhomogeneous media is direct and uninvolved. Simply set

$$f = n(1 + y'^2 + z'^2)^{1/2}$$

and substitute this quantity into the appropriate expressions.

From Eq. (VII-23), the canonical variables are

$$u = \partial f/\partial y' = ny'/(1 + y'^2 + z'^2)^{1/2},$$

$$v = \partial f/\partial z' = nz'/(1 + y'^2 + z'^2)^{1/2},$$

from which it is a simple matter to determine that

$$(1 + y'^2 + z'^2)^{1/2} = n/(n^2 - u^2 - v^2)^{1/2}.$$

Solving the two expressions for the canonical variables for y' and z', thus obtaining the expressions for P and Q in the manner indicated by Eq. (VII-24), we obtain

$$y' = P = u/(n^2 - u^2 - v^2)^{1/2},$$

$$z' = Q = v/(n^2 - u^2 - v^2)^{1/2},$$

as well as the fact that

$$f = n^2/(n^2 - u^2 - v^2)^{1/2}.$$

From Eq. (VII-25), we may now write the Hamiltonian

$$H = Pu + Qv - f$$

$$= -(n^2 - u^2 - v^2)^{1/2}.$$

Its partial derivatives are

$$\partial H/\partial y = -n/(n^2 - u^2 - v^2)^{1/2}(\partial n/\partial y),$$

$$\partial H/\partial z = -n/(n^2 - u^2 - v^2)^{1/2}(\partial n/\partial z),$$

$$\partial H/\partial u = u/(n^2 - u^2 - v^2)^{1/2},$$

$$\partial H/\partial v = v/(n^2 - u^2 - v^2)^{1/2},$$

which yield the canonical equations (VII-26)

$$-n/(n^2 - u^2 - v^2)^{1/2}(\partial n/\partial y) = du/dx,$$

$$-n/(n^2 - u^2 - v^2)^{1/2}(\partial n/\partial z) = dv/dx,$$

$$u/(n^2 - u^2 - v^2)^{1/2} = dy/dx,$$

$$v/(n^2 - u^2 - v^2)^{1/2} = dz/dx.$$

From the expression for the Hamilton–Jacobi differential equation, Eq. (VII-28), we obtain

$$(\partial\phi/\partial x) - [n^2 - (\partial\phi/\partial y)^2 - (\partial\phi/\partial z)^2]^{1/2} = 0,$$

which, on elimination of the radical, becomes

$$(\partial\phi/\partial x)^2 + (\partial\phi/\partial y)^2 + (\partial\phi/\partial z)^2 = n^2,$$

which we have already encountered as the eikonal equation, Eq. (VII-20),

$$(\nabla\phi)^2 = n^2.$$

THE THEOREM OF MALUS

It is intuitively obvious that light from a point source has associated with it a spherical wavefront in a homogeneous medium and that this wavefront, although it may no longer be spherical, persists after it passes through a refracting surface. We can also expect that the wave persists as it passes through an inhomogeneous medium. These observations were first made by Malus about the turn of the nineteenth century and are known as Malus' theorem. One rarely encounters a proof of Malus' theorem mainly because the proof usually shown, based on rather clumsy geometrical constructions, is hardly worth the effort. However, we can produce a simple proof rather easily using the formalism we have developed. (See also Young, 1969, pp. 32–34, 274–275.)

Expressed in our terms, Malus' theorem states that an orthotomic system of rays remains so after refraction. In an inhomogeneous medium in which the refractive index function n is continuous, if a system of rays is orthotomic in a portion of the medium, then it is orthotomic throughout that medium.

We treat the latter statement first. Consider any system of rays in an inhomogeneous medium. These must satisfy the ray equation

$$(d/ds)(n\mathbf{P}') = \nabla n.$$

The curl of this expression is

$$(d/ds)\nabla \times (n\mathbf{P}') = \mathbf{0},$$

and $\nabla \times (n\mathbf{P}')$ must be independent of s, and therefore if it should vanish in some portion of the medium it must vanish throughout the medium.

We now consider the case of the refracting surface. Let $\phi = 0$ be the equation of the refracting surface on which n is discontinuous. We have seen that in this case, from Eq. (V-18),

$$n_1\mathbf{P}_1' - n_2\mathbf{P}_2' = k\mathbf{N},$$

where $\mathbf{N}$ is the normal to the refracting surface, $\phi = 0$, and therefore may be regarded as being essentially the gradient of ϕ. In taking the curl of this expression, the right-hand member vanishes, and we are left with

$$\nabla \times (n_1\mathbf{P}_1') = \nabla \times (n_2\mathbf{P}_2').$$

Clearly, if a system of rays is orthotomic on one side of a refracting surface, then it must be orthotomic after passing through that surface.

REFERENCES

Bateman, H. (1966). "Differential Equations." Chelsea, Bronx, New York.

Bliss, G. A. (1946). "Lectures on the Calculus of Variations." Univ. of Chicago Press, Chicago.

Born, M., and Wolf, E. (1970). "Principles of Optics," 4th ed. Pergamon, Oxford.

Forsyth, A. R. (1929). "A Treatise on Differential Equations," 6th ed. Macmillan, London.

Rektorys, K. (ed.) (1969). Ordinary differential equations, *in* "Survey of Applicable Mathematics." MIT Press, Cambridge, Massachusetts.

Rund, H. (1966). "The Hamilton–Jacobi Theory in the Calculus of Variations." Van Nostrand-Reinhold, Princeton, New Jersey.

Young, L. C. (1969). "Lectures on the Calculus of Variations and Optimal Control Theory." Saunders, Philadelphia, Pennsylvania.

VIII

Wavefronts

In the last chapter we examined orthotomic systems of rays. These have the property that there exists a surface that intersects each ray of the system orthogonally. Although our justification for doing so was perhaps inadequate, we decided to call these orthogonal surfaces *wavefronts*. If such a surface is represented by the equation

$$\phi(x, y, z) = \text{const.},$$

then the function ϕ is a solution to the eikonal equation, Eq. (VII-20), $(\nabla\phi)^2 = n^2$, where n represents the refractive index of the medium in which the rays lie and is, in general, a function of position.

Our goal in this chapter is to derive from the eikonal equation information about the structure of wavefronts. We would like data on the geometry of wavefronts that do not depend on some particular orthotomic system of rays. The eikonal equation, since it contains no explicit referent to a system of rays, is therefore ideally suited to this purpose.

Our approach is, as usual, from the general to the specific. The eikonal equation is a nonlinear, first-order partial differential equation; therefore we will look at nonlinear, first-order partial differential equations in general and develop at the same time some results and notations that subsequently will serve us well in a different context.

We follow more or less Forsyth (1929, Chapter VIII; 1959, Chapters III, IV, VI, and VIII). Shorter, somewhat different accounts can be found in Bateman (1966), John (1970), and Rektorys (1969).

We will begin at a more elementary level and consider *linear*, first-order partial differential equations such as

$$A(\partial\phi/\partial x) + B(\partial\phi/\partial y) + C(\partial\phi/\partial z) = 0,$$

where A, B, and C are functions of x, y, and z. The right-hand member is zero, making this expression a *homogeneous*, linear, first-order partial differential equation. Note that in a vector notation this expression may be written as

$$\nabla\phi \cdot \mathbf{V} = 0,$$

where $\mathbf{V} = (A, B, C)$, and can be interpreted as describing a surface $\phi(x, y, z)$ that is perpendicular to the vector field $\mathbf{V}$.

The "open sesame" to this equation is, just as in the case of the total differential equation, the expression for the total differential of a function ϕ,

$$d\phi = (\partial\phi/\partial x)\, dx + (\partial\phi/\partial y)\, dy + (\partial\phi/\partial z)\, dz.$$

Comparing this with the given partial differential equation, we see that the two are equivalent if $d\phi = 0$ and if the coefficients of the partial derivatives are proportional, that is, if

$$dx/A = dy/B = dz/C. \qquad \text{(VIII-1)}$$

These are the so-called *characteristic equations*. From these, two total differential equations are extracted (any two will do as long as they are linearly independent and, of course, integrable) that, when integrated, yields two functions, say $a(x, y, z)$ and $b(x, y, z)$. These are called the *characteristic functions* (not to be confused with the characteristic functions encountered in optics and mechanics) or, more simply, *characteristics*. Let ψ be any differentiable function in two variables. Then the general solution of the given differential equation is $\phi = \psi(a, b)$. This general procedure is known as the method of Lagrange.

Just as the general solution of a first-order ordinary differential equation depends on one arbitrary constant, the general solution of a first-order partial differential equation involves one arbitrary function, which here is represented as ψ. As in the case of an ordinary differential equation where, say, a boundary condition determines the value of the constant of integration, the arbitrary function is determined by the imposition of a set of boundary conditions.

There is one particularly useful technique in working with characteristic equations. Note that, if α, β, and γ are any three functions, then

$$dx/A = dy/B = dz/C = \frac{\alpha\, dx + \beta\, dy + \gamma\, dz}{\alpha A + \beta B + \gamma C}. \qquad \text{(VIII-2)}$$

If, moreover, α, β, and γ can be chosen in such a way that

$$\alpha A + \beta B + \gamma C = 0,$$

then the numerator of the right-hand member must also be zero, and one can obtain quite painlessly the total differential equation

$$\alpha\, dx + \beta\, dy + \gamma\, dz = 0.$$

Now suppose we face an *inhomogeneous*, linear partial differential equation,

$$A(\partial\phi/\partial x) + B(\partial\phi/\partial y) + C(\partial\phi/\partial z) = D,$$

where the right-hand member is not zero. It turns out that this can always be reduced to the homogeneous case. We seek a function u of the three independent variables, x, y, and z, that satisfies this equation. However, the relationsip between u and x, y, and z need not be expressed explicitly but may be defined implicitly by means of an unknown function of four variables, say ψ, set equal to a constant:

$$\psi(\phi, x, y, z) = \text{const.}$$

It is a simple matter to show that

$$\begin{aligned} \partial\phi/\partial x &= -(\partial\psi/\partial x)/(\partial\psi/\partial\phi), \\ \partial\phi/\partial y &= -(\partial\psi/\partial y)/(\partial\psi/\partial\phi), \\ \partial\phi/\partial z &= -(\partial\psi/\partial z)/(\partial\psi/\partial\phi). \end{aligned}$$

When these relations are substituted into the original inhomogeneous differential equation in three variables, we obtain a homogeneous equation in four variables,

$$A(\partial\psi/\partial x) + B(\partial\psi/\partial y) + C(\partial\psi/\partial z) + D(\partial\psi/\partial\phi) = 0.$$

This we know how to solve. The characteristic equations are formed, *three* characteristic functions are found, and the general solution ψ is expressed as an arbitrary function of the three characteristic functions. This, when set equal to a constant, provides the implicit functional relation between u and x, y, and z.

This discussion of the method of Lagrange or, as it is sometimes called, the method of characteristics is deliberately sketchy. This material is usually covered in a first course in partial differential equations and should be somewhat familiar to even the unsophisticated reader.

THE NONLINEAR, FIRST-ORDER EQUATION

Consider a partial differential equation in the form

$$F(x, y, z; \phi; \partial\phi/\partial x, \partial\phi/\partial y, \partial\phi/\partial z) = 0, \tag{VIII-3}$$

where F is a given function in the seven variables x, y, z, the unknown function ϕ, and its three partial derivatives. Of course it is assumed that F is nonlinear. Our problem is to find an unknown function $\phi(x, y, z)$ that, when substituted into F, causes F to vanish. We will discuss a technique for obtaining a solution known as the method of Lagrange–Charpit or as the method of Jacobi. Strictly speaking, the method of Lagrange–Charpit refers to the case where there are only two independent variables, the method of Jacobi being an extension to the general case of n independent variables.

The technique is conceptually simple. First we obtain two additional partial differential equations F_1 and F_2, which share a solution with F. Next we solve algebraically the system of equations

$$\begin{aligned} F(x, y, z; \phi; \partial\phi/\partial x, \partial\phi/\partial y, \partial\phi/\partial z) &= 0, \\ F_1(x, y, z; \phi; \partial\phi/\partial x, \partial\phi/\partial y, \partial\phi/\partial z) &= 0, \\ F_2(x, y, z; \phi; \partial\phi/\partial x, \partial\phi/\partial y, \partial\phi/\partial z) &= 0, \end{aligned} \tag{VIII-4}$$

for the three partial derivatives and obtain

$$\begin{aligned} \partial\phi/\partial x &= P(x, y, z, \phi), \\ \partial\phi/\partial y &= Q(x, y, z, \phi), \\ \partial\phi/\partial z &= R(x, y, z, \phi). \end{aligned} \tag{VIII-5}$$

When these are substituted into the expression for the total derivative of the unknown function u, we obtain

$$d\phi = P\,dx + Q\,dy + R\,dz.$$

The solution of this is called the *complete integral.* It is nowhere near our final destination, however. The complete integral will generally depend on three constants of integration rather than on the arbitrary function that consideration of the linear case would lead us to expect. Generalization of the complete integral is possible and leads us in two directions: to a *singular integral* and to a *general integral.*

Although the procedures outlined here are conceptually simple, their application becomes particularly hairy. The critical steps are (1) finding the two additional differential equations that have a solution in common with the given partial differential equation and (2) obtaining an assurance that the resulting total differential equation is exact. These problems will be treated next.

THE BRACKET

Suppose we have a pair of nonlinear partial differential equations of the type described above,

$$a(x, y, z; \phi; \xi, \eta, \zeta) = 0, \qquad b(x, y, z; \phi; \xi, \eta, \zeta) = 0, \qquad \text{(VIII-6)}$$

in which, for convenience, we denote

$$\xi = \partial\phi/\partial x, \qquad \eta = \partial\phi/\partial y, \qquad \zeta = \partial\phi/\partial z.$$

Suppose in addition that they have a solution in common,

$$\bar{\phi} = \bar{\phi}(x, y, z).$$

That is to say, when $\bar{\phi}$ and its derivatives are substituted into a and b, the two resulting expressions considered as functions of x, y, and z vanish identically. If this is the case, their derivatives with respect to x, y, and z must also vanish identically, and we get, denoting partial derivatives by subscripts, the six expressions

$$\begin{aligned}
a_x + a_\phi\bar{\xi} + a_\xi\bar{\xi}_x + a_\eta\bar{\eta}_x + a_\zeta\bar{\zeta}_x &= 0,\\
a_y + a_\phi\bar{\eta} + a_\xi\bar{\xi}_y + a_\eta\bar{\eta}_y + a_\zeta\bar{\zeta}_y &= 0,\\
a_z + a_\phi\bar{\zeta} + a_\xi\bar{\xi}_z + a_\eta\bar{\eta}_z + a_\zeta\bar{\zeta}_z &= 0,\\
b_x + b_\phi\bar{\xi} + b_\xi\bar{\xi}_x + b_\eta\bar{\eta}_x + b_\zeta\bar{\zeta}_x &= 0,\\
b_y + b_\phi\bar{\eta} + b_\xi\bar{\xi}_y + b_\eta\bar{\eta}_y + b_\zeta\bar{\zeta}_y &= 0,\\
b_z + b_\phi\bar{\zeta} + b_\xi\bar{\xi}_z + b_\eta\bar{\eta}_z + b_\zeta\bar{\zeta}_z &= 0.
\end{aligned}$$

From the way they are defined, $d\phi = \bar{\xi}\, dx + \bar{\eta}\, dy + \bar{\zeta}\, dz$, and from the condition for exactness, Eq. (VII-3),

$$\bar{\xi}_y = \bar{\eta}_x\,, \qquad \bar{\xi}_z = \bar{\zeta}_x\,, \qquad \bar{\eta}_z = \bar{\zeta}_y\,. \tag{VIII-7}$$

Using these three relationships, we eliminate the derivatives of $\bar{\xi}$, $\bar{\eta}$, and $\bar{\zeta}$ from the six equations above and obtain

$$(a_x b_\xi - a_\xi b_x) + (a_y b_\eta - a_\eta b_y) + (a_z b_\zeta - a_\zeta b_z) \\ + (a_\phi b_\xi - a_\xi b_\phi)\bar{\xi} + (a_\phi b_\eta - a_\eta b_\phi)\bar{\eta} + (a_\phi b_\zeta - a_\zeta b_\phi)\bar{\zeta} = 0. \tag{VIII-8}$$

The entire left-hand member of this expression is called a *bracket* and is denoted by

$$[a, b].$$

It will turn out to be useful to us in another context. What we have shown is that a necessary condition for the differential equations a and b to have a solution in common is that

$$[a, b] = 0. \tag{VIII-9}$$

It turns out that this condition is also sufficient.

The definition of the bracket leads to a further definition. Suppose we have any number of partial differential equations, say n of them,

$$f_1 = 0, \qquad f_2 = 0, \qquad \ldots, \qquad f_n = 0,$$

with the property that

$$[f_i\,, f_j] = 0, \qquad i, j = 1, 2, \ldots, n.$$

Then the equations of the system have a common solution. Such a system is called a *complete system.*

The bracket notation as well as the idea of a complete system of equations will prove useful in a subsequent chapter. We use it here to develop the so-called complete integral to the single partial differential equation we are considering.

THE COMPLETE INTEGRAL

Returning now to the nonlinear, first-order partial differential equation

$$F(x, y, z; \phi; \xi, \eta, \zeta) = 0,$$

our first problem is to find two other differential equations, F_1 and F_2, that have a solution in common with F. From Eq. (VIII-9), they must satisfy

$$[F, F_1] = 0, \qquad [F, F_2] = 0, \qquad [F_1, F_2] = 0.$$

The first of these, by the definition of the bracket, Eq. (VIII-8), is

$$\begin{aligned}[][F_x(\partial F_1/\partial\xi) - F_\xi(\partial F_1/\partial x)] + [F_y(\partial F_1/\partial\eta) - F_\eta(\partial F_1/\partial y)]&\\ + [F_z(\partial F_1/\partial\zeta) - F_\zeta(\partial F_1/\partial z)] + [F_\phi(\partial F_1/\partial\xi) - F_\xi(\partial F_1/\partial\phi)\xi]&\\ + [F_\phi(\partial F_1/\partial\eta) - F_\eta(\partial F_1/\partial\phi)\eta] + [F_\phi(\partial F_1/\partial\zeta) - F_\zeta(\partial F_1/\partial\phi)\zeta] &= 0,\end{aligned}$$

or, rearranging terms,

$$\begin{aligned}(F_x + \xi F_\phi)(\partial F_1/\partial\xi) + (F_y + \eta F_\phi)(\partial F_1/\partial\eta)&\\ + (F_z + \zeta F_\phi)(\partial F_1/\partial\zeta) - (\xi F_\xi + \eta F_\eta + \zeta F_\zeta)(\partial F_1/\partial\phi)&\\ - F_\xi(\partial F_1/\partial x) - F_\eta(\partial F_1/\partial y) - F_\zeta(\partial F_1/\partial z) &= 0,\end{aligned}$$

a linear, first-order partial differential equation in F_1, in seven independent variables. To this we apply the method of Lagrange, extending Eq. (VIII-1) to a statement involving seven variables, and get the characteristic equations

$$\begin{aligned}d\xi/(F_x + \xi F_\phi) &= d\eta/(F_y + \eta F_\phi) = d\zeta/(F_z + \zeta F_\phi)\\ &= -d\phi/(\xi F_\xi + \eta F_\eta + \zeta F_\zeta) = -dx/F_\xi = -dy/F_\eta = -dz/F_\zeta.\end{aligned} \tag{VIII-10}$$

As we have seen, the most general form of F_1 is given by an arbitrary function of six characteristics obtained from six independent total differential equations derived from the above set of characteristic equations. But we do not need the most general solution! All we need is one. We may therefore extract any convenient total differential equation out of the collection of characteristic equations, integrate it, and designate its integral by F_1. We need only be certain that F_1 contains at least one of the derivatives of u: ξ, η, or ζ. We also must be sure that it is equivalent to neither the expression for the total derivative of u nor to the original partial differential equation. Thus, F_1 is found.

Finding F_2 is only slightly more difficult. Not only must it satisfy $[F, F_2] = 0$, but it also must satisfy $[F_1, F_2] = 0$. The first of these leads to a linear partial differential equation in F_2 identical in form to that already obtained for F_1, and therefore we are led to exactly the same set of characteristic equations. Any second characteristic, distinct from F_1, obtained from this set of equations will do. Again, it must contain explicitly the derivatives of ϕ and must reduce to neither the expression for the differential of ϕ nor to the original differential equation. Finally, we determine whether or not it satisfies the third equation. This rarely leads to any difficulties, but if it does, a third characteristic is tried.

We now have the triad of equations,

$$F(x, y, z; \phi; \xi, \eta, \zeta) = 0,$$
$$F_1(x, y, z; \phi; \xi, \eta, \zeta) = 0,$$
$$F_2(x, y, z; \phi; \xi, \eta, \zeta) = 0,$$

which we solve for ξ, η, and ζ:

$$\xi = P(x, y, z, \phi), \qquad \eta = Q(x, y, z, \phi), \qquad \zeta = R(x, y, z, \phi).$$

When these are substituted into the expression for the total differential of u, we get

$$d\phi = P\,dx + Q\,dy + R\,dz.$$

Now we know that this is exact. In deriving the fact that $[a, b] = 0$ is a condition for a and b to have a solution in common, we introduced the condition in Eq. (VIII-7) that $\xi_y = \eta_x$, $\xi_z = \zeta_x$, and $\eta_z = \zeta_y$, which is exactly the condition required.

We are now in a position to obtain the complete integral of the eikonal equation.

THE EIKONAL EQUATION

The simplest case of the eikonal equation (VII-20), the one in which the refractive index n is constant, will now be treated:

$$F = \xi^2 + \eta^2 + \zeta^2 - n^2 = 0,$$

where ξ, η, and ζ are the partial derivatives of ϕ, the unknown, with respect to x, y, and z, respectively. The derivatives of F are

$$F_x = 0, \quad F_y = 0, \quad F_z = 0, \quad F_\phi = 0, \quad F_\xi = 2\xi, \quad F_\eta = 2\eta, \quad F_\zeta = 2\zeta.$$

Substituting these into Eq. (VIII-10), we get the characteristic equations

$$\begin{aligned} d\xi/0 = d\eta/0 = d\zeta/0 &= d\phi/2(\xi^2 + \eta^2 + \zeta^2) \\ &= -dx/2\xi = -dy/2\eta = -dz/2\zeta. \end{aligned}$$

The easiest way out is to take as the two total differential equations

$$d\xi = 0, \qquad d\eta = 0,$$

so that the two new equations are

$$F_1 = \xi - u, \qquad F_2 = \eta - v,$$

where u and v are constants of integration. The three equations that are solved simultaneously are

$$\begin{aligned} F &= \xi^2 + \eta^2 + \zeta^2 - n^2 = 0, \\ F_1 &= \xi - u \qquad\qquad\quad = 0, \\ F_2 &= \eta - v \qquad\qquad\quad = 0, \end{aligned}$$

which yield

$$\xi = u, \qquad \eta = v, \qquad \zeta = (n^2 - u^2 - v^2)^{1/2}. \tag{VIII-11}$$

This, when substituted into the expression for the total differential of ϕ, yields

$$d\phi = u\, dx + v\, dy + (n^2 - u^2 - v^2)^{1/2}\, dz.$$

When this is integrated, we get the complete integral

$$\phi = ux + vy + (n^2 - u^2 - v^2)^{1/2}\, z + w, \tag{VIII-12}$$

where w is another constant of integration. The reader may feel a bit cheated—that this is a puny result for such a long stream of preliminary calculations. But this is by no means the terminus of our travels, only a way station.

THE SINGULAR AND GENERAL INTEGRALS

What has been developed thus far is a reasonable step-by-step procedure for obtaining a complete integral for a given nonlinear partial differential equation. We have seen that the complete integral involves constants of integration; in fact, in the class of differential equations that we have been considering (those involving three independent variables) we found three constants of integration. However, when the complete integral is compared with the general solution obtained for the case of the linear equation, which involved arbitrary functions rather than constants of integration, we see that the complete integral suffers in comparison from a lack of generality.

The complete integral can be used as the starting point in generating a more general solution. We began with our nonlinear equation,

$$F(x, y, z; \phi; \xi, \eta, \zeta) = 0$$

(where ξ, η, and ζ represent the derivatives of the unknown function ϕ), from which the characteristic equations (VIII-10) were obtained. This yielded a pair of total differential equations, say, $d\alpha = 0$, $d\beta = 0$, whose integrals are $\alpha = u$, $\beta = v$, where u and v are constants of integration. We used these together with F to form the complete system:

$$\begin{aligned} F &= 0, \\ F_1 = \alpha - u &= 0, \\ F_2 = \beta - v &= 0. \end{aligned}$$

This was then solved algebraically for the three derivatives of ϕ, and we got

$$\xi = P(x, y, z; \phi; u, v), \quad \eta = Q(x, y, z; \phi; u, v), \quad \zeta = R(x, y, z; \phi; u, v). \tag{VIII-13}$$

Note that these depend not only on the independent variables x, y, and z but also on the two constants of integration, u and v. Inserting these expressions into the formula for the total differential of ϕ, we have

$$d\phi = P\,dx + Q\,dy + R\,dz,$$

which is guaranteed to be exact. Integration produces our complete integral

$$\phi = \phi(x, y, z; u, v, w), \tag{VIII-14}$$

which now depends on three constants of integration, u, v, and w, the latter of which was introduced at the last step.

The complete integral has an interesting property. If Eq. (VIII-14) is differentiated with respect to x, y, and z, we get three expressions,

$$\xi = \bar{P}(x, y, z; u, v, w), \quad \eta = \bar{Q}(x, y, z; u, v, w), \quad \zeta = \bar{R}(x, y, z; u, v, w), \tag{VIII-15}$$

similar to Eqs. (VIII-11) but depending on u, v, and w rather than on ϕ, u, and v. If we eliminate the three constants of integration from the four equations in (VIII-12) and (VIII-13), we obtain an expression depending on ϕ and its derivatives that in general is identical to the original differential equation.

The first step in generalizing the complete integral is to replace the three constants of integration with three functions of x, y, and z,

$$u = u(x, y, z), \qquad v = v(x, y, z), \qquad w = w(x, y, z),$$

so that the integral is now, from Eq. (VIII-14),

$$\phi = \phi(x, y, z; u(x, y, z), v(x, y, z), w(x, y, z)). \tag{VIII-16}$$

However, we impose a condition on these new functions. We require that the three partial derivatives of ϕ with these functions substituted be identical to those obtained from the complete integral itself. This ensures that Eq. (VIII-16) will be a solution to the original differential equation. Moreover, it ensures that, if the three functions are eliminated from the algebraic system consisting of the solution and its three derivatives, we will obtain the partial differential equation we started with.

Differentiating Eq. (VIII-16) with respect to x produces

$$\xi + (\partial\phi/\partial u)(\partial u/\partial x) + (\partial\phi/\partial v)(\partial v/\partial x) + (\partial\phi/\partial w)(\partial w/\partial x),$$

which we have agreed must equal $\bar{P}$ in Eq. (VIII-15). For this to happen, it must be that

$$(\partial\phi/\partial u)(\partial u/\partial x) + (\partial\phi/\partial v)(\partial v/\partial x) + (\partial\phi/\partial w)(\partial w/\partial x) = 0.$$

Performing the differentiation with respect to y and z produces two additional equations. Thus we obtain the system of equations that u, v, and w must satisfy in order to meet our requirements:

$$\begin{aligned}
(\partial\phi/\partial u)(\partial u/\partial x) + (\partial\phi/\partial v)(\partial v/\partial x) + (\partial\phi/\partial w)(\partial w/\partial x) &= 0,\\
(\partial\phi/\partial u)(\partial u/\partial y) + (\partial\phi/\partial v)(\partial v/\partial y) + (\partial\phi/\partial w)(\partial w/\partial y) &= 0,\\
(\partial\phi/\partial u)(\partial u/\partial z) + (\partial\phi/\partial v)(\partial v/\partial z) + (\partial\phi/\partial w)(\partial w/\partial z) &= 0.
\end{aligned} \tag{VIII-17}$$

It is convenient to express this in matrix form:

$$\begin{pmatrix} \partial u/\partial x & \partial v/\partial x & \partial w/\partial x \\ \partial u/\partial y & \partial v/\partial y & \partial w/\partial y \\ \partial u/\partial z & \partial v/\partial z & \partial w/\partial x \end{pmatrix} \begin{pmatrix} \partial\phi/\partial u \\ \partial\phi/\partial v \\ \partial\phi/\partial w \end{pmatrix} = \begin{pmatrix} 0 \\ 0 \\ 0 \end{pmatrix}. \tag{VIII-18}$$

Recall that ϕ is the complete integral and is therefore a known function of x, y, z, u, v, and w.

We can look at Eqs. (VIII-17) or (VIII-18) as a linear system of three homogeneous equations where the three unknowns are $\partial\phi/\partial u$, $\partial\phi/\partial v$, and $\partial\phi/\partial w$. Then two possibilities arise. Either the determinant of coefficients is not zero, in which case only the trivial solution is possible and therefore $\partial\phi/\partial u = \partial\phi/\partial v = \partial\phi/\partial w = 0$; or the determinant of coefficients does vanish and nontrivial solutions exist.

Consider the first of these possibilities first. The determinant of coefficients differs from zero and

$$\partial\phi/\partial u = 0, \qquad \partial\phi/\partial v = 0, \qquad \partial\phi/\partial w = 0.$$

Now ϕ is certainly known, and it depends on u, v, and w as well as on x, y, and z. It may be possible to solve these for u, v, and w and obtain them as explicit functions of x, y, and z. When these are substituted into the expression for the complete integral, Eq. (VIII-16), we get a solution to the differential equation. This solution depends on neither an arbitrary function not any constants of integration. It falls into a special category of solution and is called the *singular integral.*

The second possibility does give rise to a solution involving an arbitrary function. In this case the determinant of coefficients vanishes:

$$\begin{vmatrix} \partial u/\partial x & \partial v/\partial x & \partial w/\partial x \\ \partial u/\partial y & \partial v/\partial y & \partial w/\partial y \\ \partial u/\partial z & \partial v/\partial z & \partial w/\partial z \end{vmatrix} = 0.$$

Now this determinant is nothing more than the Jacobian of the three functions u, v, and w, and its vanishing implies that a functional relationship among these functions exists. This means, simply, that one of them, say w, can be represented as a function of the other two,

$$w = w(u, v).$$

Now we return to Eqs. (VIII-17). We multiply the first by dx, the second by dy, the third by dz, then add, obtaining

$$(\partial\phi/\partial u)\, du + (\partial\phi/\partial v)\, dv + (\partial\phi/\partial w)\, dw = 0.$$

But now $dw = (\partial w/\partial u)\, du + (\partial w/\partial v)\, dv$; substituting this relationship into the above results in

$$\{\partial\phi/\partial u + (\partial\phi/\partial w)(\partial w/\partial u)\}\, du + \{\partial\phi/\partial v + (\partial\phi/\partial w)(\partial w/\partial v)\}\, dv = 0.$$

Since the differentials in the above expression are independent, we may set the two quantities within the braces equal to zero.

We have obtained the *general integral*, which consists of the following three expressions:

$$\phi = \phi(x, y, z; u, v, w(u, v)),$$
$$\partial\phi/\partial u + (\partial\phi/\partial w)(\partial w/\partial u) = 0, \qquad \partial\phi/\partial v + (\partial\phi/\partial w)(\partial w/\partial v) = 0. \tag{VIII-19}$$

This, the general integral, depends on w, an arbitrary function of two variables. It is not as transparent as the solution to the linear case in that the arbitrary function enters in a rather complicated way. In fact, it is more of an algorithm than a proper formal solution and its application consists of the "you give me an I'll give you" ritual so often encountered in mathematics. If you give me any arbitrary function w (differentiable, of course) of u and v, then I can plug that function into the three formulas (VIII-19), eliminate u and v, and obtain a relationship between ϕ and x, y, and z that constitutes a solution. You give me a different function for w and I'll give you a different solution. We can regard the totality of all solutions obtained in this way as the general solution of the differential equation.

THE GENERAL INTEGRAL FOR HOMOGENEOUS MEDIA

We have already the complete integral of the eikonal equation (VIII-12) for the case of the homogeneous medium,

$$\phi = ux + vy + z(n^2 - u^2 - v^2)^{1/2} + w.$$

The first step in the procedure is to replace u, v, and w by three functions of x, y, and z that are subject to the conditions given by Eqs. (VIII-17) or (VIII-18). These involve the derivatives of ϕ with respect to u, v, and w, which when calculated and substituted into Eq. (VIII-18) give us

$$\begin{pmatrix} \partial u/\partial x & \partial v/\partial x & \partial w/\partial x \\ \partial u/\partial y & \partial v/\partial y & \partial w/\partial y \\ \partial u/\partial z & \partial v/\partial z & \partial w/\partial z \end{pmatrix} \begin{pmatrix} (x - uz)/(n^2 - u^2 - v^2)^{1/2} \\ (y - vz)/(n^2 - u^2 - v^2)^{1/2} \\ 1 \end{pmatrix} = \begin{pmatrix} 0 \\ 0 \\ 0 \end{pmatrix}.$$

Clearly, there is no trivial solution to this algebraic equation, and therefore there is no singular integral to the differential equation. The determinant of coefficients must vanish, implying the existence of a functional relationship among u, v, and w, which we express as $w = w(u, v)$. Applying this and Eq. (VIII-12) to Eqs. (VIII-19), we obtain a general integral for the eikonal equation,

$$\begin{aligned} \phi &= ux + vy + z(n^2 - u^2 - v^2)^{1/2} + w(u, v), \\ x &- uz/(n^2 - u^2 - v^2)^{1/2} + \partial w/\partial u = 0, \\ y &- vz/(n^2 - u^2 - v^2)^{1/2} + \partial w/\partial v = 0. \end{aligned} \tag{VIII-20}$$

As indicated in the last section, once a function w is provided, u and v can be eliminated, resulting in a function ϕ of x, y, and z.

The equation for a wavefront is obtained by setting ϕ equal to a constant. Thus, rearranging terms, Eqs. (VIII-20) become

$$\begin{aligned} \phi &\equiv ux + vy + z(n^2 - u^2 - v^2)^{1/2} + w(u, v) = c, \\ uz &= (n^2 - u^2 - v^2)^{1/2}(x + \partial w/\partial u), \\ vz &= (n^2 - u^2 - v^2)^{1/2}(y + \partial w/\partial v), \end{aligned} \tag{VIII-21}$$

where c is a constant.

This can be interpreted in another way. Instead of eliminating u and v between the equations (VIII-21), we treat them as parameters in the representation of the wavefront $\phi = c$. Further, we define the vectors

$$\mathbf{S} = (u, v, (n^2 - u^2 - v^2)^{1/2}), \qquad \mathbf{W} = (x, y, z).$$

The vector $\mathbf{W}$ is now, of course, a function of the parameters u and v and represents a point on the wavefront. The first equation of Eqs. (VIII-21) then becomes

$$\phi \equiv \mathbf{S} \cdot \mathbf{W} + w(u, v) = c. \tag{VIII-22}$$

Note that $\nabla\phi = \mathbf{S}$ and that $\mathbf{S}^2 = n^2$. Clearly the eikonal equation is satisfied. Note also that $\nabla\phi$ is a normal vector to the wavefront and may therefore be considered as a vector in the direction of a ray. Since its magnitude is equal to n, the components of $\mathbf{S}$ are the reduced direction cosines of a ray.

We next construct a second wavefront a distance s from the first. If $\overline{\mathbf{W}}$ represents a point on that wavefront, then

$$\overline{\mathbf{W}} = \mathbf{W} + (s/n)\mathbf{S}, \tag{VIII-23}$$

where $(1/n)\mathbf{S}$ is a unit normal vector. This is shown in Fig. VIII-1.

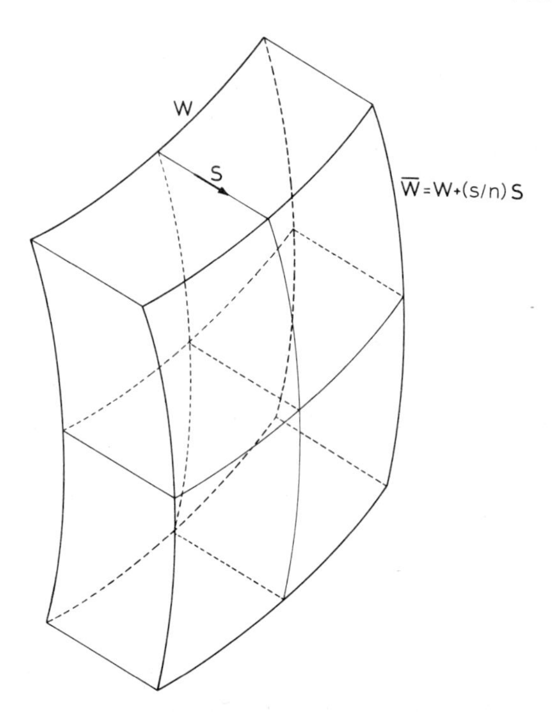

FIG. VIII-1. Wavefront propagation.

W represents the wavefront's initial position; **S** is the wavefront normal and ray direction; n is the refractive index; $\mathbf{S}^2 = n^2 \cdot \overline{\mathbf{W}}$ is the translated position of the wavefront.

In scalar form, Eq. (VIII-23) is written

$$\bar{x} = x + (su/n),$$

$$\bar{y} = y + (sv/n),$$

$$\bar{z} = z + (s/n)(n^2 - u^2 - v^2)^{1/2}.$$

This wavefront must also satisfy the eikonal equation, and we can therefore expect it to have the sort of representation given by Eq. (VIII-22),

$$\bar{\phi} \equiv \mathbf{S} \cdot \overline{\mathbf{W}} + \bar{w}(u, v) = \bar{c}.$$

Substituting Eq. (VIII-23) into this and applying Eq. (VIII-22) yields

$$ns + \bar{w}(u, v) - w(u, v) + c - \bar{c} = 0. \qquad \text{(VIII-24)}$$

Now applying Eq. (VIII-23) to the first of the two remaining equations of Eqs. (VIII-21), we get

$$\begin{aligned} u\bar{z} &- (n^2 - u^2 - v^2)^{1/2}(\bar{x} + \partial\bar{w}/\partial u) \\ &= uz - x(n^2 - u^2 - v^2)^{1/2} - (n^2 - u^2 - v^2)^{1/2}\partial\bar{w}/\partial u \\ &= (n^2 - u^2 - v^2)^{1/2}(\partial w/\partial u - \partial\bar{w}/\partial u) = 0. \end{aligned}$$

Thus, $\partial w/\partial u = \partial\bar{w}/\partial u$ and, following the same argument, using the third equation of Eqs. (VIII-21), $\partial w/\partial v = \partial\bar{w}/\partial v$. It follows that w and $\bar{w}$ differ by at most a constant. Absorbing this constant in those already present in Eq. (VIII-24), we obtain $ns + c - \bar{c} = 0$. The constant c appearing in Eqs. (VIII-21) and (VIII-22) is therefore intimately connected with the optical path length along a ray separating two successive wavefronts. We can now designate a wavefront by its distance s from some fixed wavefront. Equation (VIII-22) then becomes

$$\phi_s \equiv \mathbf{S} \cdot \mathbf{W} + w(u, v) = c_0 + ns.$$

We can further regard $\mathbf{W}$ as a function of s as well as of u and v.

We can continue this kind of reasoning still further. It will be convenient to take w as a scalar function of the vector $\mathbf{S}$ and denote its derivatives with respect to the components of $\mathbf{S}$ by w_1, w_2, and w_3. We also define the vector

$$\tilde{\nabla} w = (w_1, w_2, w_3).$$

It follows that

$$\begin{aligned} \partial w/\partial u &= w_1 + u/(n^2 - u^2 - v^2)^{1/2} w_3, \\ \partial w/\partial v &= w_2 + v/(n^2 - u^2 - v^2)^{1/2} w_3. \end{aligned}$$

Substituting these relations in the second two equations of Eqs. (VIII-21) gives us

$$\begin{aligned} uz - x(n^2 - u^2 - v^2)^{1/2} &= w_1(n^2 - u^2 - v^2) - uw_3, \\ vz - y(n^2 - u^2 - v^2)^{1/2} &= w_2(n^2 - u^2 - v^2) - vw_3, \end{aligned}$$

whence

$$vx - uy = w_2 u - w_1 v.$$

These are very easily represented in vector form,

$$\mathbf{W} \times \mathbf{S} = \mathbf{S} \times \tilde{\nabla} w.$$

To recapitulate, the general integral of the eikonal equation can be written as the pair of vector equations

$$\begin{aligned} \mathbf{W} \cdot \mathbf{S} + w(\mathbf{S}) &= c_0 + ns, \\ \mathbf{W} \times \mathbf{S} &= \mathbf{S} \times \tilde{\nabla} w, \end{aligned} \qquad \text{(VIII-25)}$$

where $\mathbf{W}$ can be regarded as a function of u, v, and s. It must be emphasized that this conclusion is valid only for homogeneous media in which the refractive index is constant.

We can continue a little further. The second equation of Eqs. (VIII-25) can be written in the form

$$(\mathbf{W} + \tilde{\nabla} w) \times \mathbf{S} = \mathbf{0},$$

which shows that the vectors $(\mathbf{W} + \tilde{\nabla} w)$ and $\mathbf{S}$ are parallel and that there exists a factor δ such that

$$\mathbf{W} + \tilde{\nabla} w = \delta \mathbf{S} \qquad \text{or} \qquad \mathbf{W} = \delta \mathbf{S} - \tilde{\nabla} w. \qquad \text{(VIII-26)}$$

The vector $\mathbf{W}$, which represents the points on a wavefront, is now given as an explicit function of the two parameters u and v.

We need to find δ. Multiplying Eqs. (VIII-26) by $\mathbf{S}$ results in

$$\mathbf{W} \cdot \mathbf{S} = n^2 \delta - \tilde{\nabla} w \cdot \mathbf{S}.$$

Comparing this with the first equation of Eqs. (VIII-25) results in

$$n^2 \delta = c_0 + ns + \tilde{\nabla} w \cdot \mathbf{S} - w, \qquad \text{(VIII-27)}$$

giving δ as a function of the arc length parameter s in addition to the parameters u and v. For our purposes, this provides the most convenient form of the general integral of the eikonal equation for a homogeneous medium.

A rather interesting special case occurs. If w happens to be a homogeneous function in its three variables, then, from Eq. (II-13),

$$\tilde{\nabla} w \cdot \mathbf{S} = w,$$

a form of the differential equation of the general homogeneous function. In that case,

$$n^2 \delta = c_0 + ns, \qquad \text{(VIII-28)}$$

and the factor δ depends only on the arc length parameter s.

We must emphasize again that this result is valid only for homogeneous media. For inhomogeneous media, the characteristic equations (VIII-10) involve the derivatives of the refractive index.

What we have shown here is that an entire orthotomic system of rays in a homogeneous medium is completely determined by the function $w(\mathbf{S})$. Suppose we have a lens of the usual sort, consisting of a sequence of surfaces separating media of constant refractive index. Suppose we consider all rays from a given object point. These form an orthotomic system. The wavefronts within each medium will be given by expressions in the form of Eqs. (VIII-25) and will be completely fixed by $w(\mathbf{S})$. We can recognize $\mathbf{S}$ as the reduced direction cosines of the rays that comprise the system. There is much more to be said on this subject, but first we will need the results of the next chapter.

REFERENCES

Bateman, H. (1966). "Differential Equations." Chelsea, Bronx, New York.

Forsyth, A. R. (1929). "A Treatise on Differential Equations," 6th ed. Macmillan, New York.

Forsyth, A. R. (1959). "Theory of Differential Equations. Part IV. Partial Differential Equations," Vol. V. Dover, New York.

John, F. (1970). Partial differential equations, *in* "Mathematics Applied to Physics" (É. Roubine, ed.). Springer-Verlag, Berlin and New York.

Rektorys, K. (ed.) (1969). Partial differential equations, *in* "Survey of Applicable Mathematics." MIT Press, Cambridge, Massachusetts.

IX

Surfaces

Wavefronts are incredibly intricate things. A converging wavefront at a great distance from its focus may appear to be quite smooth and manageable, yet as it is translated toward that focus it becomes wild and intractable. Two types of wavefront that remain uncomplicated are the plane and the sphere. Indeed, the classical solutions to the Maxwell equations had to presume either a plane or a spherical wavefront for precisely this reason. Any other shape quickly translates itself into something that is difficult to visualize and virtually impossible to analyze.

Nevertheless, the study of the geometrical properties of wavefronts has intrigued some of the best people in optics since the time of Hamilton. It should be obvious that a detailed understanding of the properties of wavefronts would be of immense technological value. One of the great unsolved problems in this area is, given any wavefront incident on any refracting surface, find the refracted wavefront in all its detail. This we will call the *in-the-large problem*, about which we will have something to say later.

A more modest problem, which we will call the *local problem*, is solvable. Suppose we look not at the entire wavefront but only at a very small region of it, which we can visualize as a neighborhood of a ray traced through an optical system. The small region of the wavefront will propagate through the optical system along the ray and always perpendicular to the ray. As the wavefront progresses from surface to surface, undergoing refraction at each surface, its shape changes continually. Our task here is to find ways to analyze these changes of shape.

But before we can analyze changes of shape, we will need to know what *shape* is, how it can be defined, and how it is treated analytically. Our first business is therefore some basic facts concerning the differential geometry of surfaces. For a more thorough treatment the reader is referred to one of the standard texts such as Blaschke (1930), Stoker (1969), Struik (1961), or Weatherburn (1927). Of particular relevance to the subject of this book is Hermann (1968).

Just as a space curve can be described as a vector function of a single parameter, say $\mathbf{P}(s)$, a surface can be described as a vector function of two variables, say $\mathbf{W}(u, v)$. In such a representation we must assume that the parameters u and v are chosen in such a way that $\mathbf{W}$ is indeed a surface. We say that u and v are *essential parameters*; i.e., there is no *a priori* relation between them that would enable one, by means of some suitable transformation, to reduce the dependence of $\mathbf{W}$ to a single parameter. This condition can be stated quite simply:

$$\partial\mathbf{W}/\partial u \times \partial\mathbf{W}/\partial v \neq 0.$$

Of course, there is a tacit assumption that $\mathbf{W}$ possesses partial derivatives to whatever order is required.

If we hold v fixed and allow u to vary, then $\mathbf{W}(u, v_0)$ describes a curve on the surface $\mathbf{W}$. Such a curve is called a *parametric curve*. Its tangent vector is $\partial\mathbf{W}(u, v_0)/\partial u$, which must also be tangent to the surface $\mathbf{W}$. In general it is not a unit vector. By holding u fixed and allowing v to vary, we get a second parametric curve $\mathbf{W}(u_0, v)$, whose tangent vector is $\partial\mathbf{W}(u_0, v)/\partial v$.

These two parametric curves will intersect on the surface at the point $\mathbf{W}(u_0, v_0)$. Both $\partial\mathbf{W}/\partial u$ and $\partial\mathbf{W}/\partial v$ are tangent vectors to the surface at that point. Therefore, the normal to the surface at that point will be given by their vector product. We can therefore define a *unit normal vector* to the surface by

$$\mathbf{N} = (\mathbf{W}_u \times \mathbf{W}_v)/[(\mathbf{W}_u \times \mathbf{W}_v)^2]^{1/2}. \tag{IX-1}$$

All of this is shown in Fig. IX-1. Here we switch to a subscript notation for partial derivatives.

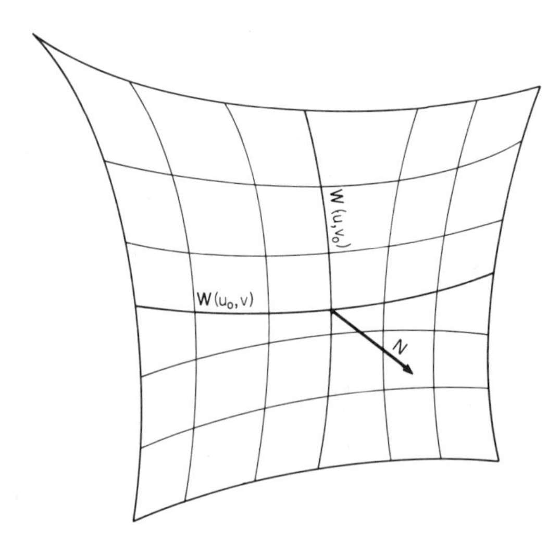

FIG. IX-1. Parametric curves and the surface normal.

A rather vital point needs be made. The choice of the parameters describing the surface **W** has no relation to the shape, i.e., to the structure of that surface. Indeed, any transformation from one system of parameters leaves the surface unchanged. The unit normal vector to the surface, being an intrinsic property of that surface, must be invariant under such a parameter transformation.

A curve on the surface **W** can be specified by replacing the two parameters u and v with a pair of functions of a new parameter, say t. We can write the expression for such a curve as

$$\mathbf{P}(t) = \mathbf{W}(u(t), v(t)). \tag{IX-2}$$

The tangent vector to this curve is

$$d\mathbf{P}/dt = (\partial\mathbf{W}/\partial u)(du/dt) + (\partial\mathbf{W}/\partial v)(dv/dt), \tag{IX-3}$$

a linear combination of the two tangent vectors to the two parametric curves. The two derivatives du/dt and dv/dt can be thought of as measures of the angles between $\mathbf{P}(t)$ and the tangent vectors of the parametric curves.

The differential of arc length is calculated in the usual way:

$$\begin{aligned} ds^2 &= (d\mathbf{P}/dt)^2\, dt^2 \\ &= (\mathbf{W}_u\, du/dt + \mathbf{W}_v\, dv/dt)^2\, dt^2 \\ &= E\, du^2 + 2F\, du\, dv + G\, dv^2, \end{aligned} \tag{IX-4}$$

where we define

$$E = \mathbf{W}_u^2, \qquad F = \mathbf{W}_u \cdot \mathbf{W}_v, \qquad G = \mathbf{W}_v^2. \tag{IX-5}$$

The expression

$$(H\,dt)^2 = E\,du^2 + 2F\,du\,dv + G\,dv^2$$

is called the *first fundamental form*, and E, F, and G are called the *first fundamental quantities*. Note that E, F, and G are properties of the surface $\mathbf{W}$ and not of the curve $\mathbf{P}(t)$. Now that E, F, and G have been defined, we may write the unit normal vector as

$$\mathbf{N} = (\mathbf{W}_u \times \mathbf{W}_v)/(EG - F^2)^{1/2} \tag{IX-6}$$

by expanding the expression in the denominator of Eq. (IX-1) and applying Eq. (IX-5). The quantity $(EG - F^2)^{1/2}$ is clearly an element of area.

An expression for the *unit* tangent vector to the curve $\mathbf{P}$ can now be written

$$\begin{aligned} \mathbf{P}' &= d\mathbf{P}/ds = (dt/ds)(d\mathbf{P}/dt) \\ &= (\mathbf{W}_u u_t + \mathbf{W}_v v_t)/H, \end{aligned} \tag{IX-7}$$

where u_t and v_t are the derivatives of u and v with respect to t and where H is as defined above.

We can now calculate the normal vector to $\mathbf{P}$:

$$\begin{aligned} \mathbf{P}'' &= d\mathbf{P}'/ds = (dt/ds)(d\mathbf{P}'/dt) \\ &= (1/H)(d/dt)[(\mathbf{W}_u u_t + \mathbf{W}_v v_t)/H] \\ &= (\mathbf{W}_{uu} u_t^2 + 2\mathbf{W}_{uv} u_t v_t + \mathbf{W}_{vv} v_t^2 + \mathbf{W}_u u_{tt} + \mathbf{W}_v v_{tt})/H^2 \\ &\quad + [(\mathbf{W}_u u_t + \mathbf{W}_v v_t)/H](d/dt)(1/H). \end{aligned}$$

Now $\mathbf{N} \cdot \mathbf{W}_u = \mathbf{N} \cdot \mathbf{W}_v = 0$, so that

$$\mathbf{N} \cdot \mathbf{P}'' = (Lu_t^2 + 2Mu_t v_t + Nv_t^2)/(Eu_t^2 + 2Fu_t v_t + Gv_t^2), \tag{IX-8}$$

where $L = \mathbf{N} \cdot \mathbf{W}_{uu}$, $M = \mathbf{N} \cdot \mathbf{W}_{uv}$, and $N = \mathbf{N} \cdot \mathbf{W}_{vv}$. The quantity $L\,du^2 + 2M\,du\,dv + N\,dv^2$ is called the *second fundamental form*, and L, M, and N are called the *second fundamental quantities*.

Now $\mathbf{N}$ is a unit vector normal to the surface $\mathbf{W}$, and, according to the second equation in Eqs. (III-2), $\mathbf{P}'' = (1/\rho)\mathbf{n}$, where $\mathbf{n}$ is the unit normal vector to the curve $\mathbf{P}$ and $1/\rho$ is its curvature. Therefore

$$\mathbf{N} \cdot \mathbf{P}'' = (\cos\theta)/\rho,$$

where θ is the angle between the surface normal **N** and the curve normal **n**. Therefore we have

$$(\cos\theta)/\rho = (Lu_t^2 + 2Mu_tv_t + Nv_t^2)/(Eu_t^2 + 2Fu_tv_t + Gv_t^2). \quad \text{(IX-9)}$$

THE THEOREMS OF MEUSNER AND GAUSS

Select a point on the surface **W**, and through it pass a plane. The plane and the surface **W** will intersect in a curve **P** passing through the chosen point. Its tangent vector **P**′ lies in the plane, and its normal vector, which also lies in the plane, makes an angle θ with the surface normal **N**. Its curvature is $1/\rho$. These quantities we have seen in Eqs. (IX-7) and (IX-9):

$$\mathbf{P}' = (u_t/H)\,\mathbf{W}_u + (v_t/H)\,\mathbf{W}_v\,, \quad \text{(IX-10)}$$

$$(\cos\theta)/\rho = (u_t/H)^2L + 2(u_t/H)(v_t/H)M + (v_t/H)^2N, \quad \text{(IX-11)}$$

where $H^2 = Eu_t^2 + 2Fu_tv_t + Gv_t^2$, as defined previously. Holding the chosen point fixed, there are two motions that we can impart to the plane:

1. We can hold the direction of **P**′ fixed and vary θ, thus imparting a rocking motion to the plane—this leads to Meusner's theorem—or
2. we can hold θ fixed and allow the direction of **P**′ to vary so that the plane will roll around the normal to the surface—this leads to Gauss' theorem.

We shall rock first and roll second.

We hold **P**′ fixed and allow θ to vary. For each value of θ we get a new pair of functions $u(t)$ and $v(t)$. However, since **P**′ is fixed, the ratios u_t/H and v_t/H remain unchanged. This can be seen from Eq. (IX-10). Now, looking at Eq. (IX-11), we can see that the entire right-hand member is constant. The quantities L, M, and N are quantities associated with the surface and have nothing to do with the direction of **P**′, and their coefficients are fixed. Thus

$$(\cos\theta)/\rho = \text{const.}$$

The curvature $1/\rho$ is inversely proportional to the cosine of the angle between the normal to the curve and the normal to the surface. Therefore, the curvature achieves its minimum value (and the radius of curvature attains its maximum) when θ is zero. This is exactly the case when the intersecting plane is normal to the surface at the chosen point; the surface normal **N** lies in this plane.

In this case, when $\theta = 0$, the curve produced is called a *normal section*; its curvature is given by

$$1/\rho = (Lu_t^2 + 2Mu_t v_t + Nv_t^2)/(Eu_t^2 + 2Fu_t v_t + Gv_t^2). \qquad \text{(IX-12)}$$

These results constitute the theorem of Meusner, which is illustrated in Fig. IX-2.

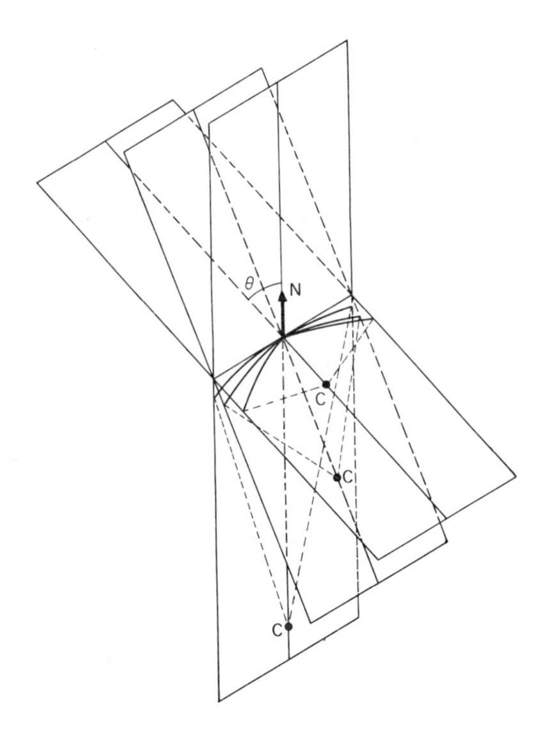

FIG. IX-2. Meusner's theorem.

The plane rocks on a tangent to the surface. **N** is the surface normal. The C's are positions of the center of curvature of the intersection of plane and surface. The theorem states that C is farthest from the surface when $\theta = 0$.

Gauss' theorem is a little more complicated. Here we set θ equal to zero and find what happens to the curvature of the normal section as the plane rotates around the surface normal. It was noted earlier that u_t and v_t were measures of the angles between the tangent vector $\mathbf{P}'$ and the tangent vectors to the parametric curves. Suppose we let $\lambda = v_t/u_t$ so that Eqs. (IX-10) and (IX-12) become

$$\mathbf{P}' = (\mathbf{W}_u + \lambda \mathbf{W}_v)/H, \tag{IX-13}$$

$$1/\rho = (L + 2M\lambda + N\lambda^2)/H^2, \tag{IX-14}$$

where $H^2 = E + 2F\lambda + G\lambda^2$. Clearly, each value of λ corresponds to a unique direction for $\mathbf{P}'$ as well as a distinct value for the curvature $1/\rho$.

Note that, if the normal section is tangent to the parametric curve in the u direction, then $\lambda = 0$ and the curvature of the normal section is L/E. If, however, the normal section is tangent to the parametric curve in the v direction, then $\lambda = \infty$ and the curvature is N/G. In other words

$$1/\rho_u = L/E, \qquad 1/\rho_v = N/G. \tag{IX-15}$$

We have an expression for the curvature of a normal section in terms of the parameter λ. We now find that value of λ that corresponds to the maximum and minimum values of the curvature as the plane rolls around the normal vector $\mathbf{N}$. We do this by differentiating Eq. (IX-14) and setting the result equal to zero:

$$\begin{aligned}(d/d\lambda)(1/\rho) &= (2/H^4)[(E + 2F\lambda + G\lambda^2)(M + N\lambda) - (L + 2M\lambda + N\lambda^2)(F + G\lambda)] \\ &= (2/H^4)[(E + F\lambda)(M + N\lambda) - (L + M\lambda)(F + G\lambda)].\end{aligned}$$

Setting this equal to zero gives us several results. First, we have a quadratic in λ,

$$(FN - GM)\lambda^2 + (EN - GL)\lambda + (EM - FL) = 0. \tag{IX-16}$$

Let λ_1 and λ_2 be the two roots of this quadratic. Then from the two statements above and Eq. (IX-14), we have

$$\begin{aligned}1/\rho_i &= (L + 2M\lambda_i + N\lambda_i^2)/(E + 2F\lambda_i + G\lambda_i^2) \\ &= (M + N\lambda_i)/(F + G\lambda_i) \\ &= (L + M\lambda_i)/(E + F\lambda_i), \qquad i = 1, 2.\end{aligned} \tag{IX-17}$$

Moreover, from a well-known relationship between the roots of a polynomial and its coefficients (see, e.g., Vilhelm, 1969, p. 76), we obtain

$$\begin{aligned}\lambda_1 + \lambda_2 &= -(EN - GL)/(FN - GM), \\ \lambda_1\lambda_2 &= (EM - FL)/(FN - GM).\end{aligned} \tag{IX-18}$$

Thus we have obtained two solutions of the quadratic Eq. (IX-16), Substituting the two roots λ_1 and λ_2 into Eq. (IX-17), we obtain the maximum and minimum values of the curvature of a normal section. These two values are called the *principal curvatures* of the surface at the chosen point. Substituting the two roots into Eq. (IX-13) gives us the two values of $\mathbf{P}'$ corresponding to the two directions in which the curvature of the normal section achieves its maximum and minimum values. These two directions are termed the *principal directions.* It turns out that the two principal directions are perpendicular to each other. This can be seen from the following. From Eq. (IX-13) we may write these as

$$\mathbf{P}_i' = (\mathbf{W}_u + \mathbf{W}_v\lambda_i)/H_i\,, \qquad i = 1, 2, \tag{IX-19}$$

where $H_i^2 = E + 2F\lambda_i + G\lambda_i^2$. Then

$$\begin{aligned}\mathbf{P}_1' \cdot \mathbf{P}_2' &= (\mathbf{W}_u + \mathbf{W}_v\lambda_1)(\mathbf{W}_u + \mathbf{W}_v\lambda_2)/H_1H_2 \\ &= [E + (\lambda_1 + \lambda_2)F + \lambda_1\lambda_2 G]/H_1H_2\,.\end{aligned}$$

When the expressions for the sum and the product of the roots, Eqs. (IX-18), are substituted, the above vanishes. Gauss' theorem is illustrated in Fig. IX-3.

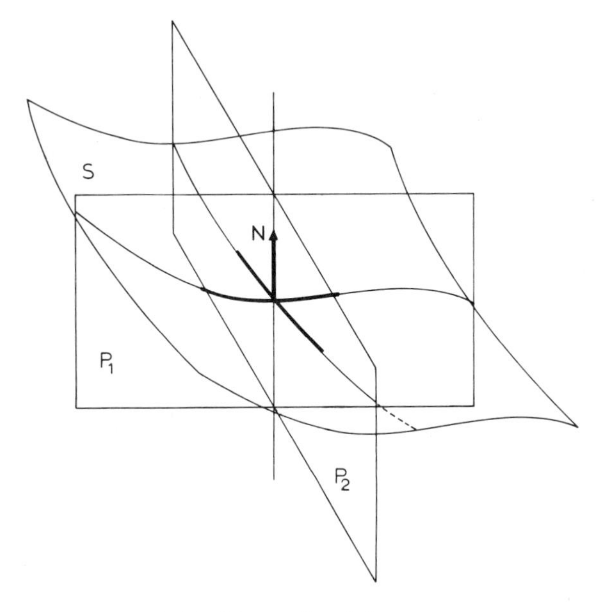

FIG. IX-3. Gauss' theorem.

Shown are the surface normal **N** and two normal sections formed by the intersection of the surface S and the planes P_1 and P_2. The theorem states that, when the curvatures of the normal sections assume their maximum and minimum values, planes P_1 and P_2 are perpendicular.

THE WEINGARTEN EQUATIONS

In this section the derivatives of the surface normals are calculated. The second fundamental quantities have been defined as

$$L = \mathbf{N} \cdot \mathbf{W}_{uu}, \qquad M = \mathbf{N} \cdot \mathbf{W}_{uv}, \qquad N = \mathbf{N} \cdot \mathbf{W}_{vv}.$$

Recall that, from the definition of **N**, Eq. (IX-1),

$$\mathbf{N} \cdot \mathbf{W}_u = 0, \qquad \mathbf{N} \cdot \mathbf{W}_v = 0.$$

Differentiating both of these expressions with respect to u and v, we obtain

$$\mathbf{N} \cdot \mathbf{W}_{uu} + \mathbf{N}_u \cdot \mathbf{W}_u = 0, \qquad \mathbf{N} \cdot \mathbf{W}_{vv} + \mathbf{N}_v \cdot \mathbf{W}_v = 0,$$

$$\mathbf{N} \cdot \mathbf{W}_{uv} + \mathbf{N}_u \cdot \mathbf{W}_v = 0, \qquad \mathbf{N} \cdot \mathbf{W}_{uv} + \mathbf{N}_v \cdot \mathbf{W}_u = 0,$$

providing us with an alternative way of writing the second fundamental quantities:

$$\begin{aligned} \mathbf{N}_u \cdot \mathbf{W}_u &= -L, \qquad & \mathbf{N}_v \cdot \mathbf{W}_v &= -N, \\ \mathbf{N}_u \cdot \mathbf{W}_v &= -M, \qquad & \mathbf{N}_v \cdot \mathbf{W}_u &= -M. \end{aligned} \tag{IX-20}$$

Moreover, since **N** is a unit vector, $\mathbf{N}^2 = 1$. Differentiating this results in $\mathbf{N} \cdot \mathbf{N}_u = \mathbf{N} \cdot \mathbf{N}_v = 0$. Therefore, $\mathbf{N}_u$ and $\mathbf{N}_v$, being perpendicular to **N**, can be written as linear combinations of $\mathbf{W}_u$ and $\mathbf{W}_v$:

$$\mathbf{N}_u = \alpha \mathbf{W}_u + \beta \mathbf{W}_v, \qquad \mathbf{N}_v = \gamma \mathbf{W}_u + \delta \mathbf{W}_v.$$

The scalar product of these two equations with $\mathbf{W}_u$ and $\mathbf{W}_v$ results in

$$\alpha E + \beta F = -L, \qquad \gamma E + \delta F = -M,$$

$$\alpha F + \beta G = -M, \qquad \gamma F + \delta G = -N.$$

Solving for α, β, γ, and δ and substituting back into the original expression gives us

$$\begin{aligned} \mathbf{N}_u &= [(FM - GL)\,\mathbf{W}_u + (FL - EM)\,\mathbf{W}_v]/(EG - F^2), \\ \mathbf{N}_v &= [(FN - GM)\,\mathbf{W}_u + (FM - EN)\,\mathbf{W}_v]/(EG - F^2). \end{aligned} \tag{IX-21}$$

These are known as the *Weingarten equations.*

A rather important special case needs to be considered. When the two parametric curves intersect at a right angle, their tangent vectors are perpendicular and therefore

$$F = \mathbf{W}_u \cdot \mathbf{W}_v = 0.$$

Then the derivatives of the normal vector are

$$\begin{aligned} \mathbf{N}_u &= -(L/E)\,\mathbf{W}_u - (M/G)\,\mathbf{W}_v\,, \\ \mathbf{N}_v &= -(M/E)\,\mathbf{W}_u - (N/G)\,\mathbf{W}_v\,. \end{aligned} \qquad \text{(IX-22)}$$

TRANSFORMATION OF THE PARAMETERS

The parameters used in representing a surface may be changed at will without altering the shape and structure of the surface in any way. This can be done by setting the original parameters equal to two functions of two variables, the variables being the new parameters

$$u = u(\xi, \eta), \qquad v = v(\xi, \eta),$$

so that the vector representation of the surface becomes

$$\mathbf{W}(\xi, \eta) = \mathbf{W}(u(\xi, \eta), v(\xi, \eta)).$$

If the original parameters are essential parameters, then the new set is essential if the Jacobian of the transform is not zero, i.e.,

$$J = \begin{vmatrix} u_\xi & u_\eta \\ v_\xi & v_\eta \end{vmatrix} = u_\xi v_\eta - u_\eta v_\xi \neq 0. \qquad \text{(IX-23)}$$

The nonvanishing Jacobian also ensures the existence of the inverse transform.

The transformation induces a new set of parametric curves, the tangent vectors of which are given by

$$\mathbf{W}_\xi = \mathbf{W}_u u_\xi + \mathbf{W}_v v_\xi\,, \qquad \mathbf{W}_\eta = \mathbf{W}_u u_\eta + \mathbf{W}_v v_\eta\,.$$

If we define the first fundamental quantities relating to the new system of parameters,

$$e = \mathbf{W}_\xi^2, \qquad f = \mathbf{W}_\xi \cdot \mathbf{W}_\eta\,, \qquad g = \mathbf{W}_\eta^2,$$

then

$$
\begin{aligned}
e &= Eu_\xi^2 + 2Fu_\xi v_\xi + Gv_\xi^2 , \\
f &= Eu_\xi u_\eta + F(u_\xi v_\eta + u_\eta v_\xi) + Gv_\xi v_\eta , \\
g &= Eu_\eta^2 + 2Fu_\eta v_\eta + Gv_\eta^2.
\end{aligned} \tag{IX-24}
$$

The corresponding inverse transform is

$$
\begin{aligned}
E &= [ev_\eta^2 - 2fv_\xi v_\eta + gv_\xi^2]/J^2 , \\
F &= -[eu_\eta v_\eta - f(u_\xi v_\eta + u_\eta v_\xi) + gu_\xi v_\xi]/J^2 , \\
G &= [eu_\eta^2 - 2fu_\xi u_\eta + gu_\xi^2]/J^2.
\end{aligned} \tag{IX-25}
$$

The reader might like to show that

$$
\mathbf{N} = (\mathbf{W}_u \times \mathbf{W}_v)/(EG - F^2)^{1/2} = (\mathbf{W}_\xi \times \mathbf{W}_\eta)/(eg - f^2)^{1/2},
$$

underlining the fact mentioned previously that the geometrical properties of the surface are invariant under a transformation of the parameters.

The second derivatives are

$$
\begin{aligned}
\mathbf{W}_{\xi\xi} &= \mathbf{W}_{uu} u_\xi^2 + 2\mathbf{W}_{uv} u_\xi v_\xi + \mathbf{W}_{vv} v_\xi^2 + \mathbf{W}_u u_{\xi\xi} + \mathbf{W}_v v_{\xi\xi} , \\
\mathbf{W}_{\xi\eta} &= \mathbf{W}_{uu} u_\xi u_\eta + \mathbf{W}_{uv}(u_\xi v_\eta + u_\eta v_\xi) + \mathbf{W}_{vv} v_\xi v_\eta + \mathbf{W}_u u_{\xi\eta} + \mathbf{W}_v v_{\xi\eta} , \\
\mathbf{W}_{\eta\eta} &= \mathbf{W}_{uu} u_\eta^2 + 2\mathbf{W}_{uv} u_\eta v_\eta + \mathbf{W}_{vv} v_\eta^2 + \mathbf{W}_u u_{\eta\eta} + \mathbf{W}_v v_{\eta\eta} .
\end{aligned}
$$

The scalar product of these with **N** produce

$$
\begin{aligned}
l &= Lu_\xi^2 + 2Mu_\xi v_\xi + Nv_\xi^2, \\
m &= Lu_\xi u_\eta + M(u_\xi v_\eta + u_\eta v_\xi) + Nv_\xi v_\eta , \\
n &= Lu_\eta^2 + 2Mu_\eta v_\eta + Nv_\eta^2,
\end{aligned} \tag{IX-26}
$$

where, of course,

$$
l = \mathbf{N} \cdot \mathbf{W}_{\xi\xi} , \qquad m = \mathbf{N} \cdot \mathbf{W}_{\xi\eta} , \qquad n = \mathbf{N} \cdot \mathbf{W}_{\eta\eta} .
$$

The inverse transformation is

$$
\begin{aligned}
L &= [lv_\eta^2 - 2mv_\xi v_\eta + nv_\xi^2]/J^2 , \\
M &= -[lu_\eta v_\eta - m(u_\xi v_\eta + u_\eta v_\xi) + nu_\xi v_\xi]/J^2 , \\
N &= [lu_\eta^2 - 2mu_\xi u_\eta + nu_\xi^2]/J^2.
\end{aligned} \tag{IX-27}
$$

Here J is the Jacobian given by Eq. (IX-23).

Now let us apply these to the results of the preceding section. Let the ξ and η directions be the two principal directions defined by the two roots, λ_1 and λ_2, of Eq. (IX-16). Let $v_\xi = \lambda_1 u_\xi$ and let $v_\eta = \lambda_2 u_\eta$. Substituting these into Eqs. (IX-24) and (IX-26) yields

$$
\begin{aligned}
e &= u_\xi^2(E + 2F\lambda_1 + G\lambda_1^2), \\
f &= u_\xi u_\eta[E + F(\lambda_1 + \lambda_2) + G\lambda_1\lambda_2], \\
g &= u_\eta^2(E + 2F\lambda_2 + G\lambda_2^2), \\
l &= u_\xi^2(L + 2M\lambda_1 + N\lambda_1^2), \\
m &= u_\xi u_\eta[L + M(\lambda_1 + \lambda_2) + N\lambda_1\lambda_2], \\
n &= u_\eta^2(L + 2M\lambda_2 + N\lambda_2^2).
\end{aligned}
\tag{IX-28}
$$

We have already seen that $f = 0$; by substituting Eq. (IX-18) into the above expression for m, we find that it also vanishes. We conclude that normal sections in the two principal directions not only are perpendicular but are also characterized by $m = 0$. Moreover, from Eq. (IX-17),

$$
1/\rho_\xi = l/e, \qquad 1/\rho_\eta = n/g. \tag{IX-29}
$$

Next, let us specialize these results slightly. Suppose that the u and v directions are orthogonal; that is to say, let $F = 0$. Then Eqs. (IX-24) and (IX-25) become

$$
\begin{aligned}
e &= Eu_\xi^2 + Gv_\xi^2, & E &= (ev_\eta^2 + gv_\xi^2)/J^2, \\
g &= Eu_\eta^2 + Gv_\eta^2, & G &= (eu_\eta^2 + gu_\xi^2)/J^2, \\
0 &= Eu_\xi u_\eta + Gv_\xi v_\eta, & 0 &= eu_\eta v_\eta + gu_\xi v_\xi.
\end{aligned}
$$

Now we can regard one system as being obtained from the other by means of a rotation. Suppose the angle between $\mathbf{W}_\xi$ and $\mathbf{W}_u$ (and therefore between $\mathbf{W}_\eta$ and $\mathbf{W}_v$) is θ. Then

$$
(\mathbf{W}_\xi/e^{1/2}) \cdot (\mathbf{W}_u/E^{1/2}) = \cos\theta, \qquad (\mathbf{W}_\eta/g^{1/2}) \cdot (\mathbf{W}_v/G^{1/2}) = \cos\theta.
$$

Using the facts that $F = 0$ and that

$$
\mathbf{W}_\xi = \mathbf{W}_u u_\xi + \mathbf{W}_v v_\xi, \qquad \mathbf{W}_\eta = \mathbf{W}_u u_\eta + \mathbf{W}_v v_\eta,
$$

we obtain

$$
\begin{aligned}
\mathbf{W}_\xi \cdot \mathbf{W}_u &= Eu_\xi = (eE)^{1/2}\cos\theta, \\
\mathbf{W}_\eta \cdot \mathbf{W}_v &= Gv_\eta = (gG)^{1/2}\cos\theta, \\
\mathbf{W}_\xi \cdot \mathbf{W}_v &= Gv_\xi = (eG)^{1/2}\sin\theta, \\
\mathbf{W}_\eta \cdot \mathbf{W}_u &= Eu_\eta = -(gE)^{1/2}\sin\theta,
\end{aligned}
$$

whence

$$u_\xi = (e/E)^{1/2} \cos\theta, \qquad u_\eta = -(g/E)^{1/2} \sin\theta,$$
$$v_\xi = (e/G)^{1/2} \sin\theta, \qquad v_\eta = (g/G)^{1/2} \cos\theta,$$
$$J = (eg/EG)^{1/2}.$$

Substituting these relations into Eq. (IX-27) results in

$$\begin{aligned} L/E &= (l/e)\cos^2\theta + (n/g)\sin^2\theta, \\ 2M/(EG)^{1/2} &= (l/e - n/g)\sin 2\theta, \\ N/G &= (l/e)\sin^2\theta + (n/g)\cos^2\theta. \end{aligned} \tag{IX-30}$$

Expressed in slightly different form, using Eqs. (IX-15) and (IX-29), we have Euler's theorem for normal curvatures:

$$1/\rho_u = (1/\rho_\xi)\cos^2\theta + (1/\rho_\eta)\sin^2\theta,$$
$$1/\rho_v = (1/\rho_\xi)\sin^2\theta + (1/\rho_\eta)\cos^2\theta,$$

where $1/\rho_\xi$ and $1/\rho_\eta$ are the two principal curvatures in the x and y directions and where $1/\rho_u$ and $1/\rho_v$ are the curvatures of two orthogonal normal sections in the u and v directions. The u direction makes an angle θ with the x direction.

Adding the two above equations gives the theorem of Dupin,

$$1/\rho_u + 1/\rho_v = 1/\rho_\xi + 1/\rho_\eta\,; \tag{IX-31}$$

i.e., the sum of the normal curvatures in two orthogonal directions is constant and equal to the sum of the two principal curvatures.

From Eq. (IX-26) we obtain the reciprocal relationships to Eq. (IX-30),

$$l/e = (L/E)\cos^2\theta + [2M/(EG)^{1/2}]\cos\theta\sin\theta + (N/G)\sin^2\theta,$$
$$0 = -(L/E - N/G)\sin 2\theta + [2M/(EG)^{1/2}]\cos 2\theta,$$
$$n/g = (L/E)\sin^2\theta - [2M/(EG)^{1/2}]\cos\theta\sin\theta + (N/G)\cos^2\theta;$$

from the second of these we are provided with an expression for the angle θ,

$$\tan 2\theta = \frac{2M/(EG)^{1/2}}{N/G - L/E}.$$

Now to summarize our results. First we introduce a new symbol

$$1/\sigma = M/(EG)^{1/2}, \tag{IX-32}$$

which, as we will see later, is related to the torsion of a geodesic curve on a surface. Recalling that the principal curvatures are $1/\rho_\xi$ and $1/\rho_\eta$, the curvatures and σ associated with a pair of normal sections making an angle θ with the principal directions are given by

$$\begin{aligned} 1/\rho_u &= (1/\rho_\xi)\cos^2\theta + (1/\rho_\eta)\sin^2\theta, \\ 1/\rho_v &= (1/\rho_\xi)\sin^2\theta + (1/\rho_\eta)\cos^2\theta, \\ 2/\sigma &= (1/\rho_\xi - 1/\rho_\eta)\sin 2\theta. \end{aligned} \tag{IX-33}$$

The inverse relationships are as follows. Given the curvatures of two orthogonal normal sections and the quantity σ, the principal curvatures and θ, the angle between the principal directions and the normal sections, are given by

$$\begin{aligned} 1/\rho_\xi &= (1/\rho_u)\cos^2\theta + (1/\rho_v)\sin^2\theta + (2/\sigma)\cos\theta\sin\theta, \\ 1/\rho_\eta &= (1/\rho_u)\sin^2\theta + (1/\rho_v)\cos^2\theta - (2/\sigma)\cos\theta\sin\theta, \\ \tan 2\theta &= (2/\sigma)/(1/\rho_v - 1/\rho_u). \end{aligned} \tag{IX-34}$$

GENERALIZED RAY TRACING

The main purpose of the preceding portion of this chapter is to set the stage for the work we are to do here. We have looked rather extensively at some of the properties of surfaces. At this point we apply these results to the study of wavefronts. We will call the results of our work *generalized ray tracing*.

We have already seen how ray tracing works. A surface $\mathbf{R}$ separates two media of constant refractive index. An incident ray, with a direction vector $\mathbf{N}$, intercepts the refracting surface at some point $\mathbf{R}_0$, giving rise to a refracted ray whose direction vector is $\mathbf{N}'$. The relation between incident and refracted ray and $\bar{\mathbf{N}}$, the unit normal to the refracting surface at $\mathbf{R}_0$, is given by Snell's law, Eq. (V-19),

$$\mathbf{N}' \times \bar{\mathbf{N}} = \mu(\mathbf{N} \times \bar{\mathbf{N}}),$$

where μ is the ratio of the two refractive indices. Indeed, we have seen that the direction vector of the refracted ray can be expressed as a linear combination of the incident ray vector and the surface normal,

$$\mathbf{N}' = \mu\mathbf{N} + \gamma\bar{\mathbf{N}},$$

where γ can be written in several ways,

$$\gamma = -\mu \cos i + (1 - \mu^2 \sin^2 i)^{1/2}$$

$$= -(\mathbf{N} \cdot \bar{\mathbf{N}}) + [1 - \mu^2(\mathbf{N} \times \bar{\mathbf{N}})^2]^{1/2}$$

$$= -\mu \cos i + \cos i'.$$

Here i and i' are the angles of incidence and refraction, respectively. This equation coupled with a transfer equation enables one to chart the course of a ray through any optical system consisting of an arrangement of lenses made of material having constant refractive index.

However, we may regard each ray traced through an optical system as being associated with a wavefront that propagates through the optical system. This wavefront in the neighborhood of the ray will have well-defined geometric properties; it will have, in general, two principal directions and two principal curvatures at the point where the ray intercepts the wavefront. The ray vector $\mathbf{N}$ will of course be the unit normal to the wavefront at the point $\mathbf{R}_0$.

The question we pose is this: Given an incident ray and the principal directions and curvatures of the wavefront associated with that ray at the point where the ray meets the wavefront, can we determine the principal directions and curvatures of that wavefront at some subsequent location? In other words, can we tell what happens to the principal directions and principal curvatures of the wavefront after transfer and after refraction. Figure IX-4 illustrates this process. The answer is an unequivocal "yes." The equations were worked out by Gullstrand (1906) in the early part of this century and again by Kneisly (1964) during the past decade using the more compact vector notation. We will follow Kneisly's derivation.

The simplest part of the problem is the matter of transfer. We saw at the end of the last chapter that a family of wavefronts can be represented in vector form by $\mathbf{W}(u, v, s)$, where u and v are surface parameters for an individual wavefront and s represents the arc length parameter along a ray. The unit normal vector to the wavefront is, as always, given by

$$\mathbf{N} = (\mathbf{W}_u \times \mathbf{W}_v)/(EG - F^2)^{1/2},$$

which also serves as the ray direction vector. However, since a ray is a straight line in a homogeneous medium, $\mathbf{N}$ will not depend on the arc length parameter s. In addition, as the ray vector, $\mathbf{N}$ can also be written as the partial derivative of $\mathbf{W}$ with respect to s: $\mathbf{N} = \partial\mathbf{W}/\partial s$. That $\mathbf{N}$

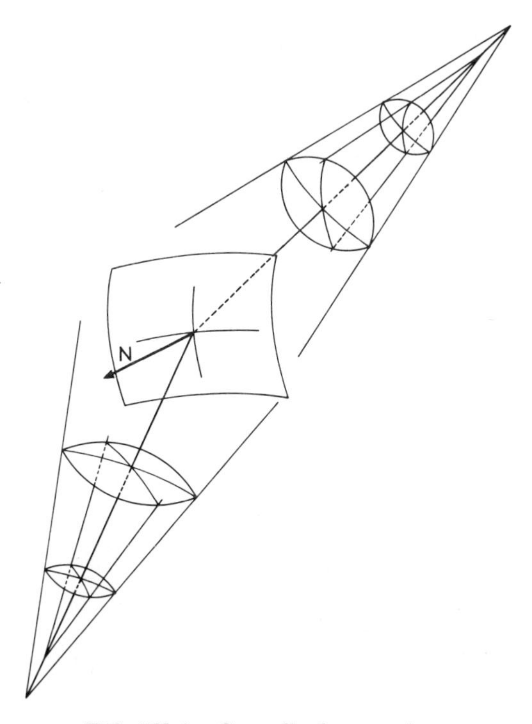

FIG. IX-4. Generalized ray tracing.

The incident ray and wavefronts, showing the two centers of curvature, the refracting surface and its normal vector **N**, and the refracted ray and its two centers of curvature.

does not depend on s can be seen from the ray equation (II-19) for a medium of constant refractive index, in which $\nabla n = 0$,

$$(d/ds)(n\mathbf{P}') = n(\partial/\partial s)(\partial \mathbf{W}/\partial s) = n(\partial/\partial s)\mathbf{N} = 0.$$

Now suppose we examine two positions of the wavefront,

$$\mathbf{W}(u, v, s_0) \qquad \text{and} \qquad \mathbf{W}(u, v, s_0 + s).$$

We may write the relationship between these two as

$$\mathbf{W}(u, v, s_0 + s) = \mathbf{W}(u, v, s_0) + s\mathbf{N}(u, v)$$

or for brevity $\mathbf{W}' = \mathbf{W} + s\mathbf{N}$. We will assume that in what follows the parametric curves are in the principal directions and that, therefore, $f = 0$ and $m = 0$. To be consistent, we will use the ξ, η parametrization to denote that these directions are principal directions.

Differentiation gives us

$$\mathbf{W}_\xi' = \mathbf{W}_\xi + s\mathbf{N}_\xi, \qquad \mathbf{W}_\eta' = \mathbf{W}_\eta + s\mathbf{N}_\eta .$$

Now using the formulas for the derivatives of **N**, the Weingarten equations, Eqs. (IX-22), we get

$$\begin{aligned} \mathbf{W}_\xi' &= (1 - sl/e)\,\mathbf{W}_\xi, \\ \mathbf{W}_\eta' &= (1 - sn/g)\,\mathbf{W}_\eta, \end{aligned} \tag{IX-35}$$

from which we calculate

$$e' = (1 - sl/e)^2 e, \qquad f' = 0, \qquad g' = (1 - sn/g)^2 g. \tag{IX-36}$$

The second derivatives are

$$\begin{aligned} \mathbf{W}'_{\xi\xi} &= (1 - sl/e)\,\mathbf{W}_{\xi\xi} + (1 - sl/e)_\xi \mathbf{W}_\xi, \\ \mathbf{W}'_{\xi\eta} &= (1 - sl/e)\,\mathbf{W}_{\xi\eta} + (1 - sl/e)_\eta \mathbf{W}_\xi \\ &= (1 - sn/g)\,\mathbf{W}_{\xi\eta} + (1 - sn/g)_\xi \mathbf{W}_\eta, \\ \mathbf{W}'_{\eta\eta} &= (1 - sn/g)\,\mathbf{W}_{\eta\eta} + (1 - sn/g)_\eta \mathbf{W}_\eta, \end{aligned}$$

whence

$$l' = (1 - sl/e)l, \qquad m' = 0, \qquad n' = (1 - sn/g)n. \tag{IX-37}$$

These equations make it suite clear that the principal directions of a wavefront are unchanged by transfer. Concerning the principal curvatures,

$$\begin{aligned} 1/\rho_\xi' &= l'/e' = (1 - sl/e)l/(1 - sl/e)^2 e \\ &= 1/(1 - s/\rho_\xi)\,\rho_\xi \\ &= 1/(\rho_\xi - s), \end{aligned} \tag{IX-38}$$

$$1/\rho_\eta' = 1/(\rho_\eta - s),$$

exactly and precisely the result that one would expect; the centers of curvature are fixed. They remain the same for each position of the wavefront.

The problem of refraction is considerably more involved. Let **W** and **W**′ represent the incident and refracted wavefronts, respectively. Let **N** and **N**′ represent the unit normal vectors or, equivalently, the direction vectors of an incident and a refracted ray. Let $\mathbf{R}_0$ be the point of intersection of these rays on the refracting surface **R**, and let the normal to the refracting surface at that point be designated by $\bar{\mathbf{N}}$. Then Snell's law applies and the ray tracing formula,

$$\mathbf{N}' = \mu\mathbf{N} + \gamma\bar{\mathbf{N}},$$

is valid. We also need the scalar form of Snell's law,

$$\sin i' = \mu \sin i.$$

We next define a unit vector. Let

$$\mathbf{P} = (\mathbf{N} \times \bar{\mathbf{N}})/\sin i. \tag{IX-39}$$

Now note that, from the vector and scalar forms of Snell's law, Eqs. (V-19) and (V-20),

$$(\mathbf{N} \times \bar{\mathbf{N}})/\sin i = (\mathbf{N}' \times \bar{\mathbf{N}})/\sin i',$$

so that $\mathbf{P}$ is perpendicular to $\mathbf{N}$, $\bar{\mathbf{N}}$, and $\mathbf{N}'$. In other words, $\mathbf{P}$ is perpendicular to the plane of incidence. Moreover, since $\mathbf{N}$ is perpendicular to $\mathbf{W}$, $\bar{\mathbf{N}}$ to $\bar{\mathbf{R}}$, and $\mathbf{N}'$ to $\mathbf{W}'$, $\mathbf{P}$ is a tangent vector to the two wavefronts and to the refracting surface.

Finally, we define three additional unit vectors,

$$\mathbf{Q} = \mathbf{N} \times \mathbf{P}, \qquad \bar{\mathbf{Q}} = \bar{\mathbf{N}} \times \mathbf{P}, \qquad \mathbf{Q}' = \mathbf{N}' \times \mathbf{P}, \tag{IX-40}$$

each perpendicular to $\mathbf{P}$ and each tangent to the appropriate surface.

We now have three orthogonal unit vectors associated with each surface:

$$\mathbf{W}: \ \mathbf{N}, \mathbf{P}, \mathbf{Q}; \qquad \mathbf{R}: \ \bar{\mathbf{N}}, \mathbf{P}, \bar{\mathbf{Q}}; \qquad \mathbf{W}': \ \mathbf{N}', \mathbf{P}, \mathbf{Q}',$$

which we will regard as providing a local coordinate system for the incident wavefront, the refracting surface, and the refracted wavefront. These are shown in Fig. IX-5. Each set of vectors obeys the following relationships:

$$\mathbf{Q} = \mathbf{N} \times \mathbf{P}, \qquad \mathbf{N} = \mathbf{P} \times \mathbf{Q}, \qquad \mathbf{P} = \mathbf{Q} \times \mathbf{N}. \tag{IX-41}$$

Moreover, the vector product of $\mathbf{P}$ with the ray-tracing formula yields

$$\mathbf{Q}' = \mu\mathbf{Q} + \gamma\bar{\mathbf{Q}}.$$

These three systems of unit vectors provide us with a means of inducing a convenient parametrization on the three surfaces. The vector $\mathbf{P}$ is tangent to all three surfaces and we may therefore define on

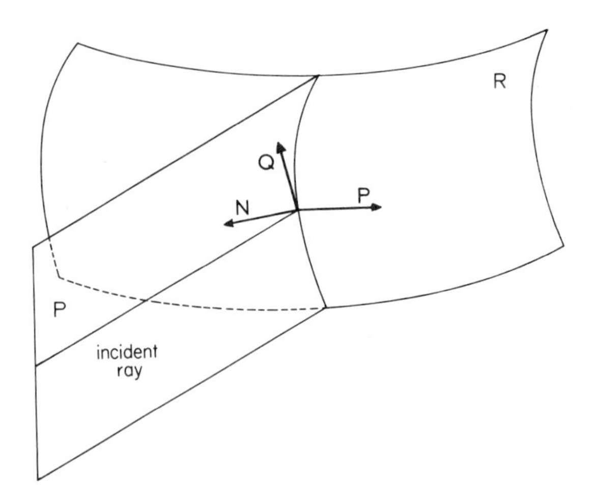

FIG. IX-5. The **N**, **P**, **Q** *coordinate system.*

N is the unit normal vector to the refracting surface R. **P** is a unit vector perpendicular to **N** and the incident ray and therefore perpendicular to P, the plane of incidence. **Q** is a unit vector perpendicular to **N** and **P**.

each of the surfaces a parametric curve that has **P** as its unit tangent vector:

$$\mathbf{W}_u/E^{1/2} = \mathbf{R}_u/\bar{E}^{1/2} = \mathbf{W}_u'/E'^{1/2} = \mathbf{P}.$$

However, more can be said here. Each of the three quantities, $E^{1/2}$, $\bar{E}^{1/2}$, and $E'^{1/2}$, is the derivative of arc length with respect to the parameter u. These three curves have a high degree of contact with each other, and therefore their arc lengths, considered as functions of u, in a region sufficiently close to $\mathbf{R}_0$ are equal. Therefore, $E = \bar{E} = E'$ and

$$\mathbf{W}_u = \mathbf{R}_u = \mathbf{W}_u' = E^{1/2}\mathbf{P}. \tag{IX-42}$$

We do the same sort of thing with the **Q** vectors. The plane of incidence intersects each of the three surfaces in a curve that is tangent to the appropriate **Q** vector. The unit tangent vector in each case is equal to the **Q** vector associated with that surface. Therefore

$$\mathbf{W}_v/G^{1/2} = \mathbf{Q}, \qquad \mathbf{R}_v/\bar{G}^{1/2} = \bar{\mathbf{Q}}, \qquad \mathbf{W}_v'/G'^{1/2} = \mathbf{Q}'. \tag{IX-43}$$

In this case the derivatives of arc length are not the same. Kneisly obtained relationships between G, $\bar{G}$, and G' and the arc length derivatives. However, this step is unnecessary; these relationships will fall out of our final result.

The method used to induce the parametrization on the three surfaces ensures that the parametric curves at $\mathbf{R}_0$ are orthogonal. As a consequence, $F = \bar{F} = F' = 0$.

The crucial step in this development is the differentiation of the ray tracing equation with respect to the newly defined parameters. To the result we apply the formulas for the derivatives of the surface normal. Differentiating

$$\mathbf{N}' - \mu\mathbf{N} - \gamma\bar{\mathbf{N}} = \mathbf{0}$$

with respect to u results in

$$\mathbf{N}_u' - \mu\mathbf{N}_u - \gamma\bar{\mathbf{N}}_u - \gamma_u\bar{\mathbf{N}} = \mathbf{0}.$$

The vector product of this and $\bar{\mathbf{N}}$ gives us

$$(\mathbf{N}_u' - \mu\mathbf{N}_u - \gamma\bar{\mathbf{N}}_u) \times \bar{\mathbf{N}} = \mathbf{0}. \qquad \text{(IX-44)}$$

In exactly the same way we obtain

$$(\mathbf{N}_v' - \mu\mathbf{N}_v - \gamma\bar{\mathbf{N}}_v) \times \bar{\mathbf{N}} = \mathbf{0}. \qquad \text{(IX-45)}$$

Now, using the expressions for the derivatives of $\mathbf{N}$,

$$\begin{aligned}
\mathbf{N}_u' \times \bar{\mathbf{N}} &= -[(L'/E')\,\mathbf{W}_u' + (M'/G')\,\mathbf{W}_v'] \times \bar{\mathbf{N}} \\
&= -[(L'/E'^{1/2})\mathbf{P} + (M'/G'^{1/2})\,\mathbf{Q}'] \times \bar{\mathbf{N}} \\
&= (L'/E'^{1/2})\bar{\mathbf{Q}} + (M'/G'^{1/2})[\bar{\mathbf{N}} \times (\mathbf{N}' \times \mathbf{P})] \\
&= (L'/E'^{1/2})\bar{\mathbf{Q}} - [(M' \cos i')/G'^{1/2}]\mathbf{P}.
\end{aligned}$$

We have used Eqs. (IX-41), (IX-42), and (IX-43). By the same methods we can show that

$$\begin{aligned}
\mathbf{N}_u \times \bar{\mathbf{N}} &= (L/E^{1/2})\bar{\mathbf{Q}} - [(M \cos i)/G^{1/2}]\mathbf{P}, \\
\bar{\mathbf{N}}_u \times \bar{\mathbf{N}} &= (\bar{L}/\bar{E}^{1/2})\bar{\mathbf{Q}} - (\bar{M}/\bar{G}^{1/2})\mathbf{P}, \\
\mathbf{N}_v' \times \bar{\mathbf{N}} &= (M'/E'^{1/2})\bar{\mathbf{Q}} - [(N' \cos i')/G'^{1/2}]\mathbf{P}, \\
\mathbf{N}_v \times \bar{\mathbf{N}} &= (M/E^{1/2})\bar{\mathbf{Q}} - [(N \cos i)/G^{1/2}]\mathbf{P}, \\
\bar{\mathbf{N}}_v \times \bar{\mathbf{N}} &= (\bar{M}/\bar{E}^{1/2})\bar{\mathbf{Q}} - (\bar{N}/\bar{G}^{1/2})\mathbf{P}.
\end{aligned}$$

Substituting the first three of these into Eq. (IX-44) and collecting terms results in

$$\begin{aligned}
&(L'E'^{1/2} - \mu L/E^{1/2} - \gamma\bar{L}/\bar{E}^{1/2})\bar{\mathbf{Q}} - (M' \cos i'/G'^{1/2} \\
&\quad -\mu M \cos i/G^{1/2} - \gamma\bar{M}/\bar{G}^{1/2})\mathbf{P} = \mathbf{0}.
\end{aligned}$$

Similarly, by substituting the last three of the above equations into Eq. (IX-45), we obtain

$$(M'/E'^{1/2} - \mu M/E^{1/2} - \gamma \bar{M}/\bar{E}^{1/2})\bar{\mathbf{Q}} - (N' \cos i'/G'^{1/2} \\ -\mu N \cos i/G^{1/2} - \gamma \bar{N}/\bar{G}^{1/2})\mathbf{P} = \mathbf{0}.$$

Now, since $\bar{\mathbf{Q}}$ and $\mathbf{P}$ are orthogonal, the coefficients of the above two expressions must be zero; hence

$$\begin{aligned} L'/E'^{1/2} &= \mu L/E^{1/2} + \gamma \bar{L}/\bar{E}^{1/2}, \\ M' \cos i'/G'^{1/2} &= \mu M \cos i/G^{1/2} + \gamma \bar{M}/\bar{G}^{1/2}, \\ M'/E'^{1/2} &= \mu M/E^{1/2} + \gamma \bar{M}/\bar{E}^{1/2}, \\ N' \cos i'/G'^{1/2} &= \mu N \cos i/G^{1/2} + \gamma \bar{N}/\bar{G}^{1/2}. \end{aligned}$$

But we have already shown that $E = \bar{E} = E'$, so the first and third of the above equations become

$$L' = \mu L + \gamma \bar{L}, \qquad M' = \mu M + \gamma \bar{M}.$$

Upon eliminating M' between the appropriate equations, we get

$$\mu M\left(\frac{\cos i}{G^{1/2}} - \frac{\cos i'}{G'^{1/2}}\right) + \gamma \bar{M}\left(\frac{1}{\bar{G}^{1/2}} - \frac{\cos i'}{G'^{1/2}}\right) = 0.$$

But M and $\bar{M}$ are associated with the properties of the incident wavefront and the refracting surface, respectively, and therefore are independent of each other. This expression then implies that their coefficients vanish, yielding

$$(G'/G)^{1/2} = \cos i'/\cos i, \qquad (G'/\bar{G})^{1/2} = \cos i',$$

or, alternatively, introducing k, a constant of proportionality,

$$G^{1/2} = k \cos i, \qquad \bar{G}^{1/2} = k, \qquad G'^{1/2} = k \cos i'. \tag{IX-46}$$

Applying these relations to the equations for L', M', and N' gives us

$$L' = \mu L + \gamma \bar{L}, \qquad M' = \mu M + \gamma \bar{M}, \qquad N' = \mu N + \gamma \bar{N},$$

a nice, simple set of expressions. However, these are not in the form we want. By using Eq. (IX-46) again, we can get

$$\begin{aligned} L'/E' &= \mu(L/E) + \gamma(\bar{L}/\bar{E}), \\ [M'/(E'G')^{1/2}] \cos i' &= \mu[M/(EG)^{1/2}] \cos i + \gamma \bar{M}/(\bar{E}\bar{G})^{1/2}, \\ (N'/G') \cos^2 i' &= \mu(N/G) \cos^2 i + \gamma(\bar{N}/\bar{G}), \end{aligned} \tag{IX-47}$$

in which form we can easily recognize the expressions for the normal curvatures of the three surfaces. Using Eqs. (IX-15) and $1/\sigma = M/(EG)^{1/2}$ from Eq. (IX-32), the above become

$$\begin{aligned} 1/\rho_u' &= \mu/\rho_u + \gamma/\bar{\rho}_u , \\ (\cos i)/\sigma' &= (\mu \cos i)/\sigma + \gamma/\bar{\sigma} , \\ (\cos^2 i')/\rho_v' &= (\mu \cos^2 i)/\rho_v + \gamma/\bar{\rho}_v . \end{aligned} \tag{IX-48}$$

These quantities are now in a form for use with Eqs. (IX-33) and (IX-34), which we so foresightedly prepared.

We now have all the tools needed to perform what we have called a generalized ray trace. The mechanics still need to be worked out and will be covered in Chapter X.

CAUSTIC SURFACES

In an earlier chapter we touched on the matter of caustic surfaces. Little was said on the subject for the simple reason that we were not yet equipped to handle the material properly. We are now so equipped.

The caustic surface was defined as an envelope to a system of rays, where it was tacitly assumed that the system of rays constituted an orthotomic system. There is, therefore, a family of wavefronts associated with the rays that we may express formally by the equation

$$\mathbf{W}(\xi, \eta, s) = \mathbf{W}(\xi, \eta) + s\mathbf{N}(\xi, \eta).$$

We assume throughout this discussion that the parametric curves are in the principal directions.

Since the caustic surface is an envelope of the system of rays, each ray must be tangent to the caustic at some point. We construct an equation for the caustic in the following way. Each pair of values for ξ and η corresponds to a point on $\mathbf{W}$ and determines a ray passing through that point whose direction vector is $\mathbf{N}(\xi, \eta)$. Define $\delta(\xi, \eta)$ as the distance, along that ray, between the point on $\mathbf{W}$ and the point where the ray is tangent to the caustic. If we denote the vector function describing the caustic by $\mathbf{C}$, then

$$\mathbf{C}(\xi, \eta) = \mathbf{W}(\xi, \eta) + \delta(\xi, \eta)\mathbf{N}(\xi, \eta).$$

Note that the parametrization of $\mathbf{W}$ has been induced on the caustic.

The tangent vectors to the caustic must be

$$\partial \mathbf{C}/\partial \xi = \partial \mathbf{W}/\partial \xi + \delta(\partial \mathbf{N}/\partial \xi) + (\partial \delta/\partial \xi)\mathbf{N},$$
$$\partial \mathbf{C}/\partial \eta = \partial \mathbf{W}/\partial \eta + \delta(\partial \mathbf{N}/\partial \eta) + (\partial \delta/\partial \eta)\mathbf{N},$$

which, by using Eqs. (IX-22) and (IX-29), become

$$\partial \mathbf{C}/\partial \xi = (1 - \delta/\rho_\xi)\, \mathbf{W}_\xi + (\partial \delta/\partial \xi)\mathbf{N},$$
$$\partial \mathbf{C}/\partial \eta = (1 - \delta/\rho_\eta)\, \mathbf{W}_\eta + (\partial \delta/\partial \eta)\mathbf{N}.$$

From these we calculate the first fundamental quantities,

$$E = (1 - \delta/\rho_\xi)^2 e + (\partial \delta/\partial \xi)^2,$$
$$F = (\partial \delta/\partial \xi)(\partial \delta/\partial \eta),$$
$$G = (1 - \delta/\rho_\eta)^2 g + (\partial \delta/\partial \eta)^2.$$

Note that, since the parametric curves on $\mathbf{W}$ are in the principal directions, $\mathbf{W}_\xi \cdot \mathbf{W}_\eta = 0$.

The element of area is

$$EG - F^2 = (1 - \delta/\rho_\xi)^2(1 - \delta/\rho_\eta)^2\, eg + (1 - \delta/\rho_\xi)^2(\partial \delta/\partial \eta)^2 e + (1 - \delta/\rho_\eta)^2(\partial \delta/\partial \xi)^2 g,$$

which is essentially a sum of squares and cannot vanish unless each term vanishes independently. The vector product of the two partial derivatives of $\mathbf{C}$ therefore does not vanish, and we are assured that the ξ and η parameters induced on $\mathbf{C}$ are essential parameters.

A normal vector to the caustic surface is given by this vector product:

$$\begin{aligned}\partial \mathbf{C}/\partial \xi \times \partial \mathbf{C}/\partial \eta &= (1 - \delta/\rho_\xi)(1 - \delta/\rho_\eta)\, \mathbf{W}_\xi \times \mathbf{W}_\eta \\ &\quad + (\partial \delta/\partial \eta)(1 - \delta/\rho_\xi)\, \mathbf{W}_\xi \times \mathbf{N} \\ &\quad - (\partial \delta/\partial \xi)(1 - \delta/\rho_\eta)\, \mathbf{W}_\eta \times \mathbf{N} \\ &= [eg(1 - \delta/\rho_\xi)(1 - \delta/\rho_\eta)\mathbf{N} \\ &\quad - e(\partial \delta/\partial \eta)(1 - \delta/\rho_\xi)\, \mathbf{W}_\eta \\ &\quad - g(\partial \delta/\partial \xi)(1 - \delta/\rho_\eta)\, \mathbf{W}_\xi]/(eg)^{1/2}.\end{aligned}$$

Here we make repeated use of the definition of $\mathbf{N}$.

This normal to the caustic must be perpendicular to the ray vector $\mathbf{N}$,

$$(\partial \mathbf{C}/\partial \xi \times \partial \mathbf{C}/\partial \eta) \cdot \mathbf{N} = (eg)^{1/2}(1 - \delta/\rho_\xi)(1 - \delta/\rho_\eta) = 0,$$

a nice little equation with exactly two solutions,

$$\delta = \rho_\xi \qquad \text{and} \qquad \delta = \rho_\eta\ .$$

The conclusion is simple. The caustic surface consists of the locus of the principal centers of curvature of the wavefront. Figure IX-6 illustrates this.

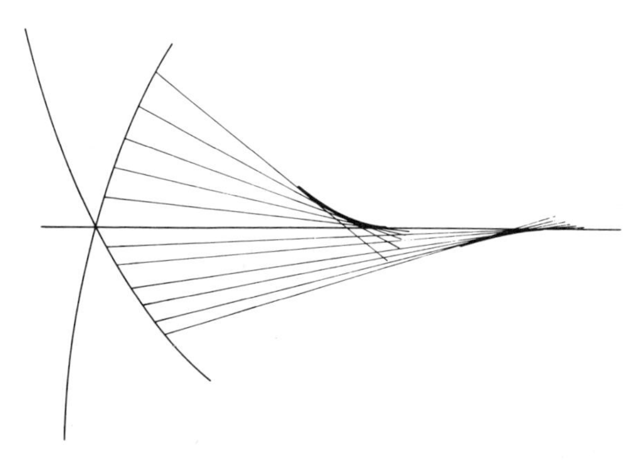

FIG. IX-6. The caustic surface.

The two curves represent normal sections on the two sheets of the caustic surface. The curves are the envelopes of the ray fans in the principal directions of the wavefront.

Since there are in general two principal centers of curvature at each point on the wavefront, the caustic surface can be treated as though it were a two-sheeted surface by setting up a separate equation for each sheet,

$$\mathbf{C}_1 = \mathbf{W} + \rho_\xi \mathbf{N}, \qquad \mathbf{C}_2 = \mathbf{W} + \rho_\eta \mathbf{N}.$$

The tangent vectors to each sheet are therefore

$$\begin{aligned}
\partial \mathbf{C}_1/\partial\xi &= (\partial\rho_\xi/\partial\xi)\mathbf{N},\\
\partial \mathbf{C}_2/\partial\xi &= (1 - \rho_\eta/\rho_\xi)\,\mathbf{W}_\xi + (\partial\rho_\eta/\partial\xi)\mathbf{N},\\
\partial \mathbf{C}_1/\partial\eta &= (1 - \rho_\xi/\rho_\eta)\,\mathbf{W}_\eta + (\partial\rho_\xi/\partial\eta)\mathbf{N},\\
\partial \mathbf{C}_2/\partial\eta &= (\partial\rho_\eta/\partial\eta)\mathbf{N}.
\end{aligned}$$

The first fundamental quantities are therefore

$$\begin{aligned}
E_1 &= (\partial\rho_\xi/\partial\xi)^2, & E_2 &= e(1 - \rho_\eta/\rho_\xi)^2 + (\partial\rho_\eta/\partial\xi)^2,\\
F_1 &= (\partial\rho_\xi/\partial\xi)(\partial\rho_\xi/\partial\eta), & F_2 &= (\partial\rho_\eta/\partial\xi)(\partial\rho_\eta/\partial\eta),\\
G_1 &= g(1 - \rho_\xi/\rho_\eta)^2 + (\partial\rho_\xi/\partial\eta)^2, & G_2 &= (\partial\rho_\eta/\partial\eta)^2,
\end{aligned}$$

and the elements of area are

$$E_1G_1 - F_1^2 = g(\partial\rho_\xi/\partial\xi)^2(1 - \rho_\xi/\rho_\eta)^2,$$
$$E_2G_2 - F_2^2 = e(\partial\rho_\eta/\partial\eta)^2(1 - \rho_\eta/\rho_\xi)^2.$$

Using these, we can show that the unit normal vectors to each sheet of the caustic are

$$\mathbf{N}_1 = -\mathbf{W}_\xi/e^{1/2}, \qquad \mathbf{N}_2 = -\mathbf{W}_\eta/g^{1/2}.$$

Therefore a normal to the caustic surface must lie in one or the other of the principal directions associated with a point on the wavefront.

REFERENCES

Blaschke, W. (1930). "Vorlesungen über Differentialgeometrie," Vol. I, 3rd ed. Springer-Verlag, Berlin and New York. (Reprinted Dover, New York, 1945.)

Gullstrand, A. (1906). Die reelle optische Abbildung. *Sv. Vetensk. Handl.* **41**, 1–119.

Hermann, R. (1968). "Differential Geometry and the Calculus of Variations." Academic Press, New York.

Kneisly, J. A., III. (1964). Local curvature of wavefronts in an optical system, *J. Opt. Soc. Amer.* **54**, 229–235.

Stoker, J. J. (1969). "Differential Geometry." Wiley (Interscience), New York.

Struik, D. J. (1961). "Lectures on Classical Differential Geometry," 2nd ed. Addison-Wesley, Reading, Massachusetts.

Vilhelm, V. (1969). Arithmetic and algebra, *in* "Survey of Applicable Mathematics" (K. Rektorys, ed.). MIT Press, Cambridge, Massachusetts.

Weatherburn, C. E. (1927). "Differential Geometry of Three Dimensions." Cambridge Univ. Press, London and New York.

X

Generalized Ray Tracing

There is nothing really new either in the material we are about to see or in the preliminary work already covered in the last chapter. As has already been mentioned, the basic work was done first by Gullstrand (1906) about fifty years ago and then by Kneisly (1964) within the past ten. Although Altrichter and Schäfer (1956) studied Gullstrand's work and placed it on a somewhat sounder basis, this method has never been used in a lens design program. This is the fate of many innovations in optical design. Lens designers are much too busy to experiment with untried methods and, for the most part, the innovators are not sufficiently skilled as designers to evaluate their methods effectively.

Prior to the advent of the digital computer there was the additional excuse that proposed new methods were too difficult to calculate. Since that time the emphasis has been in programming the tried and true methods for the digital computer. These programs are then coupled with minimization routines that adjust the optical parameters of a lens to (it is hoped) improve its performance.

No one has any reason to complain of this development. The overall result has been a phenomenal improvement in the quality of lenses designed using these techniques. But this is hardly an excuse for complacency. Certainly, a portion of our research effort ought to be expended in a quest for new software that will make the optical design process better, quicker, easier, and cheaper. With such research will come a better understanding of the image-forming process itself.

It is in this spirit that the material in these two chapters is offered. It would be foolish to claim that this is the great breakthrough that will revolutionize the industry. Think of it as a kind of fancy new screwdriver that, because it might make the handyman's life a little easier, deserves a try.

In Chapter VI we saw how ray tracing is done and that it involves two elementary operations, *refraction* and *transfer*. Beginning with some starting point and with an initial ray direction given, these two processes are applied at each refracting surface and in each space separating two refracting surfaces until the ray finally emerges into image space.

Our task here is to extend the idea of ray tracing to include the calculation of those wavefront parameters derived and defined in the preceding chapter. Again, we begin at a starting point and choose an initial ray direction. Suppose that the starting point is an object point. Then we can assume that it is the center of a train of concentric spherical wavefronts each of which is orthogonal to the given ray.

Next, let us assume that the first refracting surface is a sphere whose vertex is located a distance t from the origin of the coordinate system with which the beginning point is referred. To find the point of intersection of the ray with the sphere we use Eqs. (VI-20) to find $\bar{\lambda}$ and $\bar{\mathbf{R}}$. Now $\bar{\lambda}$ is exactly the radius of the spherical wavefront passing through $\bar{\mathbf{R}}$ and orthogonal to the given ray. We may therefore set $\rho_u = \rho_v = \bar{\lambda}$; moreover, since the wavefront is a sphere, there is no principal direction and $1/\sigma = 0$.

Since the refracting surface is also a sphere, it will also have no principal direction and $1/\bar{\sigma} = 0$. Moreover, if the sphere's radius is $r = 1/c$, we may set

$$\bar{\rho}_u = \bar{\rho}_v = r.$$

This transfer operation is shown in Fig. X-1.

Now we are in a position to perform the refraction operation. From Eq. (VI-20) we already have $\bar{\mathbf{N}}$, the normal to the refracting sphere. Then Eqs. (VI-1) and (VI-2) provide $\mathbf{N}'$, the direction vector of the refracted ray, and γ. We also need to calculate i', the angle of refraction. Next we use Eq. (IX-39) to find $\mathbf{P}$, the vector normal to the plane of incidence.

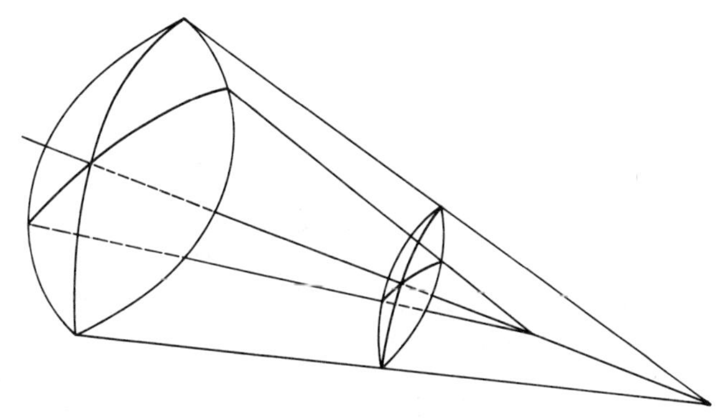

FIG. X-1. The transfer operation.
Principal directions and centers of curvature remain unchanged as the wavefront progresses in a homogeneous medium.

Now we use Eq. (IX-48) to obtain

$$1/\rho_u' = \mu/\bar{\lambda} + \gamma/r,$$

$$1/\sigma' = 0,$$

$$1/\rho_v' = \cos^2 i/\bar{\lambda}\cos^2 i' + \gamma/r\cos^2 i'.$$

Again we have a case where $1/\sigma = 0$. However, here the two curvatures are not equal, and we do have principal directions. The fact that $1/\sigma'$ is zero tells us that we are already oriented with respect to the principal directions and that the **P** vector is in the direction of one of them. This is indicated by relabeling the curvatures

$$\rho_\xi' = \rho_u', \qquad \rho_\eta' = \rho_v'.$$

To avoid confusion later on, define **T** to be a unit vector in the ξ-principal direction, so that $\mathbf{T} = \mathbf{P}$.

Now suppose the next refracting surface is also a sphere located a distance t_1 from the first surface. Again we perform the transfer operation and use Eq. (VI-20) to find $\bar{\lambda}$ and $\bar{\mathbf{R}}$. We use this value of $\bar{\lambda}$ to adjust the principal radii of curvature of the wavefront, as shown in Eq. (IX-38):

$$\rho_\xi = \rho_\xi' - \bar{\lambda}, \qquad \rho_\eta = \rho_\eta' - \bar{\lambda}.$$

The ξ-principal direction is still given by **T**.

Again, the refracting surface is a sphere and we may set

$$\bar{\rho}_u = \bar{\rho}_v = r_1 \qquad \text{and} \qquad 1/\bar{\sigma} = 0.$$

Here, r_1 is the radius of the sphere.

The normal to the refracting surface $\bar{\mathbf{N}}$ has been found from Eq. (VI-20). Then Eqs. (VI-1) and (VI-2) provide $\mathbf{N}'$ and γ, from which we can obtain i', the angle of refraction. Next, we compute $\mathbf{P}$ from Eq. (IX-39) and then $\mathbf{Q}'$ directly from the third equation of Eqs. (IX-40).

Now we need to rotate the coordinate system. First, θ must be found. This is given by $\cos\theta = \mathbf{T}\cdot\mathbf{P}$. Then the rotation is performed using Eq. (IX-33) to obtain ρ_u, ρ_v, and σ:

$$\begin{aligned} 1/\rho_u &= (1/\rho_\xi)\cos^2\theta + (1/\rho_\eta)\sin^2\theta, \\ 1/\rho_v &= (1/\rho_\xi)\sin^2\theta + (1/\rho_\eta)\cos^2\theta, \\ 2/\sigma &= (1/\rho_\xi - 1/\rho_\eta)\sin 2\theta. \end{aligned}$$

Now we are able to use Eq. (VI-43):

$$\begin{aligned} 1/\rho_u' &= \mu/\rho_u + \gamma/r_1, \\ 1/\sigma' &= (\mu\cos i)/(\sigma\cos i'), \\ 1/\rho_v' &= (\mu\cos^2 i)/(\rho_v\cos^2 i') + \gamma/(r_1\cos^2 i'). \end{aligned}$$

Finally, we need to find the principal curvatures and principal directions of the refracted wavefront. Using Eq. (IX-34), we have

$$\begin{aligned} \tan 2\theta' &= (2/\sigma')(1/\rho_v' - 1/\rho_u'), \\ 1/\rho_\xi' &= (1/\rho_u')\cos^2\theta' + (1/\rho_v')\sin^2\theta' + (2/\sigma')\cos\theta'\sin\theta', \\ 1/\rho_\eta' &= (1/\rho_u')\sin^2\theta' + (1/\rho_v')\cos^2\theta' - (2/\sigma')\cos\theta'\sin\theta'. \end{aligned}$$

The vector $\mathbf{T}$ in the x-principal direction is given by

$$\mathbf{T} = \mathbf{P}\cos\theta' + \mathbf{Q}'\sin\theta'.$$

Figure X-2 is an attempt to illustrate the refraction operation.

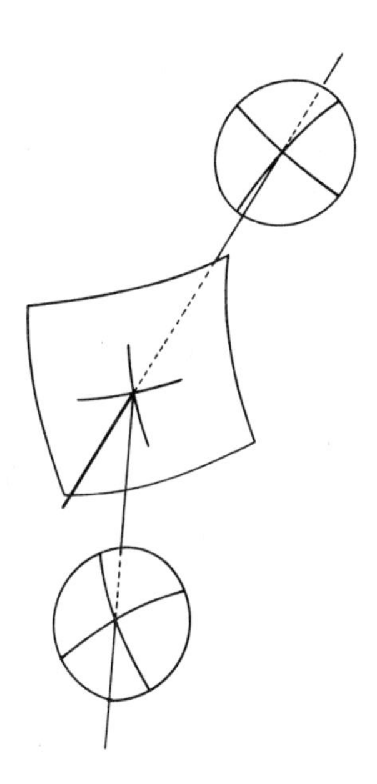

FIG. X-2. The refraction operation. Principal directions and principal curvatures of the refracted wavefront depend on principal directions and curvatures associated with the incident wavefront and the refracting surface.

SPHERICAL REFRACTING SURFACES

To complete this section we need to recapitulate and summarize only what has already been discussed. Let the design parameters of the system be as follows: (1) $r_i = 1/c_i$, the radius and curvature at each spherical surface; (2) t_i, the vertex distance between successive surfaces; (3) n_i, the refractive index between surfaces. A unit vector **A** designates the direction of the system's axis.

Consider first the transfer operation from the *i*th to the $(i + 1)$th surface. Suppose we have the following information on the *i*th surface:

- $\mathbf{R}_i$ the point of intersection of the ray with the *i*th surface,
- $\mathbf{N}_i$ the direction vector of the ray following the *i*th surface,
- $\mathbf{T}_i$ unit vector in the ξ-principal direction,
- $1/\rho_i^\xi$ principal curvature on the ξ direction,
- $1/\rho_i^\eta$ principal curvature in the η direction.

From Eqs. (VI-20), (IX-38), (IX-39), and (IX-33) the transfer equations for spherical surfaces are

$$R^2 = 1 - [(\mathbf{A} \times \mathbf{N}_i)(1 + c_{i+1}t_i) + c_{i+1}(\mathbf{N}_i \times \mathbf{R}_i)]^2, \qquad \text{(X-1a)}$$

$$U = (\mathbf{A} \cdot \mathbf{N}_i)(1 + c_{i+1}t_i) - c_{i+1}(\mathbf{N}_i \cdot \mathbf{R}_i) + R, \qquad \text{(X-1b)}$$

$$\bar{\lambda} = (1/U)[c_{i+1}(\mathbf{R}_i - t_i\mathbf{A})^2 - 2\mathbf{A}\cdot(\mathbf{R}_i - t_i\mathbf{A})], \tag{X-1c}$$

$$\mathbf{R}_{i+1} = \mathbf{R}_i - t_i\mathbf{A} + \bar{\lambda}\mathbf{N}_i \tag{X-1d}$$

$$\bar{\mathbf{N}}_{i+1} = -c_{i+1}\mathbf{R}_{i+1} + \mathbf{A}, \tag{X-1e}$$

$$\rho_{i+1}^{\xi} = \rho_i'^{\xi} - \bar{\lambda}, \qquad \rho_{i+1}^{\eta} = \rho_i'^{\eta} - \bar{\lambda}, \tag{X-1f}$$

$$\cos i = \mathbf{N}_i \cdot \bar{\mathbf{N}}_{i+1}, \tag{X-1g}$$

$$\mathbf{P} = (\mathbf{N}_i \times \bar{\mathbf{N}}_{i+1})/\sin i, \tag{X-1h}$$

$$\cos\theta = \mathbf{P}\cdot\mathbf{T}_i\,, \tag{X-1i}$$

$$1/\rho_u = (1/\rho_{i+1}^{\xi})\cos^2\theta + (1/\rho_{i+1}^{\eta})\sin^2\theta, \tag{X-1j}$$

$$1/\rho_v = (1/\rho_{i+1}^{\xi})\sin^2\theta + (1/\rho_{i+1}^{\eta})\cos^2\theta, \tag{X-1k}$$

$$1/\sigma = \tfrac{1}{2}[(1/\rho_{i+1}^{\xi}) - (1/\rho_{i+1}^{\eta})]\sin 2\theta. \tag{X-1l}$$

For refraction across the $(i + 1)$th surface the following equations are collected from Eqs. (VI-1), (VI-2), (IX-40), (IX-48), and (IX-34):

$$\mu = n_i/n_{i+1}\,, \tag{X-2a}$$

$$\sin i' = \mu\sin i, \tag{X-2b}$$

$$\gamma = -\mu\cos i + \cos i', \tag{X-2c}$$

$$\mathbf{N}_{i+1} = \mu\mathbf{N}_i + \gamma\bar{\mathbf{N}}_{i+1}\,, \tag{X-2d}$$

$$\mathbf{Q}' = \mathbf{N}_{i+1} \times \mathbf{P}, \tag{X-2e}$$

$$1/\rho_u' = \mu/\rho_u + \gamma/r_{i+1}\,, \tag{X-2f}$$

$$1/\sigma' = (\mu\cos i)/(\sigma\cos i'), \tag{X-2g}$$

$$1/\rho_v' = (\mu\cos^2 i)/(\rho_v\cos^2 i') + \gamma/(r_{i+1}\cos^2 i'), \tag{X-2h}$$

$$\tan 2\theta' = (2/\sigma')/(1/\rho_v' - 1/\rho_u'), \tag{X-2i}$$

$$1/\rho_{i+1}'^{\xi} = (1/\rho_u')\cos^2\theta' + (1/\rho_v')\sin^2\theta' + (1/\sigma')\sin 2\theta', \tag{X-2j}$$

$$1/\rho_{i+1}'^{\eta} = (1/\rho_u')\sin^2\theta' + (1/\rho_v')\cos^2\theta' - (1/\sigma')\sin 2\theta', \tag{X-2k}$$

$$\mathbf{T}_{i+1} = \mathbf{P}\cos\theta' + \mathbf{Q}'\sin\theta'. \tag{X-2l}$$

THE GENERAL ROTATIONALLY SYMMETRIC SURFACE

We are accustomed to seeing surfaces described in terms of an expression in which one coordinate is given as a function of the other two, $z = z(x, y)$. If this surface happens to be symmetric with respect to the z axis, then its equation becomes

$$z = z(x^2 + y^2),$$

a surface obtained by rotating the plane curve $z = z(x^2)$ around the z axis.

However, the equations involving the geometric properties of surfaces derived in this and the preceding chapter involve the description of a surface in terms of a vector function of two parameters, $\bar{\mathbf{R}} = \bar{\mathbf{R}}(u, v)$. Since the choice of the two parameters is arbitrary and can be assigned at will, we select in this case the two independent coordinates x and y. The vector describing our general surface of revolution then becomes

$$\bar{\mathbf{R}} = [x, y, z(x^2 + y^2)].$$

All of the equations derived previously can be applied to this vector function. We need only remember to replace u and v by x and y.

Note that z is really a function of one variable

$$z = z(u), \qquad \text{where} \quad u = \tfrac{1}{2}(x^2 + y^2),$$

and we will denote the derivative of z with respect to u by simply z'.

The derivatives of $\bar{\mathbf{R}}$ are

$$\bar{\mathbf{R}}_x = (1, 0, xz'), \qquad \bar{\mathbf{R}}_y = (0, 1, yz'),$$

$$\bar{\mathbf{R}}_{xx} = (0, 0, z' + x^2 z''), \qquad \bar{\mathbf{R}}_{xy} = (0, 0, xyz''), \qquad \bar{\mathbf{R}}_{yy} = (0, 0, z' + y^2 z'').$$

Using Eq. (IX-5), we get the first fundamental quantities,

$$\begin{aligned} E &= 1 + x^2 z'^2, \\ F &= xyz'^2, \\ G &= 1 + y^2 z'^2. \end{aligned} \tag{X-3}$$

In addition, we define

$$K^2 = EG - F^2 = 1 + 2uz'^2. \tag{X-4}$$

The unit normal vector to the surface is, from Eq. (IX-6),

$$\bar{\mathbf{N}} = -(xz', yz', -1)/K. \tag{X-5}$$

Now we use Eq. (IX-8) to get the second fundamental quantities,

$$\begin{aligned} L &= (z' + x^2 z'')/K, \\ M &= xyz''/K, \\ N &= (z' + y^2 z'')/K. \end{aligned} \tag{X-6}$$

Next the principal directions are found. We use the quadratic in Eq. (IX-16),

$$(FN - GM)\lambda^2 + (EN - GL)\lambda + (EM - FL) = 0,$$

into which we substitute the expressions for the first and second fundamental forms from Eqs. (X-3) and (X-6) to obtain

$$xy\lambda^2 + (x^2 - y^2)\lambda - xy = 0, \tag{X-7}$$

whose solutions are

$$\lambda_1 = -x/y, \qquad \lambda_2 = y/x. \tag{X-8}$$

Before going further, it must be noted that, in passing from Eq. (IX-16) to Eq. (X-7), each term of the quadratic had as a factor $z'^3 - z''$. Should this vanish, the quadratic would not possess a solution. It is a simple matter to show that this happens only when the surface is a sphere and therefore has no principal directions. Considered as a differential equation, $z'^3 - z'' = 0$, has the general solution

$$z = -(a^2 - 2u)^{1/2} + b,$$

where a and b are constants of integration. Recalling that $2u = x^2 + y^2$, this is clearly the equation of a sphere centered at $z = b$ and with radius a.

Now continuing from Eq. (X-8) and using Eq. (IX-19), the unit vectors in the principal directions are

$$\begin{aligned} \mathbf{P}_1' &= (y, -x, 0)/(x^2 + y^2)^{1/2}, \\ \mathbf{P}_2' &= (x, y, 2(x^2 + y^2)\, z')/K(x^2 + y^2)^{1/2}. \end{aligned}$$

We shall designate the first of these as the ξ direction so that

$$\bar{\mathbf{T}} = (y, -x, 0)/(x^2 + y^2)^{1/2}. \tag{X-9}$$

Note that z does not appear here; $\bar{\mathbf{T}}$ is the same for all rotationally symmetric surfaces.

Finally, the principal curvatures are calculated using Eq. (IX-17)

$$1/\rho_\xi = z'/K,$$
$$1/\rho_\eta = [z' + 2uz'']/K^3. \tag{X-10}$$

We can now include rotationally symmetric surfaces in our generalized ray tracing scheme. We must first assume that the equation for λ, Eq. (VI-8), has been solved yielding a value for $\bar{\lambda}$, the distance from the starting point to the point where the ray intercepts the surface. Then to the equations for the transfer operation, Eqs. (X-2), we need adjoin only the following:

$$K^2 = 1 + 2uz'^2, \tag{X-11a}$$
$$1/\bar{\rho}^\xi_{i+1} = z'/K, \tag{X-11b}$$
$$1/\bar{\rho}^\eta_{i+1} = [z' + 2uz'']/K^3, \tag{X-11c}$$
$$\cos\bar{\theta} = \mathbf{T}\cdot\mathbf{P}, \tag{X-11d}$$
$$1/\bar{\rho}_u = (1/\bar{\rho}^\xi_{i+1})\cos^2\bar{\theta} + (1/\bar{\rho}^\eta_{i+1})\sin^2\bar{\theta}, \tag{X-11e}$$
$$1/\bar{\rho}_v = (1/\bar{\rho}^\xi_{i+1})\sin^2\bar{\theta} + (1/\bar{\rho}^\eta_{i+1})\cos^2\bar{\theta}, \tag{X-11f}$$
$$1/\bar{\sigma} = \tfrac{1}{2}[(1/\bar{\rho}^\xi_{i+1}) - (1/\bar{\rho}^\eta_{i+1})]\sin 2\bar{\theta}. \tag{X-11g}$$

The refraction equations must also be altered. Equations (X-2f), (X-2g), and (X-2h) must be replaced by

$$1/\rho_u' = \mu/\rho_u + \gamma/\bar{\rho}_u, \tag{X-12a}$$
$$1/\sigma' = (\mu\cos i)/(\sigma\cos i') + \gamma/(\bar{\sigma}\cos i'), \tag{X-12b}$$
$$1/\rho_v' = (\mu\cos^2 i)/(\rho_v\cos^2 i') + \gamma/(\bar{\rho}_v\cos^2 i'). \tag{X-12c}$$

CONIC SURFACES

Ray tracing equations for conic surfaces now become a special case of those for the general surface of revolution. Using Eq. (VI-31),

$$z = 2cu/\{1 + [1 - 2c^2(1-\epsilon^2)u]^{1/2}\},$$

where $2u = x^2 + y^2$, we first find the derivatives

$$z' = c/[1 - 2c^2(1 - \epsilon^2)u]^{1/2},$$
$$z'' = c^3(1 - \epsilon^2)/[1 - 2c^2(1 - \epsilon^2)u]^{3/2}.$$

These are then inserted into Eq. (X-11a) to give

$$K^2 = (1 + 2c^2\epsilon^2 u)/[1 - 2c^2(1 - \epsilon^2)u].$$

The values of $\bar{\lambda}$, $\bar{\mathbf{R}}$, and $\bar{\mathbf{N}}$ are given by Eqs. (VI-35). Also, Eqs. (X-11b) and (X-11c) become

$$1/\bar{\rho}_{i+1}^{\xi} = c/(1 + 2c^2\epsilon^2 u)^{1/2},$$
$$1/\bar{\rho}_{i+1}^{\eta} = c/(1 + 2c^2\epsilon^2 u)^{3/2}.$$

To put these last three equations into vector form note that

$$2u = x^2 + y^2 = \mathbf{R}^2 - (\mathbf{R} \cdot \mathbf{A})^2 = (\mathbf{R} \times \mathbf{A})^2.$$

Then

$$K^2 = [1 + c^2\epsilon^2(\mathbf{R}_{i+1} \times \mathbf{A})^2]/[1 - c^2(1 - \epsilon^2)(\mathbf{R}_{i+1} \times \mathbf{A})^2], \qquad \text{(X-13a)}$$
$$1/\bar{\rho}_{i+1}^{\xi} = c/[1 + c^2\epsilon^2(\mathbf{R}_{i+1} \times \mathbf{A})^2]^{1/2}, \qquad \text{(X-13b)}$$
$$1/\bar{\rho}_{i+1}^{\eta} = c/[1 + c^2\epsilon^2(\mathbf{R}_{i+1} \times \mathbf{A})^2]^{3/2}. \qquad \text{(X-13c)}$$

THE PLANE REFRACTING SURFACE

A rather nice illustration of generalized ray tracing is provided by the plane refracting surface. A point object gives rise to a train of concentric spherical wavefronts centered at that point. These are refracted by the plane surface resulting in a train of more complicated wavefronts. Generalized ray tracing provides a means of analyzing these refracted wavefronts.

Locate the origin of a coordinate system on the refracting surface so that its normal is along the z axis. Then for the unit surface normal, $\bar{\mathbf{N}}_1 = \mathbf{A} = (0, 0, 1)$, and the equation of the refracting surface, relative to its coordinate system, is $\mathbf{R} \cdot \mathbf{A} = 0$.

Establish a second coordinate origin a distance t along the z axis of the first. The x, y plane of this system will be parallel to that of the first and will constitute the object reference plane. The object point $\mathbf{R}_0$ will be taken at the origin of this system so that $\mathbf{R}_0 = (0, 0, 0)$. The unit vector in the direction of the ray will be $\mathbf{N}_0$. The indices of refraction on opposite sides of the surface will be n_0 and n_1.

Now we apply the transfer equation, Eq. (VI-9) or (X-1d),

$$\bar{\mathbf{R}}_1 = \mathbf{R}_0 - t\mathbf{A} + \lambda\mathbf{N}_0$$

to the equation of the refracting surface and obtain

$$\bar{\lambda} = t/(\mathbf{N}_0 \cdot \mathbf{A}), \tag{X-14}$$

where $\bar{\lambda}$ is, of course, the distance along the ray to $\bar{\mathbf{R}}_1$, the point of intersection. The latter is given by

$$\bar{\mathbf{R}}_1 = [t/(\mathbf{N}_0 \cdot \mathbf{A})][\mathbf{N}_0 - (\mathbf{N}_0 \cdot \mathbf{A})\mathbf{A}] = [t\mathbf{A} \times (\mathbf{N}_0 \times \mathbf{A})]/(\mathbf{N}_0 \cdot \mathbf{A}). \tag{X-15}$$

Now we turn our attention to the other equations for transfer given in Eqs. (X-1). At the object point, the wavefront has a zero radius so that $\rho_0'^{\xi} = \rho_0'^{\eta} = 0$ and therefore, from Eq. (X-1f),

$$\rho_1{}^{\xi} = \rho_1{}^{\eta} = -\bar{\lambda} = -t/(\mathbf{N}_0 \cdot \mathbf{A}).$$

From Eq. (X-1g) we obtain the angle of incidence,

$$\cos i = \mathbf{N}_0 \cdot \bar{\mathbf{N}}_1 = \mathbf{N}_0 \cdot \mathbf{A},$$

which leads to the sine,

$$\sin i = [1 - (\mathbf{N}_0 \cdot \mathbf{A})^2]^{1/2} = [(\mathbf{N}_0 \times \mathbf{A})^2]^{1/2}.$$

Equation (X-1h) gives the $\mathbf{P}$ vector,

$$\mathbf{P} = (\mathbf{N}_0 \times \mathbf{A})/[(\mathbf{N}_0 \times \mathbf{A})^2]^{1/2}.$$

Because the incident wavefronts are spheres, there are no principal directions. We can ignore Eq. (X-1i) and set θ equal to 0 in Eqs. (X-1j) (X-1k), and (X-1l), obtaining from them

$$1/\rho_u = 1/\rho_1{}^{\xi} = -(\mathbf{N}_0 \cdot \mathbf{A})/t,$$

$$1/\rho_v = 1/\rho_1{}^{\eta} = -(\mathbf{N}_0 \cdot \mathbf{A})/t,$$

$$1/\sigma = 0.$$

Now we proceed to the refraction equations (X-2). The first two, Eqs. (X-2a) and (X-2b), can be disposed of quickly:

$$\mu = n_0/n_1, \qquad \sin i' = \mu[(\mathbf{N}_0 \times \mathbf{A})^2]^{1/2},$$

the second of which leads to

$$\cos i' = [1 - \mu^2(\mathbf{N}_0 \times \mathbf{A})^2]^{1/2}.$$

Next, from Eq. (X-2c) we obtain

$$\gamma = -\mu(\mathbf{N}_0 \cdot \mathbf{A}) + [1 - \mu^2(\mathbf{N}_0 \times \mathbf{A})^2]^{1/2},$$

and from Eq. (X-2d) we get the unit vector in the direction of the refracted ray:

$$\begin{aligned}\mathbf{N}_1 &= \mu\mathbf{N}_0 + \{-\mu(\mathbf{N}_0 \cdot \mathbf{A}) + [1 - \mu^2(\mathbf{N}_0 \times \mathbf{A})^2]^{1/2}\}\mathbf{A} \\ &= \mu\mathbf{A} \times (\mathbf{N}_0 \times \mathbf{A}) + [1 - \mu^2(\mathbf{N}_0 \times \mathbf{A})^2]^{1/2}\mathbf{A}. \qquad \text{(X-16)}\end{aligned}$$

Although we will not need it, we will next calculate $\mathbf{Q}'$ from Eq. (X-2e):

$$\begin{aligned}\mathbf{Q}' &= \mathbf{N}_1 \times \mathbf{P} \\ &= \{\mu\mathbf{A} \times (\mathbf{N}_0 \times \mathbf{A}) + [1 - \mu^2(\mathbf{N}_0 \times \mathbf{A})^2]^{1/2}\mathbf{A}\} \times \frac{(\mathbf{N}_0 \times \mathbf{A})}{[(\mathbf{N}_0 \times \mathbf{A})^2]^{1/2}} \\ &= \frac{\{-\mu(\mathbf{N}_0 \times \mathbf{A})^2\mathbf{A} + [1 - \mu^2(\mathbf{N}_0 \times \mathbf{A})^2]^{1/2}\mathbf{A} \times (\mathbf{N}_0 \times \mathbf{A})\}}{[(\mathbf{N}_0 \times \mathbf{A})^2]^{1/2}}.\end{aligned}$$

Since the refracting surface is a plane, its radius of curvature is infinite and therefore, in Eqs. (X-2f), (X-2g), and (X-2h), $1/r_1 = 0$. These then yield

$$\begin{aligned}1/\rho_u' &= -\mu(\mathbf{N}_0 \cdot \mathbf{A})/t, \\ 1/\sigma' &= 0, \\ 1/\rho_v' &= -\mu(\mathbf{N}_0 \cdot \mathbf{A})^3/t[1 - \mu^2(\mathbf{N}_0 \times \mathbf{A})^2].\end{aligned}$$

From Eq. (X-2i) we find that $\theta' = 0$. This tells us that the u-direction and the v-direction are the principal directions of the wavefront and that $1/\rho'_u$ and $1/\rho'_v$, which we have just calculated, are its principal curvatures. This becomes clear when we calculate Eqs. (X-2j) and (X-2k):

$$1/\rho_1'^{\xi} = -\mu(\mathbf{N}_0 \cdot \mathbf{A})/t, \qquad \text{(X-17)}$$

$$1/\rho_1'^{\eta} = -\mu(\mathbf{N}_0 \cdot \mathbf{A})^3/t[1 - \mu^2(\mathbf{N}_0 \times \mathbf{A})^2]. \qquad \text{(X-18)}$$

Finally, Eq. (X-21) gives us the unit vector in the ξ-principal direction

$$\mathbf{T}_1 = \mathbf{P} = (\mathbf{N}_0 \times \mathbf{A})/[(\mathbf{N}_0 \times \mathbf{A})^2]^{1/2}. \tag{X-19}$$

Note that the ξ-principal direction is perpendicular to the plane of incidence whereas the η-principal direction lies in it.

We are now in a position to proceed to the next surface if there happens to be one. We would repeat the steps already taken using the transfer equations (X-1) followed by the refraction equations (X-2) and obtain quantities similar to those above.

Rather than do that we shall calculate the caustic surfaces, the locus of the principal centers of curvature. Note first of all that the principal curvatures are both negative. The caustic surface must lie on the object side of the refracting surface. Therefore it must be *virtual* rather than *real.* If we could see it we would have to look back through the refracting surface toward the object point.

First we do the sheet of the caustic surface associated with the ξ-principal direction, perpendicular to the plane of incidence. We start at the point of intersection $\bar{\mathbf{R}}_1$ and measure off along the refracted ray, whose direction is given by $\mathbf{N}_1$, a distance equal to the radius of curvature $\rho_1'^{\xi}$:

$$\mathbf{C}_\xi = \bar{\mathbf{R}}_1 + \rho_1'^{\xi}\mathbf{N}_1\,.$$

Using Eqs. (X-15), (X-16), and (X-17) and reducing, this becomes

$$\mathbf{C}_\xi = -\{t[1 - \mu^2(\mathbf{N}_0 \times \mathbf{A})^2]^{1/2}/\mu(\mathbf{N}_0 \cdot \mathbf{A})\}\mathbf{A}. \tag{X-20}$$

This portion of the caustic surface is degenerate, consisting only of a straight line segment lying entirely on the z axis.

The other sheet of the caustic surface is satisfyingly more complicated. Applying Eqs. (X-15), (X-16), and (X-18) to

$$\mathbf{C}_\eta = \bar{\mathbf{R}}_1 + \rho_1'^{\eta}\mathbf{N}_1$$

and reducing, we obtain

$$\mathbf{C}_\eta = -t\left\{\frac{(1-\mu^2)(\mathbf{N}_0 \times \mathbf{A})^2}{(\mathbf{N}_0 \cdot \mathbf{A})^3}\mathbf{A} \times (\mathbf{N}_0 \times \mathbf{A}) + \frac{[1 - \mu^2(\mathbf{N}_0 \times \mathbf{A})^2]^{3/2}}{\mu(\mathbf{N}_0 \cdot \mathbf{A})^3}\mathbf{A}\right\}. \tag{X-21}$$

To analyze this surface we look at its intersection with the meridian plane. Let α be the angle between the incident ray and the z axis, so that $\mathbf{N}_0 \cdot \mathbf{A} = \cos\alpha$ and $(\mathbf{N}_0 \times \mathbf{A})^2 = \sin^2\alpha$. (In this case α is identical to i, the angle of incidence. To avoid confusing these concepts these two symbols will be kept distinct.) Next, note that the unit vector

$$\mathbf{B} = \mathbf{A} \times (\mathbf{N}_0 \times \mathbf{A})/[(\mathbf{N}_0 \times \mathbf{A})^2]^{1/2}$$

is perpendicular to the z axis and lies on the plane of incidence. Making these changes, we have

$$\mathbf{C}_\eta = -\frac{t}{\mu \cos \alpha_g} \{\mu(1-\mu^2)\sin^3\alpha\,\mathbf{B} + [1-\mu^2\sin^2\alpha]^{3/2}\mathbf{A}\}. \qquad \text{(X-22)}$$

Designating $\mathbf{B}$ as the y axis, we obtain a pair of parametric equations for the curve formed by the intersection of the caustic surface and the meridian plane:

$$y = -t(1-\mu^2)\tan^3\alpha,$$
$$z = -t(1-\mu^2\sin^2\alpha)^{3/2}/\mu\cos^3\alpha.$$

Eliminating α from these two equations results in

$$\mu^{2/3}z^{2/3} + (\mu^2-1)^{1/3}y^{2/3} = t^{2/3}.$$

When $\mu > 1$, this is the equation of the evolute of an ellipse. This case and the case where $\mu < 1$ are shown in Figs. X-3 and X-4, respectively.

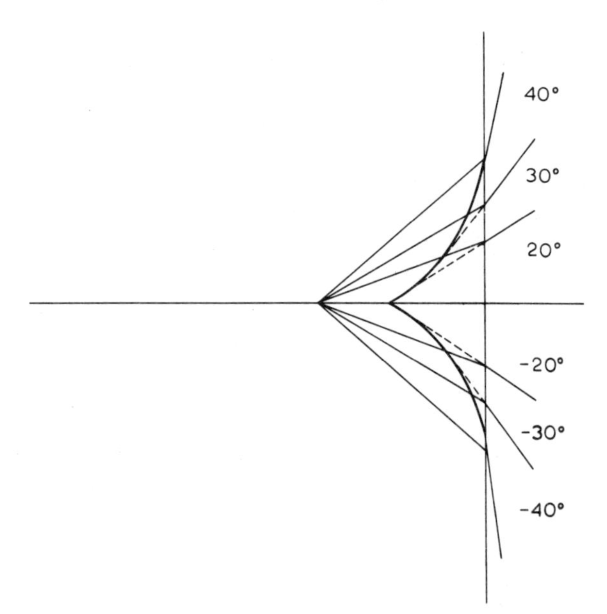

FIG. X-3. The in-the-large problem: the caustic surface with $\mu > 1$.

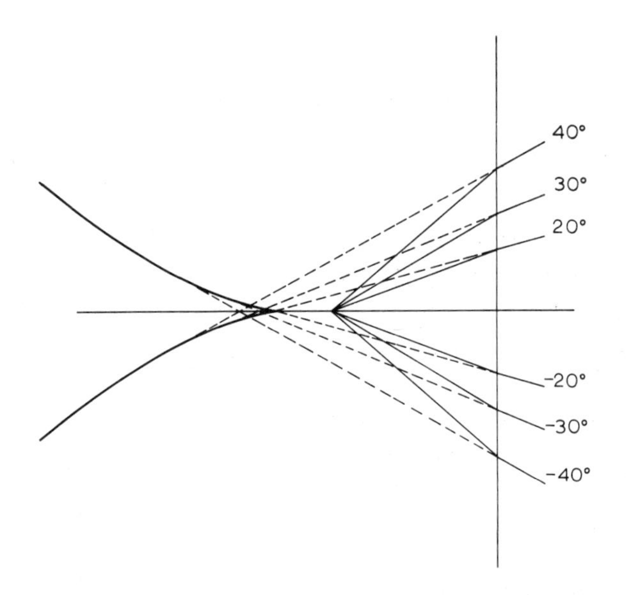

FIG. X-4. The in-the-large problem: the caustic surface with $\mu < 1$.

THE IN-THE-LARGE PROBLEM

Starting at the point of intersection $\bar{\mathbf{R}}_1$ of each ray and the refracting surface, we measure off, along each refracted ray, a distance λ' to some point **W**. The value of λ' is so determined that the points **W** all have the same optical path length from the object point. In other words, the quantity $n_0\bar{\lambda} + n_1\lambda'$, representing the total optical path length from object to **W**, is to be constant. Here $\bar{\lambda}$ is given by Eq. (X-14). All such points **W** lie on the same wavefront. In fact, the resulting equation becomes the equation for that wavefront. Let $n_1 s$ represent the total optical path length from the object to the wavefront. Thus

$$n_0\bar{\lambda} + n_1\lambda' = n_1 s, \tag{X-23}$$

where s is interpreted as an actual distance (not an optical path length) in image space. Thus

$$\lambda' = s - \mu\bar{\lambda} = s - \mu t/(\mathbf{N}_0 \cdot \mathbf{A}). \tag{X-24}$$

The equation for the wavefront is now

$$\mathbf{W} = \bar{\mathbf{R}}_1 + \lambda'\mathbf{N}_1\,, \tag{X-25}$$

which, on applying Eqs. (X-15) and (X-16), becomes

$$\mathbf{W}_s = [t/(\mathbf{N}_0 \cdot \mathbf{A})]\{(1-\mu^2)\mathbf{A} \times (\mathbf{N}_0 \times \mathbf{A}) - \mu[1-\mu^2(\mathbf{N}_0 \times \mathbf{A})^2]^{1/2}\mathbf{A}\} + s\mathbf{N}_1\,. \tag{X-26}$$

Next we set $s = 0$. In effect, this singles out the equation of a *virtual* wavefront whose optical path length from the object point is zero and which must therefore lie on the object side of the refracting surface:

$$\mathbf{W}_0 = [t/(\mathbf{N}_0 \cdot \mathbf{A})]\{(1-\mu^2)\mathbf{A} \times (\mathbf{N}_0 \times \mathbf{A}) - \mu[1-\mu^2(\mathbf{N}_0 \times \mathbf{A})^2]^{1/2}\mathbf{A}\}. \tag{X-27}$$

As in the last section, consider the plane curve resulting from the intersection of the wavefront with the meridian plane and let α be the angle between $\mathbf{N}_0$ and $\mathbf{A}$. Then

$$\mathbf{W}_0 = (t/\cos\alpha)\{(1-\mu^2)\sin\alpha\mathbf{B} - \mu[1-\mu^2\sin^2\alpha]^{1/2}\mathbf{A}\} \tag{X-28}$$

so that, in parametric form,

$$y = t(1-\mu^2)\tan\alpha, \qquad z = -(t\mu/\cos\alpha)[1-\mu^2\sin^2\alpha]^{1/2}.$$

When α is eliminated between these two expressions, we obtain

$$z^2/\mu^2t^2 - y^2/t^2(1-\mu^2) = 1.$$

This is the equation of either a hyperbola, when $\mu < 1$, or an ellipse, when $\mu > 1$. This wavefront is then either a hyperboloid or an ellipsoid depending on whether the refractive index of the second medium is greater than or less than that of the first, as shown in Figs. X-5 and X-6, respectively.

This particular wavefront is considerably simpler than all the other wavefronts in the train. Yet it contains all of the information concerning the refraction of the light from the given object point. Without stating a precise definition, let us call such surfaces *archtypical* wavefronts. Thus an *archtype* will be a member of a train of wavefronts that is distinguished from the other members of the train by some simplicity of form or expression.

Now return to the equation of the wavefront, Eq. (VIII-23),

$$\begin{aligned}\mathbf{W}_s &= \lambda'\mathbf{N}_1 + \bar{\mathbf{R}}_1 \\ &= [s - \mu t/(\mathbf{N}_0 \cdot \mathbf{A})]\,\mathbf{N}_1 + [t/(\mathbf{N}_0 \cdot \mathbf{A})]\mathbf{A} \times (\mathbf{N}_0 \times \mathbf{A}).\end{aligned} \tag{X-29}$$

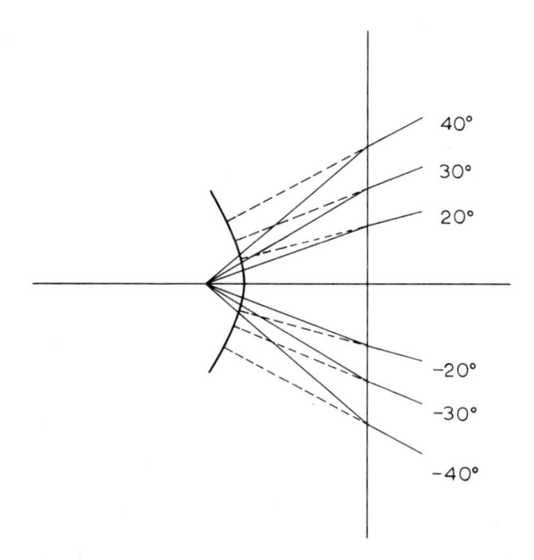

FIG. X-5. The in-the-large problem: the archtypical wavefront with $\mu < 1$.

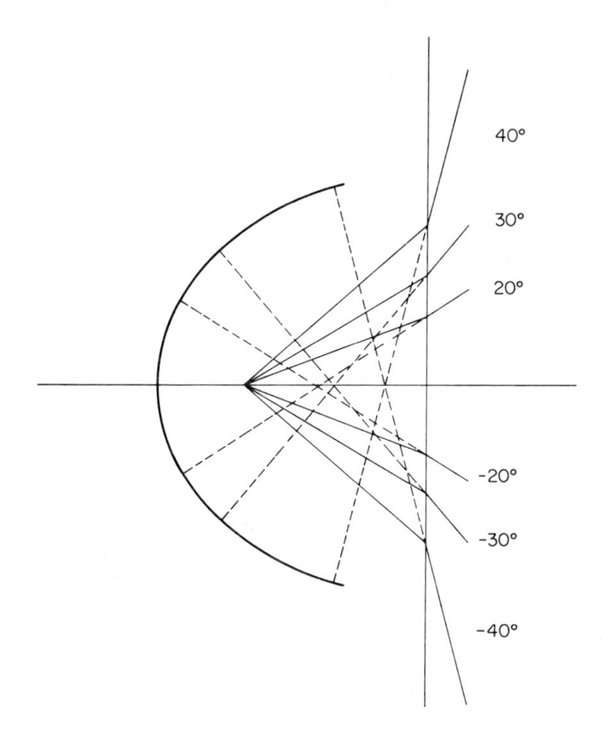

FIG. X-6. The in-the-large problem: the archtypical wavefront with $\mu > 1$.

Here we have used Eqs. (X-24) and (X-15) to make this expression depend on $\mathbf{N}_1$ rather than on $\mathbf{N}_0$. We use Snell's law, Eq. (II-19),

$$\mathbf{N}_0 \times \mathbf{A} = (1/\mu)(\mathbf{N}_1 \times \mathbf{A}),$$

whence $\mathbf{N}_0 \cdot \mathbf{A} = D/\mu$, where

$$D^2 = (\mu^2 - 1) + (\mathbf{N}_1 \cdot \mathbf{A})^2.$$

Making these substitutions, we get

$$\begin{aligned} \mathbf{W}_s\, [s - (\mu^2 t/D)]\, \mathbf{N}_1 + (t/D)\mathbf{A} \times (\mathbf{N}_1 \times \mathbf{A}) \\ = [s + t(1 - \mu^2)/D]\, \mathbf{N}_1 - [t(\mathbf{N}_1 \cdot \mathbf{A})/D]\mathbf{A}. \end{aligned} \tag{X-30}$$

Next the reduced direction vector $\mathbf{S}$ is introduced by writing $\mathbf{N}_1 = (1/n_1)\mathbf{S}_1$. Now

$$D^2 = (\mu^2 - 1) + (1/n_1^2)(\mathbf{S}_1 \cdot \mathbf{A})^2 = \Delta^2/n_1^2,$$

where

$$\Delta^2 = (n_0^2 - n_1^2) + (\mathbf{S}_1 \cdot \mathbf{A})^2$$

so that

$$\mathbf{W}_s = [(s/n_1) + t(1 - \mu^2)/\Delta]\, \mathbf{S}_1 + [t(\mathbf{S}_1 \cdot \mathbf{A})/\Delta]\mathbf{A}. \tag{X-31}$$

We can now compare this expression with Eqs. (X-27) and (X-28), the equations that constitute the general integral of the eikonal equation for a homogeneous medium, Eqs. (VIII-26) and (VIII-27),

$$\mathbf{W} = \delta\mathbf{S} - \tilde{\nabla} W, \qquad n^2\delta = c_0 + ns + \tilde{\nabla} W \cdot \mathbf{S} - W.$$

According to these,

$$\mathbf{W} \cdot \mathbf{S} = n^2\delta - \tilde{\nabla} W \cdot \mathbf{S} = c_0 + ns - W, \tag{X-32}$$

which should agree with, from Eq. (X-31),

$$\begin{aligned} \mathbf{W}_s \cdot \mathbf{S}_1 &= n_1^2\, [s/n_1 + t(1 - \mu^2)/\Delta] + t(\mathbf{S}_1 \cdot \mathbf{A})^2/\Delta \\ &= n_1 s + (t/\Delta)[(n_1^2 - n_0^2) + (\mathbf{S}_1 \cdot \mathbf{A})^2] \\ &= n_1 s + t\Delta. \end{aligned} \tag{X-33}$$

Comparing (X-32) with (X-33) we are forced to conclude that

$$W = c_0 - t\Delta \qquad \text{and therefore} \qquad \tilde{\nabla} W = -[t(\mathbf{S}_1 \cdot \mathbf{A})/\Delta]\mathbf{A},$$

which agrees exactly with Eq. (X-31). Here we have used the definition of $\tilde{\nabla}$ following Eq. (VIII-24).

We can now speculate on the nature of archtypical wavefronts. An obvious assertion is that in a train of wavefronts, given by the expression for the general integral of the eikonal equation, the archtype for this train is obtained for that value of s for which $ns + c_0 = 0$. If this is the case, then we are one step closer to the solution of the in-the-large problem, the calculation of a refracted wavefront directly from the equation of the incident wavefront.

REFERENCES

Altrichter, O., and Schäfer, G. (1956). Herleitung der Gullstrandschen Grundgleichungen für Schiefe Strahlenbüschel aus den Hauptkrümmungen der Wellenfläsche, *Optik (Stuttgart)* **13**, 241–253.

Gullstrand, A. (1906). Die reelle optische Abbildung, *Sv. Vetensk. Handl.* **41**, 1–119.

Kneisly, J. A., III. (1964). Local curvature of wavefronts in an optical system, *J. Opt. Soc. Amer.* **54**, 229–235.

XI

The Inhomogeneous Medium

In Chapter II we derived the ray equation, Eq. (II-19),

$$(d/ds)(n\mathbf{P}') = \nabla n,$$

also known in the form of Eq. (II-20),

$$n\mathbf{P}'' + (\nabla n \cdot \mathbf{P}')\,\mathbf{P}' = \nabla n,$$

whose solutions constitute the totality of ray paths in the inhomogeneous medium whose refractive index is given by the function $n(\mathbf{P})$. In Chapter VII we defined an orthotomic system of rays as an aggregate of rays associated with a wavefront; each ray in the aggregate is perpendicular to the wavefront at the point of intersection of ray and wavefront, and through each point of the wavefront there passes a ray belonging to the aggregate. We subsequently showed that associated with each orthotomic system of rays there is a one-parameter family of wavefronts.

Given an orthotomic system of rays, the differential equation, Eq. (VII-11), for the wavefront is

$$\mathbf{P}' \cdot d\mathbf{W} = 0,$$

a total differential equation. We have seen that the condition that a continuous solution should exist for this equation is Eq. (VII-12),

$$\mathbf{P}' \cdot (\nabla \times \mathbf{P}') = 0.$$

If n is an integrating factor, so that Eq. (VII-13) holds, then

$$\nabla \times (n\mathbf{P}') = \mathbf{0}.$$

Either of these can be interpreted as a condition that an aggregate of rays constitute an orthotomic system.

The expansion of Eq. (VII-13) led to Eq. (VII-14),

$$\nabla \times \mathbf{P}' = \mathbf{P}' \times \mathbf{P}'',$$

which, since the divergence of a curl is zero, led to Eq. (VII-15),

$$\nabla \cdot (\mathbf{P}' \times \mathbf{P}'') = 0,$$

and by differentiating Eq. (VII-14) we also obtained Eq. (VII-16),

$$\nabla \times \mathbf{P}'' = (d/ds)(\mathbf{P}' \times \mathbf{P}'').$$

The last two equations, we noted, resemble two of the Maxwell equations. Recall that $\mathbf{P}''$ is in the direction of the normal vector and that $\mathbf{P}' \times \mathbf{P}''$ is in the direction of the binormal vector. Both are therefore perpendicular to the direction of propagation and tangent to the wavefront just as are the $\mathbf{E}$ and $\mathbf{H}$ vectors that occur in the Maxwell equations:

$$\begin{aligned} \nabla \times \mathbf{H} &= (1/c)(d/dt)(\epsilon\mathbf{E}), & \nabla \cdot (\epsilon\mathbf{E}) &= 0, \\ \nabla \times \mathbf{E} &= (-1/c)(d/dt)(\mu\mathbf{H}), & \nabla \cdot (\mu\mathbf{H}) &= 0. \end{aligned} \tag{XI-1}$$

(See, for example, Born and Wolf, 1970, pp. 1–10.)

Here $\mathbf{E}$ is the electric field vector, $\mathbf{H}$ is the magnetic field vector, ϵ is the dielectric constant, and μ is magnetic permeability. In this version of the Maxwell equations it is assumed for convenience that the specific conductivity and the electric charge density are both zero. This pair of assumptions implies that the electric current density vector is also zero.

Clearly, Eqs. (VII-15) and (VII-16) bear a resemblance to a pair of these equations. The principal objective of this chapter is to see just how close to the Maxwell equations we can get under the assumptions of geometrical optics.

GEODESIC CURVES

Before beginning, we require some ancillary information. First we need to investigate geodesic curves on a surface. This material is a continuation of Chapter IX.

We define a *geodesic curve*, or more simply, a *geodesic* as that curve on a surface with the property that, given any pair of points on the curve, the arc length of the curve between these points is an extremum (Struik, 1961, p. 131 *et seq.*). Suppose such a curve is given by

$$\mathbf{P}(t) = \mathbf{W}(u(t), v(t)),$$

where $\mathbf{W}(u, v)$ represents the surface. Choose any pair of points, $\mathbf{P}_0 = \mathbf{P}(t_0)$ and $\mathbf{P}_1 = \mathbf{P}(t_1)$. Then

$$I = \int_{\mathbf{P}_0}^{\mathbf{P}_1} ds = \int_{t_0}^{t_1} (Eu_t^2 + 2Fu_tv_t + Gv_t^2)^{1/2}\, dt$$

must be an extremum, where we have used Eq. (IX-4). Here $u_t = du/dt$, $v_t = dv/dt$. This is now a variational problem to which we can apply the results of Chapter II.

The Euler equations in the form adopted from Eq. (II-15),

$$(d/dt)(\partial f/\partial u_t) = \partial f/\partial u, \qquad (d/dt)(\partial f/\partial v_t) = \partial f/\partial v, \tag{XI-2}$$

apply. Here $f = (Eu_t^2 + 2Fu_tv_t + Gv_t^2)^{1/2}$. Note that we may also write $f = ds/dt$. Then

$$\partial f/\partial u_t = (Eu_t + Fv_t)/f, \qquad \partial f/\partial v_t = (Fu_t + Gv_t)/f.$$

Recalling from Eq. (IX-5) that $E = \mathbf{W}_u^2$, $F = \mathbf{W}_u \cdot \mathbf{W}_v$, and $G = \mathbf{W}_v^2$, we may write

$$\begin{aligned} \partial f/\partial u_t &= (1/f)(\mathbf{W}_u u_t + \mathbf{W}_v v_t) \cdot \mathbf{W}_u \\ &= (ds/dt)(d\mathbf{P}/dt) \cdot \mathbf{W}_u = \mathbf{P}' \cdot \mathbf{W}_u . \end{aligned}$$

Similarly

$$\partial f/\partial v_t = \mathbf{P}' \cdot \mathbf{W}_v .$$

Again using Eq. (IX-5), we calculate

$$\begin{aligned}
\partial f/\partial u &= (1/2f)[(\partial E/\partial u)\, u_t^2 + 2(\partial F/\partial u)\, u_t v_t + (\partial G/\partial u)\, v_t^2] \\
&= (1/f)[\mathbf{W}_u \cdot \mathbf{W}_{uu} u_t^2 + (\mathbf{W}_{uu} \cdot \mathbf{W}_v + \mathbf{W}_u \cdot \mathbf{W}_{uv})\, u_t v_t + \mathbf{W}_v \cdot \mathbf{W}_{uv} v_t^2] \\
&= (1/f)(\mathbf{W}_u u_t + \mathbf{W}_v v_t) \cdot (\mathbf{W}_{uu} u_t + \mathbf{W}_{uv} v_t) \\
&= (dt/ds)(d\mathbf{P}/dt) \cdot d\mathbf{W}_u/dt \\
&= \mathbf{P}' \cdot d\mathbf{W}_u/dt.
\end{aligned}$$

In like manner,

$$\partial f/\partial v = \mathbf{P}' \cdot (d\mathbf{W}_v/dt).$$

Substituting these into the Euler equations, Eps. (XI-2), yields

$$\begin{aligned}
(d/dt)(\mathbf{P}' \cdot \mathbf{W}_u) &= \mathbf{P}' \cdot (d\mathbf{W}_u/dt), \\
(d/dt)(\mathbf{P}' \cdot \mathbf{W}_v) &= \mathbf{P}' \cdot (d\mathbf{W}_v/dt).
\end{aligned}$$

These in turn reduce to

$$\mathbf{P}'' \cdot \mathbf{W}_u = 0, \qquad \mathbf{P}'' \cdot \mathbf{W}_v = 0.$$

Now $\mathbf{P}''$ is a normal vector to the curve; in particular, $\mathbf{P}'' = (1/\rho)\mathbf{n}$, where $\mathbf{n}$ is the curve's unit normal vector. Moreover, $\mathbf{W}_u$ and $\mathbf{W}_v$ are tangent vectors to the surface. We conclude that the normal to a geodesic curve must be parallel to the surface normal at every point. The proof is reversible: Any curve that has the property that its normal at every point is perpendicular to the surface has the extremum property. In fact, many writers define a geodesic curve in this way—as that curve having its normal perpendicular to the surface at every point—and then deduce the extremum property (Struik, 1961, p. 131).

But the niceties of the definition do not concern us here. We know what a geodesic curve is, and from the extremum property we know that a regular point and a direction on a surface determine a geodesic curve through that point.

The unit tangent vector to the geodesic, using Eqs. (IX-3) and (IX-7), is

$$\mathbf{t} = \mathbf{P}' = (\mathbf{W}_u u_t + \mathbf{W}_v v_t)/f = \mathbf{W}_u u' + \mathbf{W}_v v', \tag{XI-3}$$

where the prime, as always, denotes the derivative with respect to the arc length parameter s.

Since the unit normal vector to the geodesic is equal to the unit surface normal, we may write

$$\mathbf{n} = \mathbf{N} = (\mathbf{W}_u \times \mathbf{W}_v)/(EG - F^2)^{1/2}. \tag{XI-4}$$

The binormal vector is therefore

$$\mathbf{b} = \mathbf{t} \times \mathbf{n} = (\mathbf{W}_u u' + \mathbf{W}_v v') \times [(\mathbf{W}_u \times \mathbf{W}_v)/(EG - F^2)^{1/2}]$$
$$= (EG - F^2)^{-1/2}[(Fu' + Gv')\,\mathbf{W}_u - (Eu' + Fv')\,\mathbf{W}_v]. \qquad \text{(XI-5)}$$

Next we calculate $\mathbf{n}'$ as the derivative of $\mathbf{N}$ with respect to s,

$$\mathbf{n}' = \mathbf{N}_u u' + \mathbf{N}_v v'.$$

Using the expressions for $\mathbf{N}_u$ and $\mathbf{N}_v$ given in the Weingarten equations, Eqs. (IX-21), this becomes

$$\mathbf{n}' = (EG - F^2)^{-1}\{[(FM - GL)\,u' + (FN - GM)\,v']\,\mathbf{W}_u$$
$$+ [(FL - EM)\,u' + (FM - EN)\,v']\,\mathbf{W}_v\}. \qquad \text{(XI-6)}$$

Now we use the second of the Frenet equations, Eqs. (III-5),

$$\mathbf{n}' = -(1/\rho)\mathbf{t} + (1/\tau)\mathbf{b},$$

which, when mutliplied by $\mathbf{t}$, leads to

$$1/\rho = -\mathbf{n}' \cdot \mathbf{t}$$
$$= -(EG - F^2)^{-1}\{[(FM - GL)\,u' + (FN - GM)\,v']\,\mathbf{W}_u$$
$$+ [(FL - EM)\,u' + (FM - EN)\,v']\,\mathbf{W}_v\} \cdot (\mathbf{W}_u u' + \mathbf{W}_v v')$$
$$= Lu'^2 + 2Mu'v' + Nv'^2.$$

Here use was made of Eqs. (XI-3) and (XI-6). This agrees with a previous result, Eq. (IX-12).

With this as reassurance, we next calculate the torsion of the geodesic. The scalar product of the second Frenet equation above with $\mathbf{b}$ gives

$$1/\tau = \mathbf{n}' \cdot \mathbf{b}.$$

When Eqs. (XI-5) and (XI-6) are substituted, the result is

$$1/\tau = (EG - F^2)^{-1/2}[(EM - FL)\,u'^2 + (EN - GL)\,u'v' + (FN - GM)\,v'^2]. \qquad \text{(XI-7)}$$

The quantity in brackets is exactly and precisely Eq. (IX-16), the quadratic equation whose solutions determine the principal directions of a surface at a point. Therefore, if u' and v' are such that the torsion of the geodesic curve vanishes, then its tangent vector is in a principal direction of the surface at that point. Conversely, if the tangent to a geodesic curve is in a principal direction, then its torsion must be zero.

We continue in this vein just a little further. First, consider the geodesic in the u direction. Then $v' = 0$ and, from Eqs. (XI-3) through (XI-7),

$$\mathbf{t}^u = \mathbf{W}_u u', \qquad \mathbf{n}^u = \mathbf{W}_u \times \mathbf{W}_v/(EG - F^2)^{1/2},$$
$$\mathbf{b}^u = (F\mathbf{W}_u - E\mathbf{W}_v)\, u'/(EG - F^2)^{1/2},$$
$$1/\rho_u = Lu'^2, \qquad 1/\tau^u = (EM - FL)\, u'^2/(EG - F^2)^{1/2}.$$

Squaring the first of these, we can see that $u'^2 = 1/E$, so the above may be written in the form

$$\mathbf{t}^u = \mathbf{W}_u/E^{1/2}, \qquad \mathbf{b}^u = (F\mathbf{W}_u - E\mathbf{W}_v)/[E(EG - F^2)]^{1/2},$$
$$1/\rho_u = L/E, \qquad 1/\tau_u = (EM - FL)/E(EG - F^2)^{1/2}. \tag{XI-8}$$

The same thing is done for the geodesic curve in the v direction. In this case, $u' = 0$ and $v'^2 = 1/G$. We have

$$\begin{aligned} \mathbf{t}^v &= \mathbf{W}_v/G^{1/2}, \qquad \mathbf{b}^v = (G\mathbf{W}_u - F\mathbf{W}_v)/[G(EG - F^2)]^{1/2}, \\ 1/\rho_v &= N/G, \qquad 1/\tau_v = (FN - GM)/G(EG - F^2)^{1/2}. \end{aligned} \tag{XI-9}$$

Now suppose that the u and v directions are perpendicular to one another. Then $F = 0$ and

$$1/\tau_u = M/(EG)^{1/2}, \qquad 1/\tau_v = -M/(EG)^{1/2}.$$

The conclusion is that the torsion of two geodesic curves passing through a point and perpendicular to each other are equal in magnitude and opposite in sign. Using the symbol introduced in Eq. (IX-32), we write the above as

$$\tau_u = \sigma \qquad \text{and} \qquad \tau_v = -\sigma, \tag{XI-10}$$

where

$$1/\sigma = M/(EG)^{1/2}. \tag{XI-11}$$

SOME VECTOR IDENTITIES

The vector notation used here is a fantastic shorthand for operations and calculations that would otherwise be cumbersome and obscured by unimportant details. Nevertheless, there comes a time when the vector notation itself becomes awkward and unclear, so much so that it seems almost to take a Dr. Strangelove to love it. So a list of the more difficult vector identities is included at this point as an aid to the reader and as a guide to the writer. No proofs are given. (For a rigorous derivation of these formulas see, e.g., Phillips, 1933, p. 38; Korn and Korn, 1968, p. 159.)

The first group of equations involves the application of the del operator to the product of a vector and a scalar:

directional derivative

$$(\mathbf{F} \cdot \nabla)(\gamma \mathbf{E}) = (\nabla \gamma \cdot \mathbf{F})\mathbf{E} + \gamma(\mathbf{F} \cdot \nabla)\mathbf{E}, \tag{XI-12a}$$

divergence

$$\nabla \cdot (\gamma \mathbf{E}) = (\nabla \gamma \cdot \mathbf{E}) + \gamma(\nabla \cdot \mathbf{E}), \tag{XI-12b}$$

curl

$$\nabla \times (\gamma \mathbf{E}) = (\nabla \gamma \times \mathbf{E}) + \gamma(\nabla \times \mathbf{E}). \tag{XI-12c}$$

The scalar product involves

gradient

$$\nabla(\mathbf{E} \cdot \mathbf{F}) = (\mathbf{E} \cdot \nabla)\mathbf{F} + (\mathbf{F} \cdot \nabla)\mathbf{E} + \mathbf{E} \times (\nabla \times \mathbf{F}) + \mathbf{F} \times (\nabla \times \mathbf{E}), \tag{XI-13}$$

and the vector product involves

divergence

$$\nabla \cdot (\mathbf{E} \times \mathbf{F}) = \mathbf{F} \cdot (\nabla \times \mathbf{E}) - \mathbf{E} \cdot (\nabla \times \mathbf{F}), \tag{XI-14a}$$

curl

$$\nabla \times (\mathbf{E} \times \mathbf{F}) = (\mathbf{F} \cdot \nabla)\mathbf{E} - (\mathbf{E} \cdot \nabla)\mathbf{F} + (\nabla \cdot \mathbf{F})\mathbf{E} - (\nabla \cdot \mathbf{E})\mathbf{F}. \tag{XI-14b}$$

A useful formula for the directional derivative is

$$2(\mathbf{E} \cdot \nabla)\mathbf{F} = \nabla \times (\mathbf{F} \times \mathbf{E}) + \nabla(\mathbf{E} \cdot \mathbf{F}) - (\nabla \cdot \mathbf{E})\mathbf{F} + (\nabla \cdot \mathbf{F})\mathbf{E}$$
$$-\mathbf{F} \times (\nabla \times \mathbf{E}) - \mathbf{E} \times (\nabla \times \mathbf{F}). \tag{XI-15}$$

The Laplacian operator is written as the square of the del operator

$$\nabla^2 = \partial^2/\partial x^2 + \partial^2/\partial y^2 + \partial^2/\partial z^2.$$

Vector identities involving second derivatives include

divergence of a gradient

$$\nabla \cdot (\nabla \gamma) = \nabla^2 \gamma, \tag{XI-16a}$$

gradient of a divergence

$$\nabla(\nabla \cdot \mathbf{E}) = \nabla^2\mathbf{E} + \nabla \times (\nabla \times \mathbf{E}),$$

curl of a curl (XI-16b)

$$\nabla \times (\nabla \times \mathbf{E}) = \nabla(\nabla \cdot \mathbf{E}) - \nabla^2\mathbf{E},$$

curl of a gradient

$$\nabla \times (\nabla\gamma) = \mathbf{0}, \tag{XI-16c}$$

divergence of a curl

$$\nabla \cdot (\nabla \times \mathbf{E}) = 0. \tag{XI-16d}$$

GEODESICS ON A WAVEFRONT

In what follows we will make extensive use of the idea of the directional derivative introduced in connection with the Frenet equations in Eq. (III-17):

$$(\mathbf{t} \cdot \nabla)\mathbf{t} = (1/\rho)\mathbf{n},$$

$$(\mathbf{t} \cdot \nabla)\mathbf{n} = -(1/\rho)\mathbf{t} + (1/\tau)\mathbf{b},$$

$$(\mathbf{t} \cdot \nabla)\mathbf{b} = -(1/\tau)\mathbf{n}.$$

We need a short lemma. Let $\mathbf{V}$ be any vector such that $\mathbf{V}^2 = \text{const.}$ Then, using the formula for the gradient of a scalar product, Eq. (XI-13),

$$\nabla(\mathbf{V}^2) = 2\mathbf{V} \times (\nabla \times \mathbf{V}) + 2(\mathbf{V} \cdot \nabla)\mathbf{V} = \mathbf{0},$$

so that

$$(\mathbf{V} \cdot \nabla)\mathbf{V} = -\mathbf{V} \times (\nabla \times \mathbf{V}). \tag{XI-17}$$

Now we return to an orthotomic system of rays. These are the solutions of the ray equation, Eqs. (II-19) or (II-20), that also satisfy the condition of integrability, Eqs. (VIII-12) or (VIII-13), and therefore are associated with a one-parameter family of wavefronts. Choose a ray and on that ray select a point. Through that point there passes a wavefront whose unit normal vector $\mathbf{N}$ at that point coincides with the unit tangent vector $\mathbf{t}$ of the ray at that point. Also associated with the ray at that point are the normal and binormal vectors $\mathbf{n}$ and $\mathbf{b}$. These are tangent to the wavefront at the chosen point and therefore determine a pair of orthogonal directions on the wavefront.

Let $\mathbf{W}(u, v)$ be the vector function describing the wavefront with the u and v parametric curves being geodesics. Moreover, we assume that the u direction is parallel to $\mathbf{n}$ and that the v direction is parallel to $\mathbf{b}$. In other words,

$$\mathbf{n} = \mathbf{W}_u/E^{1/2} = \mathbf{t}^u \quad \text{and} \quad \mathbf{b} = \mathbf{W}_v/G^{1/2} = \mathbf{t}^v.$$

Collecting the appropriate formulas from Eqs. (XI-8)–(XI-11), we have

$$\begin{aligned} \mathbf{t}^u &= \mathbf{W}_u/E^{1/2} = \mathbf{n}, & \mathbf{t}^v &= \mathbf{W}_v/G^{1/2} = \mathbf{b}, \\ \mathbf{n}^u &= \mathbf{N} = \mathbf{t}, & \mathbf{n}^v &= \mathbf{N} = \mathbf{t}, \\ \mathbf{b}^u &= -\mathbf{W}_v/G^{1/2} = -\mathbf{b}, & \mathbf{b}^v &= \mathbf{W}_u/E^{1/2} = \mathbf{n}, \\ 1/\rho_u &= L/E, & 1/\rho_v &= N/G, \\ 1/\tau_u &= M/(EG)^{1/2} = 1/\sigma, & 1/\tau_v &= -M/(EG)^{1/2} = -1/\sigma. \end{aligned} \tag{XI-18}$$

Now we need to calculate the Frenet equations for the two geodesics. In the u direction they are

$$\begin{aligned} (\mathbf{t}^u \cdot \nabla)\, \mathbf{t}^u &= (1/\rho_u)\, \mathbf{n}^u, \\ (\mathbf{t}^u \cdot \nabla)\, \mathbf{n}^u &= -(1/\rho_u)\, \mathbf{t}^u + (1/\tau_u)\, \mathbf{b}^u, \\ (\mathbf{t}^u \cdot \nabla)\, \mathbf{b}^u &= -(1/\tau_u)\, \mathbf{n}^u \end{aligned}$$

and in the v direction they are

$$\begin{aligned} (\mathbf{t}^v \cdot \nabla)\, \mathbf{t}^v &= (1/\rho_v)\, \mathbf{n}^v, \\ (\mathbf{t}^v \cdot \nabla)\, \mathbf{n}^v &= -(1/\rho_v)\, \mathbf{t}^v + (1/\tau_v)\, \mathbf{b}^v, \\ (\mathbf{t}^v \cdot \nabla)\, \mathbf{b}^v &= -(1/\tau_v)\, \mathbf{n}^v. \end{aligned}$$

We next substitute into the preceding equations the relations in Eqs. (XI-12). This results in the following, which includes for convenience the three relations that constitute the Frenet equations of the ray, Eq. (III-17):

$$\begin{gathered} (\mathbf{t} \cdot \nabla)\mathbf{t} = (1/\rho)\mathbf{n}, \quad (\mathbf{n} \cdot \nabla)\mathbf{n} = (1/\rho_u)\mathbf{t}, \quad (\mathbf{b} \cdot \nabla)\mathbf{b} = (1/\rho_v)\mathbf{t}, \\ (\mathbf{t} \cdot \nabla)\mathbf{n} = -(1/\rho)\mathbf{t} + (1/\tau)\mathbf{b}, \quad (\mathbf{n} \cdot \nabla)\mathbf{t} = -(1/\rho_u)\mathbf{n} - (1/\sigma)\mathbf{b}, \\ (\mathbf{b} \cdot \nabla)\mathbf{t} = -(1/\rho_v)\mathbf{b} - (1/\sigma)\mathbf{n}, \\ (\mathbf{t} \cdot \nabla)\mathbf{b} = -(1/\tau)\mathbf{n}, \quad (\mathbf{n} \cdot \nabla)\mathbf{b} = (1/\sigma)\mathbf{t}, \quad (\mathbf{b} \cdot \nabla)\mathbf{n} = (1/\sigma)\mathbf{t}. \end{gathered} \tag{XI-19}$$

We are now in a position to calculate the divergence and the curl of each of the vectors associated with the ray, **t**, **n**, and **b**. Using Eqs. (XI-14a), (XI-17), and the relationships between **t**, **n**, and **b** contained in Eq. (III-3), we proceed:

$$\begin{aligned}\nabla \cdot \mathbf{t} &= \nabla \cdot (\mathbf{b} \times \mathbf{n}) = \mathbf{b} \cdot (\nabla \times \mathbf{n}) - \mathbf{n} \cdot (\nabla \times \mathbf{b}) \\ &= (\mathbf{t} \times \mathbf{n}) \cdot (\nabla \times \mathbf{n}) - (\mathbf{b} \times \mathbf{t}) \cdot (\nabla \cdot \mathbf{b}) \\ &= \mathbf{t} \cdot [\mathbf{n} \times (\nabla \times \mathbf{n}) + \mathbf{b} \times (\nabla \times \mathbf{b})] \\ &= -\mathbf{t} \cdot [(\mathbf{n} \cdot \nabla)\mathbf{n} + (\mathbf{b} \cdot \nabla)\mathbf{b}].\end{aligned}$$

Now using Eq. (XI-19),

$$\nabla \cdot \mathbf{t} = -\mathbf{t} \cdot [\mathbf{t}/\rho_u + \mathbf{t}/\rho_v] = -(1/\rho_u + 1/\rho_v). \tag{XI-20}$$

This is exactly the result given by Kline and Kay (1965, p. 186), obtained by essentially the same proof. Proceeding in the same way,

$$\nabla \cdot \mathbf{n} = -1/\rho, \tag{XI-21}$$

$$\nabla \cdot \mathbf{b} = 0. \tag{XI-22}$$

Note that Eq. (XI-22) strongly resembles Eq. (VII-15).

The calculation of the curls is similar,

$$\begin{aligned}\nabla \times \mathbf{t} &= \nabla \times (\mathbf{n} \times \mathbf{b}) \\ &= (\mathbf{b} \cdot \nabla)\mathbf{n} - (\mathbf{n} \cdot \nabla)\mathbf{b} + (\nabla \cdot \mathbf{b})\mathbf{n} - (\nabla \cdot \mathbf{n})\mathbf{b} \qquad \text{(XI-23)} \\ &= (1/\rho)\mathbf{b}.\end{aligned}$$

Here we use Eqs. (XI-14b), (XI-21), and (XI-22). This result is identical to Eq. (VII-14). In what follows the same procedure is used:

$$\nabla \times \mathbf{n} = (1/\sigma - 1/\tau)\mathbf{n} - (1/\rho_u)\mathbf{b}, \tag{XI-24}$$

$$\nabla \times \mathbf{b} = -(1/\sigma + 1/\tau)\mathbf{b} + (1/\rho_v)\mathbf{n}. \tag{XI-25}$$

MORE ON THE FISH EYE

We really need to apply some of these results to an orthotomic system of rays in a particular inhomogeneous medium. In this section we will do exactly that and reconsider systems of rays in Maxwell's fish eye.

In Chapter IV we found that ray paths in Maxwell's fish eye are arcs of circles. Suppose we consider the totality of rays emanating from a fixed object point $\mathbf{P}_0$. Such a collection of rays must necessarily constitute an orthotomic system. Each ray in this system is determined uniquely by its initial direction $\mathbf{t}_0$. The vector equation for these rays is given by Eq. (IV-16):

$$\mathbf{P} = \mathbf{P}_0 + \rho \mathbf{t}_0 \sin(s/\rho) + 2\rho^2 n_0[1 - \cos(s/\rho)]\, \mathbf{t}_0 \times (\mathbf{t}_0 \times \mathbf{P}_0).$$

As usual, s denotes the arc length parameter. The value of the refractive index at the initial point $\mathbf{P}_0$ is given by n_0. The curvature of each ray is constant for that ray and therefore depends only on $\mathbf{t}_0$. This is seen from Eq. (IV-14),

$$2\rho n_0 = [(\mathbf{t}_0 \times \mathbf{P}_0)^2]^{-1/2}.$$

This aggregate of rays from the fixed object point $\mathbf{P}_0$ can be represented as a two-parameter vector function. Since each member of the family is completely determined by the initial direction vector $\mathbf{t}_0$, any parametrization we choose to apply to $\mathbf{t}_0$ imposes a parametrization on $\mathbf{t}$.

First, we define a coordinate system with the z axis passing through the center of the fish eye and through the object point. Thus $\mathbf{P}_0 = r(0, 0, 1)$ where $r = (\mathbf{P}_0{}^2)^{1/2}$ is the distance from the object point to the center. Next let

$$\mathbf{t}_0 = (\sin u \cos v, \sin u \sin v, \cos u). \tag{XI-26}$$

When this is substituted into Eq. (XI-16), $\mathbf{P}$ becomes a function of the two parameters u and v as well as of the arc length parameter s.

Note that $\mathbf{P}_0 \cdot \mathbf{t}_0 = r \cos u$. Comparing this with Eq. (IV-24), one can see that the parameter u is identical to θ, the angle between the vector to the object point $\mathbf{P}_0$ and the initial ray direction $\mathbf{t}_0$. Next calculate

$$\mathbf{t}_0 \times \mathbf{P}_0 = r \sin u\, (\sin v, -\cos u, 0), \tag{XI-27}$$

from which we obtain, using Eq. (IV-14),

$$1/\rho = 2n_0 r \sin u. \tag{XI-28}$$

Finally,

$$\mathbf{t}_0 \times (\mathbf{t}_0 \times \mathbf{P}_0) = r \sin u\, (\sin u \cos v, \sin u \sin v, -\cos v). \tag{XI-29}$$

Next we differentiate Eq. (XI-26) and obtain

$$\mathbf{t}_{0u} = (\cos u \cos v, \cos u \sin v, -\sin u),$$

$$\mathbf{t}_{0v} = -\sin u\, (\sin v, -\cos v, 0)$$

and

$$\mathbf{t}_{0uu} = -(\sin u \cos v, \sin u \sin v, \cos u),$$
$$\mathbf{t}_{0uv} = -\cos u (\sin v, -\cos v, 0),$$
$$\mathbf{t}_{0vv} = -\sin u (\cos v, \sin v, 0).$$

Applying these to Eqs. (XI-27) and (XI-29), we get

$$\mathbf{t}_0 \times \mathbf{P}_0 = -r\mathbf{t}_{0v}, \tag{XI-30}$$

$$\mathbf{t}_0 \times (\mathbf{t}_0 \times \mathbf{P}_0) = r \sin u \, \mathbf{t}_{0u}. \tag{XI-31}$$

Note also that

$$\mathbf{t}_0 \cdot \mathbf{t}_{0u} = \mathbf{t}_0 \cdot \mathbf{t}_{0v} = \mathbf{t}_{0u} \cdot \mathbf{t}_{0v} = 0,$$

$$\mathbf{t}_0^2 = \mathbf{t}_{0u}^2 = 1, \qquad \mathbf{t}_{0v}^2 = \sin^2 u.$$

In addition

$$\begin{aligned} \mathbf{t}_{0uu} &= -\mathbf{t}_0, \qquad \mathbf{t}_{0uv} = \cot u \, \mathbf{t}_{0v}, \\ \mathbf{t}_{0vv} &= (1/r) \cos u \, \mathbf{P}_0 - \mathbf{t}_0, \end{aligned} \tag{XI-32}$$

$$\begin{aligned} \mathbf{t}_0 \times \mathbf{t}_{0u} &= \csc u \, \mathbf{t}_{0v}, \qquad \mathbf{t}_0 \times \mathbf{t}_{0v} = -\sin u \, \mathbf{t}_{0u}, \\ \mathbf{t}_{0u} \times \mathbf{t}_{0v} &= \sin u \, \mathbf{t}_0. \end{aligned} \tag{XI-33}$$

Applying these to Eq. (IV-16), we get, after a considerable amount of reduction,

$$\mathbf{P} = \mathbf{P}_0 + 2\rho \sin(s/2\rho)[\mathbf{t}_0 \cos(s/2\rho) + \mathbf{t}_{0u} \sin(s/2\rho)]. \tag{XI-34}$$

For each value of u and v, this equation defines a single ray from the object point $\mathbf{P}_0$, whose initial direction is given by the vector $\mathbf{t}_0(u, v)$. Thus the vector function $\mathbf{P}$ depends on u and v as well as on the arc length parameter s. It would be nice to be able to say that the wavefronts associated with this two-parameter family of orthotomic rays are obtained by holding s fixed. This, however, is not the case. A constant s produces a surface a constant *geometrical* distance along the ray from the object point. But this is not the wavefront. What is needed is the surface whose points are a constant optical path length from $\mathbf{P}_0$.

To find this we turn to Eq. (IV-28), the equation for the optical path length along a ray in the fish eye,

$$I = \arctan\left[\frac{\sin(s/2\rho)}{(\mathbf{P}_0^2)^{1/2} \sin(s/2\rho + \theta)}\right].$$

Translated into the current context, $\theta = u$ and $(\mathbf{P}_0{}^2)^{1/2} = r$. This provides a relation between the optical path length I and the arc length parameter s. This equation can be written in the form

$$r \tan I = \sin(s/2\rho)/\sin(s/2\rho + u). \tag{XI-35}$$

Holding the left member of this equation constant gives us a relation between s and u that defines a wavefront. Solving this for s gives

$$\tan(s/2\rho) = r \tan I \sin u/[1 - r \tan I \cos u],$$

from which we get

$$\begin{aligned} \cos(s/2\rho) &= (1 - r \tan I \cos u)\, H^{-1/2}, \\ \sin(s/2\rho) &= r \tan I \sin u\, H^{-1/2}, \end{aligned} \tag{XI-36}$$

where

$$H = 1 - 2r \tan I \cos u + r^2 \tan^2 I. \tag{XI-37}$$

Substituting Eqs. (XI-36) and (XI-37) into Eq. (XI-34) gives us

$$\mathbf{P} = \mathbf{P}_0 + \tan I[\mathbf{t}_0(1 - r \tan I \cos u) + \mathbf{t}_{0u}\, r \tan I \sin u]/n_0 H. \tag{XI-38}$$

When I is held fixed, this is the parametric vector equation of a wavefront. Its optical path length from $\mathbf{P}_0$ is equal to I.

We next calculate the derivatives of $\mathbf{P}$ using Eqs. (XI-30), (XI-31), and (XI-32):

$$\mathbf{P}_u = \tan I[(H - 2r^2 \tan^2 I \sin^2 u\, \mathbf{t}_{0u} - 2r \tan I \sin u(1 - r \tan I \cos u)\, \mathbf{t}_0]/n_0 H^2,$$

$$\mathbf{P}_v = \tan I\, \mathbf{t}_{0v}/n_0 H.$$

We use Eq. (XI-32) to calculate the cross product

$$\mathbf{P}_u \times \mathbf{P}_v = \tan^2 I \sin u(H \cos 2u\, \mathbf{t}_0 + 2r \tan I \sin u(1 - r \tan I \cos u)\, \mathbf{t}_{0u}]/n_0{}^2 H^3.$$

We also calculate, using the equations preceding Eq. (XI-32),

$$E = \tan^2 I/n_0{}^2 H^2, \qquad F = 0, \qquad G = \tan^2 I \sin^2 u/n_0{}^2 H^2.$$

Using Eq. (IX-6), we obtain the unit normal vector to the surface,

$$\begin{aligned} \mathbf{N} &= (\mathbf{P}_u \times \mathbf{P}_v)/(EG - F^2)^{1/2} \\ &= [(H - 2r^2 \tan^2 I \sin^2 u\, \mathbf{t}_0 + 2r \tan I \sin u(1 - r \tan I \cos u)\, \mathbf{t}_{0u}]/H. \end{aligned}$$

Its derivatives are then calculated and are used with Eq. (IX-20) to obtain the second fundamental quantities,

$$\begin{aligned} L &= -\mathbf{N}_u \cdot \mathbf{P}_u = -\tan I(1 - r^2 \tan^2 I)/n_0 H^2, \\ M &= -\mathbf{N}_u \cdot \mathbf{P}_v = 0, \\ N &= -\mathbf{N}_v \cdot \mathbf{P}_u = -\tan I \sin^2 u(1 - r^2 \tan^2 I)/n_0 H^2. \end{aligned} \tag{XI-39}$$

From these and Eq. (IX-15) we obtain the principal curvatures of the wavefront in the fish eye

$$\begin{aligned} 1/\rho_u &= L/E = -n_0(1 - r^2 \tan^2 I)/\tan I, \\ 1/\rho_v &= N/G = -n_0(1 - r^2 \tan^2 I)/\tan I. \end{aligned} \tag{XI-40}$$

This shows that the wavefronts in the fish eye medium are spheres. When I is small, the curvature is large. As the wavefront progresses, its curvature decreases until somewhere near the midpoint of the ray, when $\tan I = 1/r$, it becomes a plane. As it passes through this region it again becomes a sphere but with the sign of its curvature reversed so that it now converges to the image point. This is illustrated in Fig. XI-1.

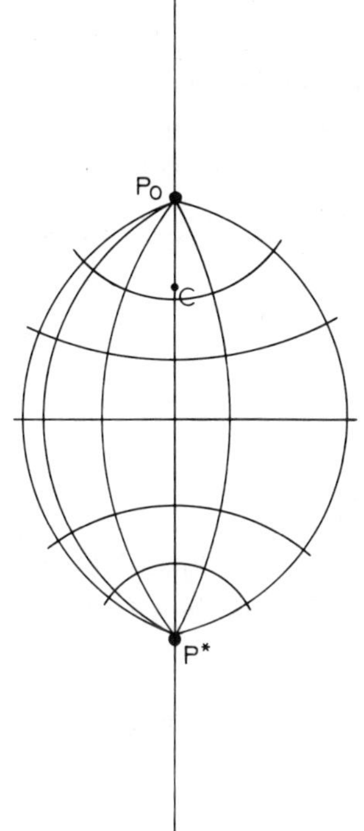

FIG. XI-1. Wavefronts in the fish eye. The object point is $\mathbf{P}_0$, its image $\mathbf{P}^*$. C is the center of symmetry of the fish eye medium. Shown are the circular ray paths connecting $\mathbf{P}_0$ and $\mathbf{P}^*$ and the spherical wavefronts that intersect the rays orthogonally.

THE PSEUDO-MAXWELL EQUATIONS

We leave the fish eye for the moment and return to Eqs. (XI-21), (XI-22), (XI-24), and (XI-25). Begin with the first of these and calculate

$$\nabla \cdot (n\mathbf{n}) = \nabla n \cdot \mathbf{n} + n\nabla \cdot \mathbf{n} = (n/\rho) - (n/\rho) = 0,$$

where we have used Eq. (XI-12b) and the ray equation (II-20). We do the same thing with Eq. (XI-22) and get

$$\nabla \cdot (n\mathbf{b}) = \nabla n \cdot \mathbf{b} + n\nabla \cdot \mathbf{b} = 0,$$

which also follows from Eq. (XI-12b) and Eq. (II-20). The two equations involving curls are treated in exactly the same way. First Eq. (XI-24):

$$\begin{aligned}
\nabla \times (n\mathbf{n}) &= \nabla n \times \mathbf{n} + n\nabla \times \mathbf{n} \\
&= \nabla n \times (\mathbf{b} \times \mathbf{t}) + n[(1/\sigma + 1/\tau)\mathbf{n} - (1/\rho_u)\mathbf{b}] \\
&= (\nabla n \cdot \mathbf{t})\mathbf{b} - (n/\tau)\mathbf{n} + (n/\sigma)\mathbf{n} - (n/\rho_u)\mathbf{b} \\
&= [(\mathbf{t} \cdot \nabla)n]\mathbf{b} + n(\mathbf{t} \cdot \nabla)\mathbf{b} + (n/\sigma)\mathbf{n} - (n/\rho_u)\mathbf{b} \\
&= (\mathbf{t} \cdot \nabla)(n\mathbf{b}) + (n/\sigma)\mathbf{n} - (n/\rho_v)\mathbf{b}.
\end{aligned}$$

In this sequence of calculations we have used Eq. (XI-12c) and the third equation of the left column in Eqs. (XI-19). The process is repeated for Eq. (XI-25):

$$\begin{aligned}
\nabla \times (n\mathbf{b}) &= \nabla n \times \mathbf{b} + n\nabla \times \mathbf{b} \\
&= \nabla n \times (\mathbf{t} \times \mathbf{n}) + n[-(1/\sigma + 1/\tau)\mathbf{b} + (1/\rho_v)\mathbf{n}] \\
&= (\nabla n \cdot \mathbf{n})\mathbf{t} - (\nabla n \cdot \mathbf{t})\mathbf{n} - (n/\tau)\mathbf{b} - (n/\sigma)\mathbf{b} + (n/\rho_v)\mathbf{n} \\
&= -[(\mathbf{t} \cdot \nabla)n]\mathbf{n} - n[-(1/\rho)\mathbf{t} + (1/\tau)\mathbf{b}] - (n/\sigma)\mathbf{b} + (n/\rho_v)\mathbf{n} \\
&= -(\mathbf{t} \cdot \nabla)(n\mathbf{n}) - (n/\sigma)\mathbf{b} + (n/\rho_v)\mathbf{n}.
\end{aligned}$$

Here we have used Eq. (XI-12c) and the second equation in the left column of Eqs. (XI-19).

Putting these all together, we have

$$\begin{aligned}
\nabla \cdot (n\mathbf{n}) &= 0, \\
\nabla \times (n\mathbf{n}) &= (\mathbf{t} \cdot \nabla)(n\mathbf{b}) + (n/\sigma)\mathbf{n} - (n/\rho_u)\mathbf{b}, \\
\nabla \cdot (n\mathbf{b}) &= 0, \\
\nabla \times (n\mathbf{b}) &= -(\mathbf{t} \cdot \nabla)(n\mathbf{n}) - (n/\sigma)\mathbf{b} + (n/\rho_v)\mathbf{n}.
\end{aligned} \qquad \text{(XI-41)}$$

Since the directional derivative $(\mathbf{t} \cdot \nabla)$ can be interpreted as a derivative in the direction of propagation, it is clear that these equations represent a step closer to the structure of the Maxwell equations in Eqs. (XI-1).

The next step is a little more difficult. There is a theorem due to Dupin concerning triply orthogonal systems of surfaces (Blaschke, 1945, pp. 98–100; Struik, 1961, pp. 99–103). The theorem states that, given three surfaces having the property that they all intersect each other at right angles, the curves of intersection are lines of curvature for each surface. As lines of curvature, their torsion must be zero.

In order to apply this theorem we have to be more careful about the wavefront. As we saw in the case of the fish eye, it is not enough to hold the arc length parameter s fixed. It is necessary to fix the optical path length I, defined in the variational integral,

$$I = \int n \, ds. \tag{XI-42}$$

We have the vector function of the three parameters s, u, and v,

$$\mathbf{W} = \mathbf{W}(s, u, v), \tag{XI-43}$$

where we have identified $\mathbf{W}_u$ with $\mathbf{n}$, $\mathbf{W}_v$ with $\mathbf{b}$, and $\mathbf{W}_s$ with $\mathbf{t}$. Here, as always, $\mathbf{t}$, $\mathbf{n}$, and $\mathbf{b}$ are the three orthogonal vectors to a ray. Strictly speaking, $\mathbf{W}$ in Eq. (XI-42) does not represent a wavefront when s is constant but does when the optical path length I is fixed. We assume that this is done—that Eq. (XI-42) is solved for s as a function of I and that this is substituted into Eq. (XI-43), giving

$$\mathbf{W} = \mathbf{W}(I, u, v). \tag{XI-44}$$

Now if I is held fixed, $\mathbf{W}$ will be the vector function of a wavefront, the parametric curves being generated by u and v. Moreover, if we hold u fixed and allow I and v to vary, we get a second surface perpendicular to the first. Finally, if we hold v fixed and allow I and u to vary, we obtain a third surface perpendicular to the first two. In this way we have produced a triply orthogonal system of surfaces. The parametric curves on the wavefront must therefore be lines of curvature, and these have zero torsion. Thus $1/\sigma = 0$ and Eqs. (XI-41) become

$$\begin{aligned} \nabla \cdot (n\mathbf{n}) &= 0, & \nabla \times (n\mathbf{n}) &= (\mathbf{t} \cdot \nabla)(n\mathbf{b}) - (1/\rho_u)(n\mathbf{b}), \\ \nabla \cdot (n\mathbf{b}) &= 0, & \nabla \times (n\mathbf{b}) &= -(\mathbf{t} \cdot \nabla)(n\mathbf{n}) + (1/\rho_v)(n\mathbf{n}). \end{aligned} \tag{XI-45}$$

There is a corollary to this argument. An additional consequence is that a ray is also a line of curvature on two of these surfaces. It must therefore also have zero torsion, which implies that the expression for the torsion of a ray path in Eq. (III-14) must always vanish. Even though the cases we have studied—the fish eye, the heated window, and the brachistochrone—all have rays with zero torsion, it would be indeed rash to say that this is always the case. What probably happens is that Eq. (III-14) vanishes when s, the arc length parameter, is replaced by I, the optical path length, and the derivatives in Eq. (III-14) are taken with respect to I. This is like twisting the metric of the medium, changing to a nonuniform system of coordinates, as it were, so that in the new coordinate system the rays are all planar.

We have encountered something like this already. Recall from Eq. (VII-13) that $\nabla \times (n\mathbf{t}) = \mathbf{0}$, which implies that rays are irrotational. Yet from this equation one derives the fact that $\nabla \times \mathbf{t} = (1/\rho)\mathbf{b}$, which states that an orthotomic system of rays has a component that rotates about the binormal vector. The difference between these two statements is only the presence or absence of n, the refractive index function, inside the curl, which results in a distortion of the vector space.

Let us return now to Eq. (XI-45), where we replace the derivative in the direction of ray propagation, $(\mathbf{t} \cdot \nabla)$, by the partial derivative operator $(\partial/\partial s)$. The two right equations in (XI-45) become

$$\nabla \times (n\mathbf{n}) = (\partial/\partial s)(n\mathbf{b}) - (1/\rho_u)(n\mathbf{b}),$$

$$\nabla \times (n\mathbf{b}) = -(\partial/\partial s)(n\mathbf{n}) + (1/\rho_v)(n\mathbf{n}).$$

Note that an integrating factor can be applied to these equations. These are, for the first and second equation, respectively,

$$\mu_u = \exp\left[-\int (1/\rho_u)\, ds\right],$$
$$\mu_v = \exp\left[-\int (1/\rho_v)\, ds\right]. \tag{XI-46}$$

Where these are used and a quadrature taken, we obtain

$$\nabla \times (n\mathbf{n}) = \mu_u^{-1}(\partial/\partial s)(n\mu_u\mathbf{b}),$$
$$\nabla \times (n\mathbf{b}) = -\mu_v^{-1}(\partial/\partial s)(n\mu_v\mathbf{n}). \tag{XI-47}$$

Finally, we change the parameter s, the arc length parameter, to the time parameter t. To do this we use the following string of identities:

$$\partial/\partial s = (dt/ds)(\partial/\partial t) = (1/v)(\partial/\partial t) = (n/c)(\partial/\partial t).$$

Here v stands for the velocity of light in the medium and c is, as always, its velocity *in vacuo*. Applying this, the previous equations become

$$\nabla \times (n\mathbf{n}) = (n/c\mu_u)(\partial/\partial t)(n\mu_u \mathbf{b}),$$

$$\nabla \times (n\mathbf{b}) = -(n/c\mu_v)(\partial/\partial t)(n\mu_v \mathbf{n}).$$

For an illustration, we refer once again to the fish eye. The equation for the curvature of the spherical wavefront, Eq. (XI-40),

$$1/\rho_u = -n_0(1 - r^2 \tan^2 I)/\tan I,$$

can be rewritten, using Eq. (XI-35), in the form

$$1/\rho_u = -n_0 r[\sin(s/2\rho + u)/\sin(s/2\rho) - \sin(s/2\rho)/\sin(s/2\rho + u)].$$

Now

$$\sin(s/2\rho + u)/\sin(s/2\rho) = \cos u + \sin u \cot(s/2\rho).$$

Using this and a similar expression for the other term, the equation reduces to

$$1/\rho_u = -(1/2\rho)[\cot(s/2\rho) + \cot(s/2\rho + u)].$$

In getting this, Eq. (XI-28) has been used. With this we calculate the integrating factors for the fish eye. From Eq. (XI-46)

$$\begin{aligned} \log \mu_u &= -\int (1/\rho_u)\, ds \\ &= -\int [\cot(s/2\rho) + \cot(s/2\rho + u)](ds/2\rho) \\ &= -\log \sin(s/2\rho) - \log \sin(s/2\rho + u). \end{aligned}$$

From this we get the fact that

$$\mu_u^{-1} = \mu_v^{-1} = \sin(s/2\rho)\sin(s/2\rho + u).$$

This surprisingly simple expression is then the integrating factor for the fish eye.

Going back to the general statement, it is convenient to define a few vectors. Since the calculations involving $\mathbf{n}$ were a little easier than those involving $\mathbf{b}$, we will call $(n\mathbf{n})$ the easy vector $\mathbf{E}$ and $(n\mathbf{b})$ the hard vector $\mathbf{H}$. Thus,

$$\mathbf{E} = -n\mathbf{n}, \qquad \mathbf{H} = n\mathbf{b}.$$

Substituting these in Eqs. (XI-47) and in the divergences appearing in Eqs. (XI-45), we get

$$\begin{aligned} \nabla \times \mathbf{H} &= (1/c)(n/\mu_v)(\partial/\partial t)(\mu_v \mathbf{E}), \qquad \nabla \cdot \mathbf{H} = 0, \\ \nabla \times \mathbf{E} &= -(1/c)(n/\mu_u)(\partial/\partial t)(\mu_u \mathbf{H}), \qquad \nabla \cdot \mathbf{E} = 0. \end{aligned} \tag{XI-48}$$

This is about as close as we can get to Maxwell's equations—which, one must admit, is close indeed. It is rather remarkable that we have come so close. Beginning with only Fermat's principle, certain theorems from the calculus of variations and classical differential geometry, and the idea of an orthotomic system of rays, we have almost re-created the fundamental system of equations that we usually regard as axiomatic to physical optics. One usually justifies geometrical optics, on the other hand, by appealing to the Maxwell equations, from which Fermat's principle, the eikonal equation, and the ray equation are derived (Luneburg, 1964, Chapter I; Kline and Kay, 1965, Chapter III).

We are tempted to jump to all sorts of conclusions. One certain implication is that the geometrical component in physical optics is greater than anyone thought and that the usual methods for extracting geometrical optics from the Maxwell equations may be too restrictive.

It is said that during prohibition one could still buy hops-flavored malt in cans. Now, the only conceivable use for hops-flavored malt is in the brewing of beer, at that time an illegal beverage. Each can, therefore, dutifully carried a warning. In big letters one was advised that under no circumstances were the contents to be mixed with 36 quarts of water, 5 pounds of sugar, and a package of yeast and be kept in a cool place for seven days, *et cetera*. Perhaps that is the best thing for us to do here. Under no circumstances are the equations in (XI-48) to be confused with the Maxwell equations, Eqs. (XI-1); nor is **E** to be confused with the electric field vector, nor is **H** to be compared with the magnetic field vector, nor shall μ_v and μ_u be regarded in any way as resembling the dielectric constant and the magnetic permeability, respectively—or an illegal beverage will result.

REFERENCES

Blaschke, W. (1930). "Vorlesungen über Differentialgeometrie," Vol. I, 3rd ed., Springer-Verlag, Berlin and New York. (Reprinted Dover, New York, 1945.)

Born, M., and Wolf, E. (1970). "Principles of Optics," 4th ed. Pergamon, Oxford.

Kline, M., and Kay, I. W. (1965). "Electromagnetic Theory and Geometrical Optics." Wiley (Interscience), New York.

Korn, G. A., and Korn, T. M. (1968). "Mathematical Handbook for Scientists and Engineers," 2nd ed. McGraw-Hill, New York.

Luneburg, R. K. (1964). "Mathematical Theory of Optics." Univ. of California Press, Berkeley.

Phillips, H. B (1933). "Vector Analysis." Wiley, New York.

Struik, D. J. (1961). "Lectures on Classical Differential Geometry," 2nd ed. Addison-Wesley, Reading, Massachusetts.

XII

Classical Aberration Theory

Many years ago, when computers were quite new, several of us consulted an applied mathematician of repute for some ideas on how to apply the computer to optical design. We laid out the problem in an abstract way: We enumerated the parameters and the rays and showed how they entered the equations; we defined as best we could a reasonable objective for a design process and a reasonable price tag that would make the process competitive with a real, live optical designer. We had hardly got through our presentation when he threw up his hands and declared the problem unreasonable—that it would take just too many hours of machine time, and even then there could be no assurance of convergence. We assured him that problems of this type were solved routinely by a class of humans known as lens designers. He was amazed and wondered whether any of *his* unsolvable problems could be cast in the form of a lens design problem and solved by this class of humans.

This is ancient history of course. Now lenses are designed routinely by computer, and the modern lens designer needs to know as much about data processing and input and output devices as he does about lenses. But whether he relies on his computer or on his art there are several basic ideas that are fundamental. We have already encountered Fermat's principle. This is the key that enables us to describe the performance and the limitations of an ideal lens and provides us with the tools for measuring how far a given real lens misses the mark. In this chapter we will look at these two aspects of lens design— the description of the ideal lens in terms of a first-order theory and a casual examination of third-order aberration theory.

In this chapter we will develop the classical aberration theory from Fermat's principle. We will lean very heavily on Steward (1928) and Synge (1937a,b,c) although ultimately this work depends on ideas that originated with Hamilton and Bruns.

The etiology of the metamorphosis of Hamilton's theory makes rather a droll story. Hamilton's bold definition of the characteristic function, based on a variational formulation of Fermat's principle, and similar to what we will do here, first saw the light of day in 1828 when "Theory of Systems of Rays" was published in *Transactions of the Royal Irish Academy*. Subsequent papers appeared in 1830 and 1837 in this journal and as relatively brief notices in *British Association Reports* in 1832 and 1833 (Conway and Synge, 1931). And that was about the extent of Hamilton's writing on this particular optical theory. At this time he was in his late twenties or early thirties. His reputation, which was considerable for such a young man, had nevertheless not yet reached the continent. Neither did, if we believe the various legends, very many copies of the *Transactions of the Royal Irish Academy*.

However, in 1834 Hamilton published in a *British Association Report* a short note entitled "On the Application to Dynamics of a General Mathematical Method Previously Applied to Optics." In the same year appeared his definitive article "On a General Method in Dynamics: by Which the Study of the Motions of All Free Systems of Attracting or Repelling Points Is Reduced to the Search and Differentiation of One Central Relation or Characteristic Function." Note the reappearance of the *characteristic function* in this more general context. This last article was published in the *Philosophic Transactions of the Royal Society*, a prestigious publication of a formidable organization with, even at that time, a global circulation (Conway and Synge, 1940).

These two titles, long as they are, tell the story in a nutshell. If Hamilton's formalism works for the dynamics of light corpuscles, then it can be extended to the dynamics of other kinds of particles.

As indeed it could; and under the aegis of Karl Jacobi it thrived and grew. Yet the optical work out of which it evolved was completely unknown on the continent. It was only a matter of time before someone would hit on the idea of applying the Hamilton–Jacobi theory of mechanics to geometrical optics. The time was 1895 and the someone was H. Bruns (1895). His formulation, according to Synge (1937b), was based directly on the theorem of Malus rather than on a statement of Fermat's principle. Nevertheless the end result was remarkably similar to Hamilton's original optical work. Bruns named his optical version of the characteristic function the *eikonal.*

This entity now had two names, *characteristic function*, the preferred term west of the English Channel (and perhaps *de rigueur* west of the Irish Channel), and *eikonal.* There is a slight difference between the two: Hamilton's characteristic function depends on six variables while Bruns' depends only on four. Yet this slight difference became the basis of (one guesses) a somewhat acrimonious debate concerning the relative importance of the contributions to optics of Hamilton and Bruns (Herzberger, 1936, 1937; Synge, 1937a, b). One gathers however, that this was only a tempest in a teapot. Steward (1928) blandly uses both terms, *characteristic function* for the "point" function and *eikonal* for the "angle" function.

I prefer *eikonal.* The term *characteristic function* is found in other areas of mathematics—partial differential equations, for example, as well as theoretical mechanics. In a subject area as mathematical as geometrical optics this can be sometimes confusing. Eikonal has not only the great advantage of being much shorter, but it is a word peculiar to geometrical optics and is therefore absolutely unambiguous. And so, with a very deep bow to Sir William, we shall use *eikonal.*

THE POINT EIKONAL

We begin with the variational integral along a ray path

$$I = \int n\, ds = I(\mathbf{P}_0, \mathbf{P}_1) \tag{XII-1}$$

calculated between a pair of fixed points $\mathbf{P}_0$ and $\mathbf{P}_1$. The location of these points places limits on the integral. In general, the location of $\mathbf{P}_0$ and $\mathbf{P}_1$ determines completely the path taken by the ray as well as the initial and terminal direction vectors, which we will denote by $\mathbf{P}_0'$ and $\mathbf{P}_1'$.

We use the identity

$$d\mathbf{P} = \mathbf{P}'\, ds$$

and note that the scalar product of it and $\mathbf{P}'$ results in

$$ds = \mathbf{P}' \cdot d\mathbf{P}. \tag{XII-2}$$

When this is substituted into the variational integral, a line integral is obtained on which we may place the limit of integration in vector form

$$I(\mathbf{P}_0, \mathbf{P}_1) = \int_{\mathbf{P}_0}^{\mathbf{P}_1} n\,\mathbf{P}' \cdot d\mathbf{P}. \tag{XII-3}$$

Define two gradient-like operators,

$$\begin{aligned} \nabla_0 &= (\partial/\partial x_0, \partial/\partial y_0, \partial/\partial z_0), \\ \nabla_1 &= (\partial/\partial x_1, \partial/\partial y_1, \partial/\partial z_1), \end{aligned} \tag{XII-4}$$

where we denote $\mathbf{P}_0 = (x_0, y_0, z_0)$, $\mathbf{P}_1 = (x_1, y_1, z_1)$. Applying these to Eq. (XII-3), we obtain

$$\nabla_0 I = -n_0 \mathbf{P}_0', \qquad \nabla_1 I = n_1 \mathbf{P}_1', \tag{XII-5}$$

where n_0 and n_1 are the refractive indices of the medium at $\mathbf{P}_0$ and $\mathbf{P}_1$, respectively. Written out in scalar form these become

$$\begin{aligned} \partial I/\partial x_0 &= -n_0\, dx_0/ds, & \partial I/\partial x_1 &= n_1\, dx_1/ds, \\ \partial I/\partial y_0 &= -n_0\, dy_0/ds, & \partial I/\partial y_1 &= n_1\, dy_1/ds, \\ \partial I/\partial z_0 &= -n_0\, dz_0/ds, & \partial I/\partial z_1 &= n_1\, dz_1/ds. \end{aligned} \tag{XII-6}$$

It can be seen that these are simply another form of the Hamilton–Jacobi equations, Eq. (VII-26), and that the sums of squares of each set result in a form of the eikonal equation, Eq. (VII-20). Here the derivatives of x_1, y_1, z_1, x_0, y_0, and z_0 with respect to s are the direction cosines of the ray at its initial and terminal points.

Next suppose that we are dealing with a conventional optical system, that is, homogeneous media separated by refracting surfaces. Then ray paths will consist of straight line segments in each medium. We will designate a point in object space by $\mathbf{P} = (x, y, z)$ and a point in image space by $\mathbf{P}^* = (x^*, y^*, z^*)$. We will use reduced direction cosines to designate the ray direction in object and image space,

$$\boldsymbol{\Xi} = (\xi, \eta, \zeta) = n_0 \mathbf{P}_0', \qquad \boldsymbol{\Xi}^* = (\xi^*, \eta^*, \zeta^*) = n_1{}^*\mathbf{E}_1',$$

where we are now denoting the refractive indices of object space and image space by n and n^*, respectively. Then Eqs. (XII-6) read

$$\begin{aligned} \partial I/\partial x &= -\xi, & \partial I/\partial x^* &= \xi^*, \\ \partial I/\partial y &= -\eta, & \partial I/\partial y^* &= \eta^*, \\ \partial I/\partial z &= -\zeta, & \partial I/\partial z^* &= \zeta^*. \end{aligned} \tag{XII-7}$$

Our goal is to obtain a functional dependence of coordinates and direction cosines of a ray in image space with those in object space; that is,

$$\begin{aligned} x^* &= x^*(x, y, z; \xi, \eta), \\ y^* &= y^*(x, y, z; \xi, \eta), \\ z^* &= z^*(x, y, z; \xi, \eta), \\ \xi^* &= \xi^*(x, y, z; \xi, \eta), \\ \eta^* &= \eta^*(x, y, z; \xi, \eta). \end{aligned} \tag{XII-8}$$

We can do this in the following way. In what follows, the subscript will denote partial differentiation. We can regard the first three of the equations in Eq. (XII-7),

$$\begin{aligned} I_x(x, y, z; x^*, y^*, z^*) &= -\xi, \\ I_y(x, y, z; x^*, y^*, z^*) &= -\eta, \\ I_z(x, y, z; x^*, y^*, z^*) &= -\zeta, \end{aligned} \tag{XII-9}$$

as a set of three simultaneous equations in the unknowns x^*, y^*, and z^* whose solutions provide the first three of the equations in Eq. (XII-8). Substituting these values into the second three equations in (XII-7) gives us the last two equations in (XII-8). A third equation for ζ^* is actually redundant because of the identity, $\xi^{*2} + \eta^{*2} + \zeta^{*2} = n^{*2}$.

There is in fact an additional redundancy. Once x, y, z, ξ, and η are chosen—that is, once a ray is specified in object space—a ray, not a point, is determined in image space. Any point on that ray will qualify for the honor of being designated by x^*, y^*, and z^*. We avoid this ambiguity by stipulating that z^* will be fixed and that the coordinates x^* and y^* represent the intersection of the ray in image space on a plane.

A shift in the plane is accomplished by a change in the value of z^*. We adjust Eqs. (XII-8) to conform to these changes:

$$
\begin{aligned}
x^* &= x^*(x, y;\ \xi, \eta;\ z, z^*),\\
y^* &= y^*(x, y;\ \xi, \eta;\ z, z^*),\\
\xi^* &= \xi^*(x, y;\ \xi, \eta;\ z, z^*),\\
\eta^* &= \eta^*(x, y;\ \xi, \eta;\ z, z^*).
\end{aligned}
\tag{XII-10}
$$

THE ANGLE EIKONAL

There remains one enormous difficulty. Suppose the two points $\mathbf{P} = (x, y, z)$ and $\mathbf{P}^* = (x^*, y^*, z^*)$ are perfect images of each other. Then every ray leaving $\mathbf{P}$ arrives at $\mathbf{P}^*$. Although the optical path length I between these two points is well defined, there is an infinite number of rays connecting them. Therefore, in Eqs. (XII-10), x^* and y^* cannot depend on the reduced direction cosines of the ray in object space, ξ, η, and ζ. Yet ξ^* and η^*, being the reduced direction cosines of the ray in image space, must. As a result the first set of relations in Eqs. (XII-7) cannot be solved for x^*, y^*, and z^* in the case where $\mathbf{P}$ and $\mathbf{P}^*$ are conjugates.

To take care of this troublesome fact we go back to the optical path length function $I = I(x, y, z;\ x^*, y^*, z^*)$ and assume for the moment that the two points $\mathbf{P}$ and $\mathbf{P}^*$ are not conjugates. Then its total differential is

$$dI = I_x\,dx + I_y\,dy + I_z\,dz + I_{x^*}\,dx^* + I_{y^*}\,dy^* + I_{z^*}\,dz^*,$$

which becomes, after applying Eqs. (XII-7),

$$
\begin{aligned}
dI &= -\xi\,dx - \eta\,dy - \zeta\,dz + \xi^*\,dx^* + \eta^*\,dy^* + \zeta^*\,dz^*\\
&= -d(x\xi + y\eta + z\zeta) + x\,d\xi + y\,d\eta + z\,d\zeta\\
&\quad + d(x^*\xi^* + y^*\eta^* + z^*\zeta^*) - x^*\,d\xi^* - y^*\,d\eta^* - z^*\,d\zeta^*.
\end{aligned}
$$

In vector form this can be expressed as

$$
\begin{aligned}
dI &= -\boldsymbol{\Xi}\cdot d\mathbf{P} + \boldsymbol{\Xi}^*\cdot d\mathbf{P}^*\\
&= -d(\boldsymbol{\Xi}\cdot\mathbf{P}) + \mathbf{P}\cdot d\boldsymbol{\Xi} + d(\boldsymbol{\Xi}^*\cdot\mathbf{P}^*) - \mathbf{P}^*\cdot d\boldsymbol{\Xi}.
\end{aligned}
$$

Therefore, if we define a new function

$$W(\xi, \eta, \zeta; \xi^*, \eta^*, \zeta^*) = I + \boldsymbol{\Xi} \cdot \mathbf{P} - \boldsymbol{\Xi}^* \cdot \mathbf{P}^*, \tag{XII-11}$$

then $dW = \mathbf{P} \cdot d\boldsymbol{\Xi} - \mathbf{P}^* \cdot d\boldsymbol{\Xi}^*$ so that

$$\begin{aligned} \partial W/\partial \xi &= x, & \partial W/\partial \xi^* &= -x^*, \\ \partial W/\partial \eta &= y, & \partial W/\partial \eta^* &= -y^*, \\ \partial W/\partial \zeta &= z, & \partial W/\partial \zeta^* &= -z^*. \end{aligned} \tag{XII-12}$$

The function W is called the *angle characteristic function* or the *angle eikonal*. Now if we set up the system of simultaneous equations

$$\begin{aligned} W_\xi(\xi, \eta, \zeta; \xi^*, \eta^*, \zeta^*) &= x, \\ W_\eta(\xi, \eta, \zeta; \xi^*, \eta^*, \zeta^*) &= y, \\ W_\zeta(\xi, \eta, \zeta; \xi^*, \eta^*, \zeta^*) &= z, \end{aligned}$$

we can, in principle, obtain as solutions

$$\begin{aligned} \xi^* &= \xi^*(x, y, z; \xi, \eta, \zeta), \\ \eta^* &= \eta^*(x, y, z; \xi, \eta, \zeta), \\ \zeta^* &= \zeta^*(x, y, z; \xi, \eta, \zeta). \end{aligned}$$

Substituting these into

$$\begin{aligned} W_{\xi^*}(\xi, \eta, \zeta; \xi^*, \eta^*, \zeta^*) &= -x^*, \\ W_{\eta^*}(\xi, \eta, \zeta; \xi^*, \eta^*, \zeta^*) &= -y^*, \\ W_{\zeta^*}(\xi, \eta, \zeta; \xi^*, \eta^*, \zeta^*) &= -z^*, \end{aligned}$$

and solving yields

$$\begin{aligned} x^* &= x^*(x, y, z; \xi, \eta, \zeta), \\ y^* &= y^*(x, y, z; \xi, \eta, \zeta), \\ z^* &= z^*(x, y, z; \xi, \eta, \zeta). \end{aligned}$$

Because object and image points do not appear as arguments in the angle eikonal, no difficulties are encountered as far as conjugate points are concerned.

It is simple enough to visualize the point eikonal as representing the optical path length between a point in object space and a point in image space. The angle eikonal is a bit more complicated. Drop a perpendicular from the coordinate origin to the ray. Let the point of intersection be $\mathbf{R}$. The segment of the ray between $\mathbf{R}$ and $\mathbf{P}$ and the vector $\mathbf{R}$ will form two sides of a right triangle whose hypotenuse is the vector $\mathbf{P}$. Let θ

represent the angle between **P** and Ξ. Then the length of the ray segment between **R** and **P** along Ξ is

$$|\mathbf{P}| \cos\theta = (1/n)\,\mathbf{P}\cdot\Xi,$$

a geometrical distance which when multiplied by n yields an optical path length. This is shown in Fig. XII-1.

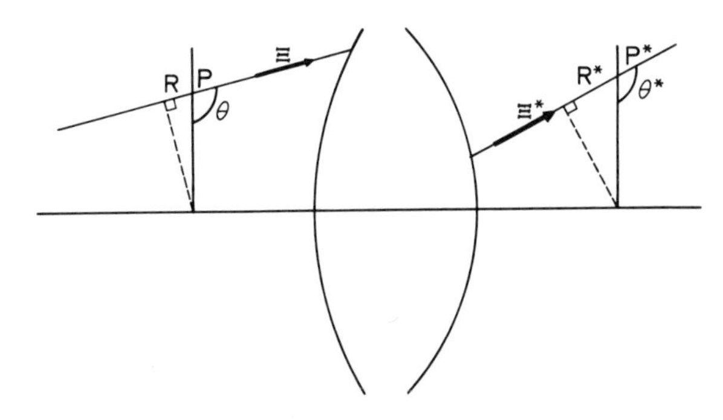

FIG. XII-1. The angle eikonal construction.

Shown is the ray connecting an object point **P** and an image point **P***. The reduced direction cosines of the ray are Ξ and Ξ^* in object and image space, respectively. The point eikonal is the optical path length from **P** to **P*** along the ray. Drop a perpendicular from the coordinate origins to the rays. The points of intersection are at **R** and **R***. The *angle eikonal* is the optical path length between these two points.

From this we can see that the angle eikonal is equal to the point eikonal with the ray segment adjoined in object space and removed in image space. More precisely, the angle eikonal is the optical path length measured from the foot of a perpendicular dropped to the ray from the coordinate origin in object space to a similar point in image space.

We can then go on to ring all the changes and define two types of mixed eikonals,

$$\begin{aligned} M^* &= M^*(\xi, \eta, \zeta; x^*, y^*, z^*) = I + \Xi\cdot\mathbf{P}, \\ M &= M(x, y, z; \xi^*, \eta^*, \zeta^*) = I - \Xi^*\cdot\mathbf{P}^*. \end{aligned} \tag{XII-13}$$

THE ROTATIONALLY SYMMETRIC SYSTEM

These rather general equations are now specialized to a rotationally symmetric lens. First of all we assume that all refracting surfaces considered here are rotationally symmetric. Then we stipulate that all of the axes of rotation shall lie on the same straight line and that this line shall coincide with the z axis of any coordinate system we may refer to.

Under these assumptions any eikonal function must be invariant under a rotation of the x and y coordinate axes. One therefore must be able to write the point eikonal as a function of three symmetric functions, each of which is invariant under a rotation of these same coordinate axes. These three symmetric functions we choose to be

$$u = \tfrac{1}{2}(x^2 + y^2), \qquad m = xx^* + yy^*, \qquad u^* = \tfrac{1}{2}(x^{*2} + y^{*2}), \tag{XII-14}$$

which do indeed possess the required invariant property.

The point eikonal then becomes $I(u, m, u^*; z, z^*)$ and its derivatives become

$$\begin{aligned} I_x &= I_u x + I_m x^* = -\xi, \qquad & I_{x^*} &= I_m x + I_{u^*} x^* = \xi^*, \\ I_y &= I_u y + I_m x^* = -\eta, \qquad & I_{y^*} &= I_m y + I_{u^*} y^* = \eta^*. \end{aligned} \tag{XII-15}$$

Here we have used Eqs. (XII-7).

Exactly the same sort of thing can be done for the angle eikonal. Define the three symmetric functions

$$w = \tfrac{1}{2}(\xi^2 + \eta^2), \qquad s = \xi\xi^* + \eta\eta^*, \qquad w^* = \tfrac{1}{2}(\xi^{*2} + \eta^{*2}). \tag{XII-16}$$

These become the arguments for the angle eikonal for the rotationally symmetric optical system. Its derivatives, with the aid of Eqs. (XII-12), yield

$$\begin{aligned} x &= W_\xi = W_w \xi + W_s \xi^*, \qquad & -x^* &= W_{\xi^*} = W_s \xi + W_{w^*} \xi^*, \\ y &= W_\eta = W_w \eta + W_s \eta^*, \qquad & -y^* &= W_{\eta^*} = W_s \eta + W y_* \eta^*. \end{aligned} \tag{XII-17}$$

From the point eikonal, Eqs. (XII-15), we can deduce an important invariant quantity associated with rotationally symmetric systems. Using the second pair of equations in Eqs. (XII-15), we calculate

$$x^*\eta^* - y^*\xi^* = I_m(x^*y - y^*x).$$

Solving the first and second equations of Eqs. (XII-15), for x^* and y^*, respectively, and substituting into the right-hand member of the above yields

$$x^*\eta^* - y^*\xi^* = x\eta - y\xi. \tag{XII-18}$$

This is called the *skewness invariant*. The *skewness* p is defined as

$$p = x\eta - y\xi \tag{XII-19}$$

and turns out to be equal to the sine of the angle between the skew ray and the meridian plane times the refractive index. We will have much more to say about the skewness and its invariance in a subsequent chapter.

The next step in this development is the formation of a power series for the rotationally symmetric angle eikonal in terms of the three symmetric functions given in Eqs. (XII-14):

$$\begin{aligned} W(w, s, w^*) = {} & J + J_w w + J_s s + J_{w^*} w^* \\ & + \tfrac{1}{2}[J_{ww} w^2 + J_{ss} s^2 + J_{w^*w^*} w^{*2} \\ & + 2J_{ws} ws + 2J_{ww^*} ww^* + 2J_{sw^*} sw^*] + \cdots . \end{aligned} \qquad \text{(XII-20)}$$

Here subscripts, as always, denote partial derivatives. The J's denote W's calculated at the point $w = s = w^* = 0$.

There is a tacit assumption needed in this step. It is important that this series be absolutely and uniformly convergent over a sufficiently large domain of the independent variables. About this we will have more to say later.

Differentiating this series and applying Eqs. (XII-12), we get

$$\begin{aligned} x &= J_w \xi + J_s \xi^* + A\xi + B\xi^*, \\ y &= J_w \eta + J_s \eta^* + A\eta + B\eta^*, \\ -x^* &= J_s \xi + J_w \xi^* + B\xi + C\xi^*, \\ -y^* &= J_s \eta + J_{w^*} \eta^* + B\eta + C\eta^*, \end{aligned} \qquad \text{(XII-21)}$$

$$\begin{aligned} A &= J_{ww} w + J_{ws} s + J_{ww^*} w^*, \\ B &= J_{ws} w + J_{ss} s + J_{sw^*} w^*, \\ C &= J_{ww^*} w + J_{sw^*} s + J_{w^*w^*} w^*. \end{aligned} \qquad \text{(XII-22)}$$

Note that in each equation of (XII-21) there are two first-degree terms and six terms of degree *three*.

THE FIRST ORDER: GAUSSIAN OPTICS

We consider now rays confined to a region so close to the axis of the optical system that second and higher powers of the ray variables can be ignored. A first consequence of this restriction is that $\zeta = n$ and

$\zeta^* = n^*$. Under this assumption, ray tracing formulas degenerate into rather simple linear relationships. When these approximations are used to trace rays outside of their domain of validity, the rays are called *paraxial rays* and represent a crude approximation to a real ray trace. This first-order optics is called *Gaussian optics.*

Under this assumption Eqs. (XII-21) degenerate into

$$\begin{aligned} x &= J_w\xi + J_s\xi^*, & -x^* &= J_s\xi + J_{w*}\xi^*, \\ -y &= J_w\eta + J_s\eta^*, & -y^* &= J_s\eta + J_{w*}\eta^*. \end{aligned} \qquad \text{(XII-23)}$$

From these equations we can obtain a number of important consequences. The first involves finding out what happens to the J's when the reference planes are translated. We use a version of the transfer equations given in Eq. (VI-5),

$$\begin{aligned} \bar{x} &= x + \bar{z}\xi/n, & \bar{x}^* &= x^* + \bar{z}^*\xi^*/n^*, \\ \bar{y} &= y + \bar{z}\eta/n, & \bar{y}^* &= y^* + \bar{z}^*\eta^*/n^*, \end{aligned}$$

in which ζ and ζ^* are replaced by n and n^* and where the barred quantities represent ray coordinates on the translated reference planes. Thus

$$\begin{aligned} \bar{x} &= x + \bar{z}\xi/n = (J_w\xi + J_s\xi^*) + \bar{z}\xi/n \\ &= (J_w + \bar{z}/n)\xi + J_s\xi^* \\ &= \bar{J}_w\xi + \bar{J}_s\xi^* \end{aligned}$$

and we have

$$\bar{J}_w = J_w + \bar{z}/n, \qquad \bar{J}_s = J_s\,. \qquad \text{(XII-24a,b)}$$

In like manner

$$\bar{J}_{w*} = J_{w*} - \bar{z}^*/n^*. \qquad \text{(XII-24c)}$$

The next step involves solving the first two equations of (XII-23) for ξ^* and η^* and substituting the result into the second pair. From this operation we obtain

$$\begin{aligned} -\bar{x}^* &= (1/J_s)[\bar{J}_{w*}\bar{x} + (J_s^2 - \bar{J}_w\bar{J}_{w*})\xi], \\ -\bar{y}^* &= (1/J_s)[\bar{J}_{w*}\bar{y} + (J_s^2 - \bar{J}_w\bar{J}_{w*})\eta]. \end{aligned} \qquad \text{(XII-25)}$$

For the two planes to be conjugates, every ray from the point $(\bar{x}, \bar{y})$ on the object side must pass through the point $(\bar{x}^*, \bar{y}^*)$. The expressions for $\bar{x}^*$ and $\bar{y}^*$ must therefore be independent of ξ and η. For this to happen

$$J_s^2 - \bar{J}_w\bar{J}_{w*} = 0. \qquad \text{(XII-26)}$$

Then Eqs. (XII-25) become

$$-\bar{x}^* = (\bar{J}_{w*}/J_s)\bar{x}, \qquad -\bar{y}^* = (\bar{J}_{w*}/J_s)\bar{y}.$$

It is natural to define the magnification $\bar{M}$ relative to these conjugate planes as

$$\bar{M} = -\bar{J}_{w*}/J_s \tag{XII-27}$$

so that Eqs. (XII-25) become

$$\bar{x}^* = \bar{M}\bar{x}, \qquad \bar{y}^* = \bar{M}\bar{y}. \tag{XII-28}$$

Next we apply Eq. (XII-27) to Eq. (XII-26) and get $1 + (\bar{J}_w/J_s)\bar{M} = 0$ so that

$$J_s/\bar{J}_w = -\bar{M}. \tag{XII-29}$$

From Eqs. (XII-24),

$$\bar{M} = M + \bar{z}^*/n^*J_s, \qquad 1/\bar{M} = 1/M - \bar{z}/nJ_s\,, \tag{XII-30}$$

where M is defined as

$$M = -J_{w*}/J_s = -J_s/J_w\,. \tag{XII-31}$$

Next, using Eq. (XII-31), the first two equations of Eqs. (XII-23) are solved for ξ^* and η^*,

$$\begin{aligned} \xi^* &= (x - J_w\xi)/J_s = \xi/M + x/J_s\,, \\ \eta^* &= (y - J_w\eta)/J_s = \eta/M + y/J_s\,. \end{aligned} \tag{XII-32}$$

Equation (XII-26), which, using Eqs. (XII-24), we write out as

$$J_s^2 - (J_w + \bar{z}/n)(J_{w*} - \bar{z}^*/n^*) = 0,$$

reduces to

$$M\bar{z}/n - \bar{z}^*/Mn^* + \bar{z}\bar{z}^*/J_snn^* = 0 \tag{XII-33}$$

when Eqs. (XII-30) are applied. It is this equation that must be satisfied for a pair of planes to be conjugate.

We use this to find the *focal* planes of the optical system these equations represent. Divide it by $\bar{z}\bar{z}^*$ to obtain

$$M/n\bar{z}^* - 1/Mn^*\bar{z} = -1/J_snn^*$$

and let $\bar{z}$ approach infinity. Then

$$\bar{z}^* = z_f^* = -MJ_sn^* = +J_{w*}n^*, \tag{XII-34a}$$

the position of the image space focal plane. When we allow $\bar{z}^*$ to become infinite we get

$$\bar{z} = z_f = nJ_s/M = -J_w n. \tag{XII-34b}$$

In deriving these formulas we have used Eq. (XII-31).

The *principal planes* are those unique conjugate planes relative to which the magnification of the lens is *unity*. Thus, by setting $\overline{M} = 1$ in Eqs. (XII-30), we obtain the appropriate values of $\bar{z}$ and $\bar{z}^*$

$$\begin{aligned} z_p^* &= (1 - M)\, n^* J_s = (J_s + J_{w*})\, n^*, \\ z_p &= -(1 - 1/M)\, nJ_s = -(J_s + J_w)n. \end{aligned} \tag{XII-35}$$

Equation (XII-31) is also used here.

Finally we come to the *nodal planes*, a pair of conjugate planes with the property that an object ray through the axial point of the object space nodal plane emerges through the axial point of the image space nodal plane at exactly the same angle. Thus, for this pair of conjugate planes, when $\bar{x} = \bar{y} = 0$,

$$\xi^*/n^* = \xi/n \qquad \text{and} \qquad \eta^*/n^* = \eta/n.$$

Referring to Eqs. (XII-32), we see that this happens when $1/\overline{M} = n^*/n$.

This is then applied to Eqs. (XII-30), which results in a pair of values of $\bar{z}$ and $\bar{z}^*$ for the nodal planes

$$\begin{aligned} z_n^* &= J_s(n - n^*M) = (nJ_s + n^*J_{w*}), \\ z_n &= -J_s(n^* - n/M) = -(n^*J_s + nJ_w). \end{aligned} \tag{XII-36}$$

Here we have also used Eq. (XII-31).

Principal planes, nodal planes, and focal planes are shown in Fig. XII-2.

The distance between a focus and a principal plane is called the *focal length* of the lens. In image space this is

$$f^* = z_p^* - z_f^* = n^*J_s\,, \tag{XII-37a}$$

and in object space

$$f = z_p - z_f = -nJ_s\,. \tag{XII-37b}$$

Note that when the refractive indices of object space and image space are the same, the two focal lengths are equal.

Combining Eq. (XII-31) and Eqs. (XII-37)

$$J_s = f^*/n^* = -f/n, \qquad J_{w*} = Mf/n, \qquad J_w = f/Mn. \tag{XII-38}$$

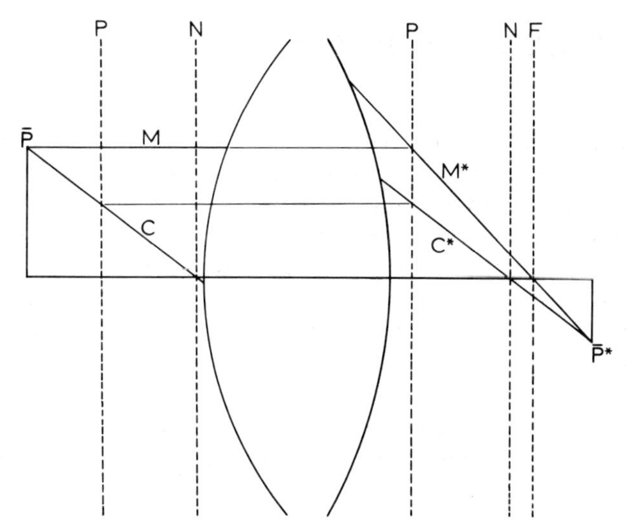

FIG. XII-2. The cardinal points of a lens.

The two conjugate planes labeled P are the principal planes that intersect the axis at the principal points. Since the magnification of the lens relative to the principal planes is unity, the ray M will pass through the two principal planes at the same height. The two conjugate planes N are the nodal planes. The ray C entering the lens through the first nodal point will emerge from the lens through the second nodal point as C^*. C and C^* are parallel. Only one focal plane F is shown. The ray M parallel to the axis will emerge from the lens as M^* and pass through the focus. This construction shows how $\mathbf{P}^*$, the image of $\mathbf{P}$, is formed.

These results can now be applied to Eqs. (XII-32) and (XII-28), which we collect in the form

$$\begin{aligned} x^* &= Mx, \qquad \xi^* = \xi/M - nx/f, \\ y^* &= My, \qquad \eta^* = \eta/M - ny/f. \end{aligned} \tag{XII-39}$$

Written as a matrix relation, this becomes

$$\begin{pmatrix} x^* \\ y^* \\ \xi^* \\ \eta^* \end{pmatrix} = \begin{pmatrix} M & 0 & 0 & 0 \\ 0 & M & 0 & 0 \\ -n/f & 0 & 1/M & 0 \\ 0 & -n/f & 0 & 1/M \end{pmatrix} \begin{pmatrix} x \\ y \\ \xi \\ \eta \end{pmatrix}. \tag{XII-40}$$

Equations (XII-39) and Eq. (XII-40) refer to a pair of conjugate planes that we will regard as being fixed. These will be referred to as the *object plane* and the *image plane*. Distances to all other planes will be measured from these.

SOME PRACTICAL FORMULAS

It is hard to believe that our abstract approach, dealing with the most general of generalities, can lead to down-to-earth formulas. From Fermat's principle alone, as interpreted by an eikonal function and an assumption of rotational symmetry, we have been able to define and describe the *cardinal points* of a lens: the two foci, the two principal points, and the two nodal points. Moreover, most of the basic formulas of practical optics are obtainable from Eqs. (XII-30), by use of Eqs. (XII-37).

For example, let u be the distance from the object plane to the object principal plane and let v be the distance from the image principal plane to the image plane. Then, from Eqs. (XII-35) and Eqs. (XII-37),

$$u = z_p = -(1 - 1/M)\,nJ_s = (1 - 1/M)f,$$

$$v = z_p{}^* = (1 - M)\,n^*J_s = (1 - M)f^*.$$

If the refractive indices of object space and image space happen to be equal, then $f = -f^*$ and the above become the useful "enlarging lens" formulas,

$$u = (1 - 1/M)f \qquad \text{and} \qquad v = -(1 - M)f.$$

Perhaps a better known formula is $v/u = M$, obtained by taking the ratio of the above.

In Eq. (XII-33) we have

$$M\bar{z}/n - \bar{z}^*/Mn^* + \overline{zz^*}/J_s nn^* = 0,$$

from which we obtain

$$M\bar{z}/f + \bar{z}^*/Mf^* = -\overline{zz^*}/ff^*. \qquad \text{(XII-41)}$$

Now, measuring distances of the object and image planes from the two principal planes on which $M = 1$, we obtain

$$u/f + v/f^* + uv/ff^* = 0,$$

whence

$$1/fv + 1/f^*u = -1/ff^*.$$

Again, if object and image spaces have the same refractive indices, then $f = -f^*$ and we are left with the familiar

$$1/v + 1/u = 1/f.$$

PUPIL PLANES

Somewhere in this lens there is a *dominant aperture*, a *diaphragm*, a *stop*—some sort of obstruction that does the work of limiting the beam size. The Gaussian image of this hole on the object side of the lens is called the *entrance pupil*; on the image side, the *exit pupil*. The pupil planes are characterized by the fact that, to the first order, a ray passing through the axial of either pupil passes through the center of the dominant aperture. Such a ray is called a *principal ray*. In the absence of aberrations, it is the center of symmetry of the bundle from the same object point that passes through the lens. A ray that just clears the dominant aperture is called a *marginal ray*. The first order approximations of these rays are referred to as the *paraxial principal ray* and the *paraxial marginal* say.

The pupil planes are conjugate planes and therefore can be represented by the methods derived above. Let the distance from the object plane to the entrance pupil plane be z_s. Then the distance from the image plane to the exit pupil plane is, from Eq. (XII-41),

$$z_s^* = -z_s M^2 f^*/(f + Mz_s). \tag{XII-42}$$

The magnification relative to the pupil planes is given by

$$M_s = Mf/(f + Mz_s). \tag{XII-43}$$

This comes from Eqs. (XII-30).

Now using the equations for a transfer, we obtain the pupil coordinates,

$$\begin{aligned} x_s &= x + z_s\xi/n, \qquad & x_s^* &= x^* + z_s^*\xi^*/n^*, \\ y_s &= y + z_s\eta/n, \qquad & y_s^* &= y^* + z_s^*\eta/n^*. \end{aligned} \tag{XII-44}$$

Applying Eqs. (XII-42) and (XII-43), we get

$$x_s^* = M_s x_s \qquad \text{and} \qquad y_s^* = M_s y_s, \tag{XII-45}$$

which is exactly as it should be.

Figure XII-3 shows a pair of marginal rays and the principal ray from a finite object point through a lens.

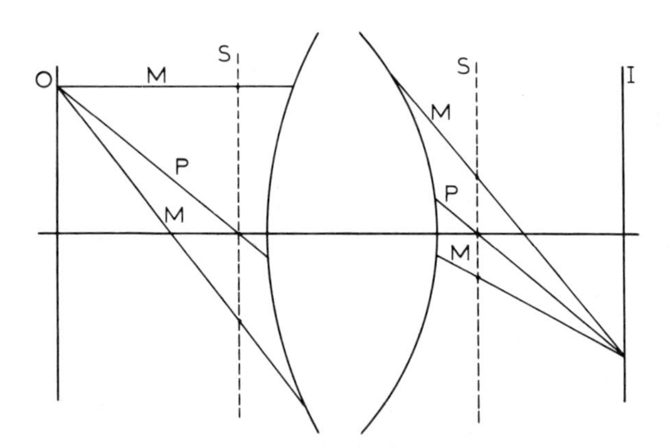

FIG. XII-3. The pupil planes: marginal and principal rays.
The two conjugate planes labeled S are the pupil planes, the Gaussian images of the physical stop. O and I represent conjugate object and image planes, respectively. The principal ray P passes through the center of the two pupil planes. The marginal rays M just miss the edge of the stop.

THE THIRD ORDER

The first-order behavior of an optical system traditionally is taken to represent the ideal behavior of a lens. Higher-order terms therefore must represent errors in a lens or the aberrations that must be exorcised. This convention is not arbitrary; the gross behavior of a lens, no matter how poor, conforms fairly faithfully to the practical formulas derived earlier. It would seem that the natural expression of a lens is first order and that all higher orders can be regarded only as defects that must be eliminated.

This is stating things a little too strongly. There are occasions when it is desirable to introduce one kind of aberration to compensate for the presence of another. These niceties, however, do not concern us here. We seek the description and the structure of the aberrations in the third-order terms of the derivatives of the eikonal function. In other words, we shall examine the so-called *Seidel aberrations*.

We will take the first order equations as exemplified by Eqs. (XII-39), as representing the coordinates and direction of an *ideal ray*. These we will designate by the "sharp" symbol $\#$. Thus

$$x^{\#} = Mx, \qquad \xi^{\#} = \xi/M - nx/f, \qquad y^{\#} = My, \qquad \eta^{\#} = \eta/M - ny/f. \tag{XII-46}$$

Let δx^*, δy^*, $\delta\xi^*$, and $\delta\eta^*$ represent the departure of the real ray from the ideal ray. In other words,

$$x^* = x^\# + \delta x^*, \qquad \xi^* = \xi^\# + \delta\xi^*,$$
$$y^* = y^\# + \delta y^*, \qquad \eta^* = \eta^\# + \delta\eta^*. \tag{XII-47}$$

These two sets of equations, (XII-46) and (XII-47), are substituted into Eqs. (XII-21) to obtain

$$(f/n)\,\delta\xi^* = (A + B/M)\xi - (n/f)\,Bx,$$
$$(f/n)\,\delta\eta^* = (A + B/M)\eta - (n/f)\,By, \tag{XII-48}$$

$$-\delta x^* = M(f/n)\,\delta\xi^* + (B + C/M)\xi - (n/f)\,Cx,$$
$$-\delta y^* = M(f/n)\,\delta\eta^* + (B + C/M)\eta - (n/f)\,Cy. \tag{XII-49}$$

Here we have used Eqs. (XII-38).

Now note that, by Eq. (XII-34a) and the second equation of Eqs. (XII-38),

$$z_f^* = J_{w*}n^* = Mn^*f/n.$$

Therefore

$$\delta x^* + (Mf/n)\,\delta\xi^* = \delta x^* + (z_f^*/n^*)\,\delta\xi^* = \delta x_f^*$$

represents a transfer from the image plane to the focal plane. We let δx_f^*, be the third-order aberration function defined in the focal plane. Then, from Eqs. (XII-49) we obtain

$$-\delta x_f^* = (B + C/M)\xi - (n/f)\,Cx,$$
$$-\delta y_f^* = (B + C/M)\eta - (n/f)\,Cy. \tag{XII-50}$$

The form of the aberration function here is such that it depends on the coordinates of the object point x, y and the ray directions. It would be more useful to convert these to pupil coordinates by means of Eq. (XII-44):

$$\xi = (x_s - x)\,n/z_s \qquad \text{and} \qquad \eta = (y_s - y)\,n/z_s\,. \tag{XII-51}$$

Substituting these into Eqs. (XII-50) results in the following expressions for the aberration functions:

$$-\delta x_f^* = [(B + C/M)\,x_s - (B + C/M_s)x]n/z_s\,,$$
$$-\delta y_f^* = [(B + C/M)\,y_s - (B + C/M_s)y]\,n/z_s\,. \tag{XII-52}$$

Here we made use of Eq. (XII-43), in which M_s is defined.

In like manner we can get

$$(f/n)\,\delta\xi^* = [(A + B/M)\,x_s - (A + B/M_s)x]\,n/z_s\,,$$
$$(f/n)\,\delta\eta^* = [(A + B/M)\,y_s - (A + B/M_s)y]\,n/z_s\,. \qquad \text{(XII-53)}$$

We now need to calculate the quadratic expressions A, B, and C given in Eqs. (XII-22), in terms of the object plane coordinates x, y and the entrance pupil plane coordinates x_s, y_s. Again we use Eqs. (XII-51), which we obtained from Eq. (XII-44) and a pair of companion equations

$$\xi^* = (x_s^* - x^*)\,n^*/z_s^* \qquad \text{and} \qquad \eta^* = (y_s^* - y^*)\,n^*/z_s^*.$$

Substituting these into Eqs. (XII-16), in which the symmetric functions were defined, we obtain

$$w = (u_s + u - \sigma)\,n^2/z_s^2,$$
$$w^* = (u_s^* + u^* - \sigma^*)\,n^{*2}/z_s^2,$$
$$s = [(x_s x_s^* + y_s y_s^*) - (x_s x^* + y_s y^*) - (x x_s^* + y y_s^*),$$
$$+(xx^* + yy^*)]\,nn^*/z_s z_s^*,$$

where

$$\sigma = xx_s + yy_s\,, \qquad \sigma^* = x^*x_s^* + y^*y_s^*.$$

We also use here the symmetric functions defined in Eqs. (XII-14). To these equations we apply Eqs. (XII-47). However, we are interested only in terms no greater than degree *two*. We therefore drop δx^* and δy^* since these involve third-degree terms. Next we apply Eqs. (XII-45) and (XII-46) and obtain

$$w = (u_s + u - \sigma)\,n^2/z_s^2,$$
$$w^* = (M_s^2 u_s + M^2 u - MM_s\sigma)\,n^{*2}/z_s^{2*},$$
$$s = [2M_s u_s + 2Mu - (M + M_s)\sigma]\,nn^*/z_s z_s^*.$$

Using Eq. (XII-42), Eq. (XII-43), and the first of Eqs. (XII-38), we get

$$n^*/z_s^* = -(n^*/f^*)(f + Mz_s)/z_s M^2 f^*$$
$$= -(n^*/f^*)f/z_s M_s M$$
$$= (n/f)f/z_s M_s M$$
$$= (n/z_s)/M_s M.$$

Substituting this into the above results in

$$w = (u_s + u - \sigma)(n/z_s)^2,$$
$$w^* = (M_s^2 u_s + M^2 u - MM_s \sigma)(n/z_s)^2/M^2 M_s^2,$$
$$s = [2M_s u_s + 2Mu - (M + M_s)\sigma](n/z_s)^2/MM_s .$$

These relations are used in conjunction with Eqs. (XII-22), (XII-52)) and (XII-53) to obtain

$$\begin{aligned} -\delta x_f^* &= (n/z_s)^3\{[(B_1 + a/M)\, u_s - (B_2 + a/M)\sigma + (B_3 + a/M)u]\, x_s \\ &\quad - [(B_2 + a/M_s)\, u_s - (B_3 + a/M_s)\sigma + (B_4 + a/M_s)u]x\}, \\ -\delta y_f^* &= (n/z_s)^3\{[(B_1 + a/M)\, u_s - (B_2 + a/M)\sigma + (B_3 + a/M)u]\, y_s \\ &\quad - [(B_2 + a/M_s)\, u_s - (B_3 + a/M_s)\sigma + (B_4 + a/M_s)u]y\}, \end{aligned} \tag{XII-54}$$

$$\begin{aligned} (f/n)\, \delta\xi^* &= (n/z_s)^3\{[(A_1 + a/M^2)\, u_s - (A_2 + a/MM_s)\sigma + (A_3 + a/M_s^2)u]\, x_s \\ &\quad - [(A_2 + a/M^2)\, u_s - (A_3 + a/MM_s)\sigma + (A_4 + a/M_s^2)u]x\}, \\ (f/n)\, \delta\eta^* &= (n/z_s)^3\{[(A_1 + a/M^2)\, u_s - (A_2 + a/MM_s)\sigma + (A_3 + a/M_s^2)u]\, y_s \\ &\quad - [(A_2 + a/M^2)\, u_s - (A_3 + a/MM_s)\sigma + (A_4 + a/M_s^2)u]y\}, \end{aligned} \tag{XII-55}$$

$$\begin{aligned} A_1 &= b + c/M, & B_1 &= c + d/M, \\ A_2 &= b + c/M_s , & B_2 &= c + d/M_s , \\ A_3 &= b_s + c_s/M, & B_3 &= c_s + d_s/M, \\ A_4 &= b_s + c_s/M_s , & B_4 &= c_s + d_s/M_s , \end{aligned} \tag{XII-56}$$

$$\begin{aligned} a &= J_{ww^*} - J_{ss} , \\ b &= J_{ww} + 2J_{ws}/M + J_{ss}/M^2, & b_s &= J_{ww} + 2J_{ws}/M_s + J_{ss}/M_s^2, \\ c &= J_{ws} + 2J_{ss}/M + J_{sw^*}/M^2, & c_s &= J_{ws} + 2J_{ss}/M_s + J_{sw^*}/M_s^2, \\ d &= J_{ss} + 2J_{sw^*}/M + J_{w^*w^*}/M^2, & d_s &= J_{ss} + 2J_{sw^*}/M_s + J_{w^*w^*}/M_s^2. \end{aligned} \tag{XII-57}$$

We can use Eqs. (XII-54) and (XII-55) to shift the aberration functions to different planes in image space. A shift can be made back to the image plane by means of the linear combination

$$\begin{aligned} \delta x^* &= \delta x_f^* - (z_f^*/n^*)\, \delta\xi^* \\ &= \delta x_f^* - M(f/n)\, \delta\xi^*. \end{aligned} \tag{XII-58}$$

Similarly, a shift can be made to the pupil plane. First note that z_s can be written as a function of M_s. From Eq. (XII-43) we can get

$$z_s = f(M - M_s)/MM_s,$$

which, when substituted into Eq. (XII-42), results in

$$z_s^* = -f^*(M - M_s).$$

Thus,

$$z_s^*/n^* = (f/n)(M - M_s),$$

using once again Eq. (XII-38), and we see that a shift of the aberration function to the pupil plane is accomplished by

$$\begin{aligned} \delta x_s^* &= \delta x^* + (z_s^*/n^*)\,\delta\xi^* \\ &= \delta x^* + (M - M_s)(f/n)\,\delta\xi^* \\ &= \delta x_f^* - M_s(f/n)\,\delta\xi^*. \end{aligned} \tag{XII-59}$$

Recall here that z_s^* represents the distance between the object plane and the pupil plane.

These arguments have been concerned with changes in the position of the plane on which the aberration functions are calculated. What this amounts to is a coarse description of the effects of focusing error on the aberration functions themselves.

The effect on the aberration functions owing to shifts in the position of the pupil planes, which is tantamount to a shift in the stop position, can also be accounted for in Eqs. (XII-54), (XII-56), and (XII-57). Such a shift will result in a change in the value of z_s and in M_s by way of Eq. (XII-43).

Finally, the effect of a shift in the object and image planes can be studied by means of transfer equations of the type given in Eq. (VI-5) as used to obtain, for example, Eqs. (XII-24). Such transfers result, of course, in altered values of M.

THE SEIDEL ABERRATIONS

Our examination of the third-order aberrations begins with the transformation Eq. (XII-58), which translates the aberration functions from the focal plane back to the image plane.

This yields

$$\begin{aligned} \delta x^* &= (\mathfrak{B}u_s - \mathfrak{F}\sigma + \mathfrak{D}u)\, x_s - (\mathfrak{F}u_s - \mathfrak{C}\sigma + \mathfrak{E}u)x, \\ \delta y^* &= (\mathfrak{B}u_s - \mathfrak{F}\sigma + \mathfrak{D}u)\, y_s - (\mathfrak{F}u_s - \mathfrak{C}\sigma + \mathfrak{E}u)y, \end{aligned} \tag{XII-60}$$

where

$$\begin{aligned} \mathfrak{B} &= [B_1 + MA_1 + 2a/M](n/z_s)^3, \\ \mathfrak{F} &= [B_2 + MA_2 + a(1/M + 1/M_s)](n/z_s)^3, \\ \mathfrak{D} &= [B_3 + MA_3 + a(1/M + M/M_s^2)](n/z_s)^3, \\ \mathfrak{C} &= [B_3 + MA_3 + 2a/M_s](n/z_s)^3, \\ \mathfrak{E} &= [B_4 + MA_4 + a(1/M_s + M/M_s^2)](n/z_s)^3. \end{aligned} \tag{XII-61}$$

The symbols here are defined in Eqs. (XII-56) and Eqs. (XII-57).

These five coefficients, $\mathfrak{B}$, $\mathfrak{F}$, $\mathfrak{D}$, $\mathfrak{C}$, and $\mathfrak{E}$, are the *third-order aberration coefficients* or the *Seidel aberration coefficients* and refer to the aberrations associated with the image plane. Had we used the transformation given by Eq. (XII-59), translating the aberration functions back to the exit pupil plane, we would have obtained aberration coefficients associated with that plane.

The aberration functions, δx^* and δy^*, given in Eqs. (XII-60), describe the distribution of the rays on the image plane about the ideal image point. To a fixed object point, determined by fixed values of x and y, there will correspond a fixed ideal image point determined by Eqs. (XII-46). Each ray from the object point through a point on the entrance pupil plane, whose coordinates are x_s and y_s, should go through this same ideal image point. The aberration functions δx^* and δy^* give the distances by which they miss the mark.

The five aberration coefficients given in Eqs. (XII-61) contribute distinctive geometrical characteristics to the structure of the image whose appearances agree rather closely with what is seen in the laboratory. The next task is to find out what these structures look like.

With no loss of generality, we can restrict object points to the y axis on the image plane so that $x = 0$. We also write the coordinates of the ray on the entrance pupil plane in polar coordinates:

$$\begin{aligned} x_s &= (2u_s)^{1/2} \cos\theta, \\ y_s &= (2u_s)^{1/2} \sin\theta. \end{aligned}$$

Then $u = y^2/2$, $\sigma = y(2u_s)^{1/2} \sin\theta$, and Eqs. (XII-60) become

$$\begin{aligned}\delta y^* &= 2^{1/2}\,\mathfrak{B}u_s^{3/2}\sin\theta - \mathfrak{F}yu_s(2\sin^2\theta + 1)\\ &\quad + 2^{1/2}(\tfrac{1}{2}\mathfrak{D} + \mathfrak{C})\,y^2u_s^{1/2}\sin\theta - \tfrac{1}{2}\mathfrak{E}y^3, \qquad \text{(XII-62)}\\ \delta x^* &= 2^{1/2}\,\mathfrak{B}u_s^{3/2}\cos\theta - 2\mathfrak{F}yu_s\sin\theta\cos\theta + 2(\tfrac{1}{2}\mathfrak{D})\,y^2u_s^{1/2}\cos\theta.\end{aligned}$$

We will be concerned with rays through concentric circles on the entrance pupil whose centers all lie on the axial point which identifies the principal ray. The Gaussian image of these circles will also be circles on the exit pupil. We will assume that the aberrations associated with the pupil planes are negligible compared with the magnitude of the circles we are considering. Of course the magnification between entrance and exit pupil planes must be taken into consideration.

We begin with $\mathfrak{B}$, the coefficient associated with *spherical aberration*, and set all other aberration coefficients equal to zero. Then Eqs. (XII-62) reduce to

$$\delta y^* = 2^{1/2}\mathfrak{B}u_s^{3/2}\sin\theta,$$
$$\delta x^* = 2^{1/2}\mathfrak{B}u_s^{3/2}\cos\theta,$$

a one-parameter family of concentric circles centered on the ideal image point. Each of these circles consists of all the rays from the object point that pass through a circle on the entrance pupil plane. The relationship between the radii of these circles is cubic. If r_i is the radius of the circle in the image plane and if $r_s = (2u_s)^{1/2}$ is the radius of the corresponding circle on the entrance pupil plane, then

$$r_i = \tfrac{1}{2}\mathfrak{B}r_s{}^3.$$

Figure XII-4 shows *lateral spherical aberration*, which has been described here, and *longitudinal spherical aberration*.

Next is $\mathfrak{F}$, the coefficient associated with *coma*:

$$\delta y^* = -\mathfrak{F}yu_s(2\sin^2\theta + 1), \qquad \delta x^* = -2\mathfrak{F}yu_s\sin\theta\cos\theta.$$

These expressions can be written as

$$\delta y^* = -2\mathfrak{F}yu_s + \mathfrak{F}yu_s\cos 2\theta, \qquad \delta x^* = -\mathfrak{F}yu_s\sin 2\theta. \qquad \text{(XII-63)}$$

Assume that the object point is fixed so that y is constant. For a fixed circle in the entrance pupil plane the figure is a circle of radius $\mathfrak{F}yu_s$ with its center at $-2\mathfrak{F}yu_s$. Note that a point making a single circuit of the circle on the entrance pupil goes twice around the counterpart of that

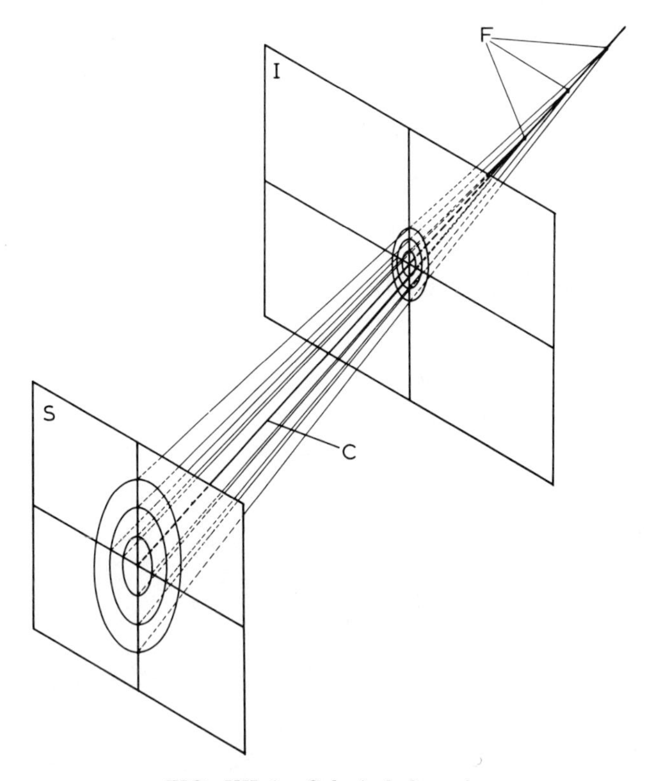

FIG. XII-4. Spherical aberration.

S is the exit pupil, I the Gaussian image plane. The ideal image point is at the intersection of C, the principal ray, and I. Rays through concentric circles on the exit pupil will pass through different points F on the principal ray. The spread of these points is described as *longitudinal spherical aberration.* The system of concentric circles on the image plane is described as *lateral spherical aberration.*

circle on the image plane. The relationship between the radius of the entrance pupil circle and the circle on the image plane is quadratic:

$$r_i = \tfrac{1}{2}\mathfrak{F}yr_s{}^2.$$

The envelope of this one-parameter family of circles is found in the following way. Eliminate θ between the equations of Eqs. (XII-63) and obtain

$$(\delta y^* + 2\mathfrak{F}yu_s)^2 + (\delta x^*)^2 = \mathfrak{F}^2y^2u_s{}^2.$$

Differentiating this with respect to the parameter u_s yields

$$2(\delta y^* + 2\mathfrak{F}yu_s) = \mathfrak{F}yu_s\,.$$

Between these two equations we eliminate u_s and obtain

$$(\delta x^*)^2 - (\delta y^*)^2/3 = 0,$$

the equation of a pair of straight lines

$$\delta y^* = \pm 3^{1/2} \delta x^*$$

intersecting at a 60° angle at the ideal image point. Coma is shown in Fig. XII-5.

The next term in Eqs. (XII-62) involves both $\mathfrak{D}$ and $\mathfrak{C}$. Setting all other aberration coefficients equal to zero yields

$$\delta y^* = (\tfrac{1}{2}\mathfrak{D} + \mathfrak{C})\, y^2 (2u_s)^{1/2} \sin\theta, \qquad \delta x^* = \tfrac{1}{2}\mathfrak{D} y^2 (2u_s)^{1/2} \cos\theta,$$

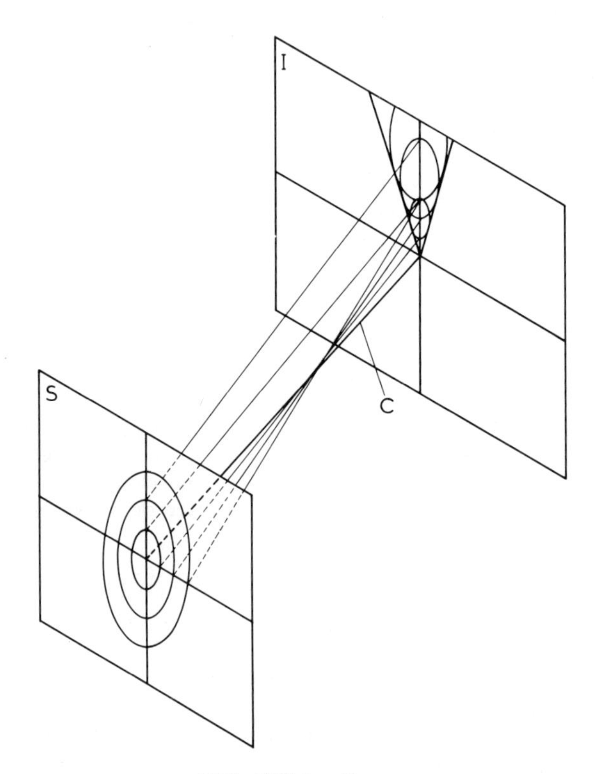

FIG. XII-5. Coma.

S is the exit pupil, I the Gaussian image plane. The ideal image point is at the intersection of C, the principal ray, and I. Rays through concentric circles on the exit pupil will pass through circles on the exit pupil centered on the intersection of I and the meridian plane. The envelope of these circles is a pair of straight lines intersecting at a 60° angle at the ideal image point. A point making a single circuit of a circle on the exit pupil goes twice around its counterpart on the ideal image plane.

the equation for an ellipse. Thus, the rays from the fixed object point, passing through a circle on the pupil plane, must also pass through this ellipse on the image plane. The relationship between the radius of the circle on the entrance pupil plane and the semiaxes of the ellipse is linear.

Let the circle on the entrance pupil plane be fixed and let $\theta = \pm\pi/2$, thus determining two meridian rays on opposite sides of the circle and lying, of course, on the plane determined by the δy^* axis and the principal ray. These two rays pass on opposite sides of the ellipse on an axis at a distance proportional to $\mathfrak{D} + 2\mathfrak{C}$. These two rays are symmetrical with respect to the principal ray and will intersect it at the same point. It turns out that all meridian rays will pass through this point, which is called the *meridian focus* or the *tangential focus*.

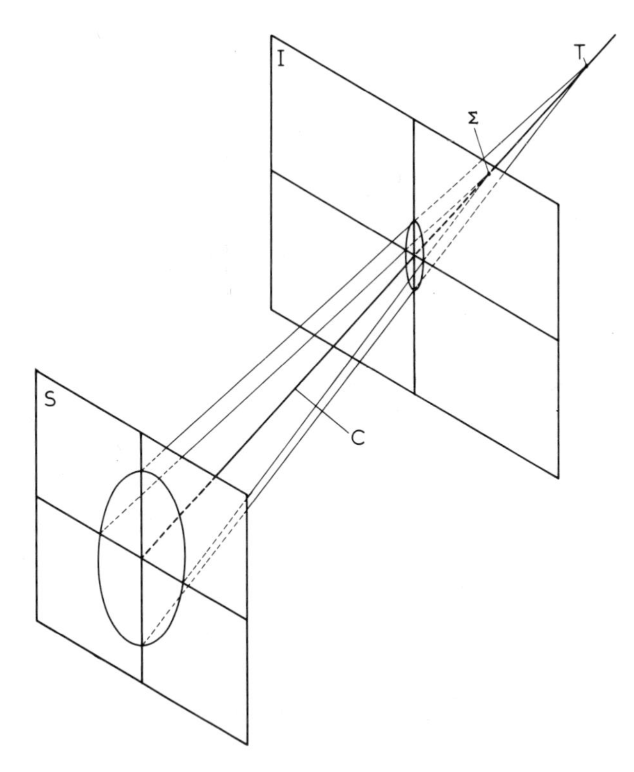

FIG. XII-6. Astigmatism.

S is the exit pupil, *I* the Gaussian image plane. The ideal image point is at the intersection of *C*, the principal ray, and *I*. Rays through a circle on the exit pupil will pass through an ellipse on the image plane, then converge to two foci on *C*, the principal ray. Meridian rays pass through the *tangential focus*, *T*. The *sagittal focus* is the point at which a fan of sagittal rays converges.

A second symmetric pair of rays is determined by $\theta = 0$ and $\theta = \pi$. These skew rays, called *sagittal rays*, also pass through opposite sides of both the circle on the exit pupil and the ellipse on the image plane, where their distance from the ideal image point along an axis of the ellipse will be proportional to $\mathfrak{D}$. These rays will intersect at a second point on the principal ray, in general distinct from the first. Any fan of sagittal rays from the fixed object point will converge to this, the *sagittal focus*. See Fig. XII-6.

These, the tangential focus and the sagittal focus, are the two *astigmatic foci*. Note that when $\mathfrak{C}$ is zero these two foci coincide and there

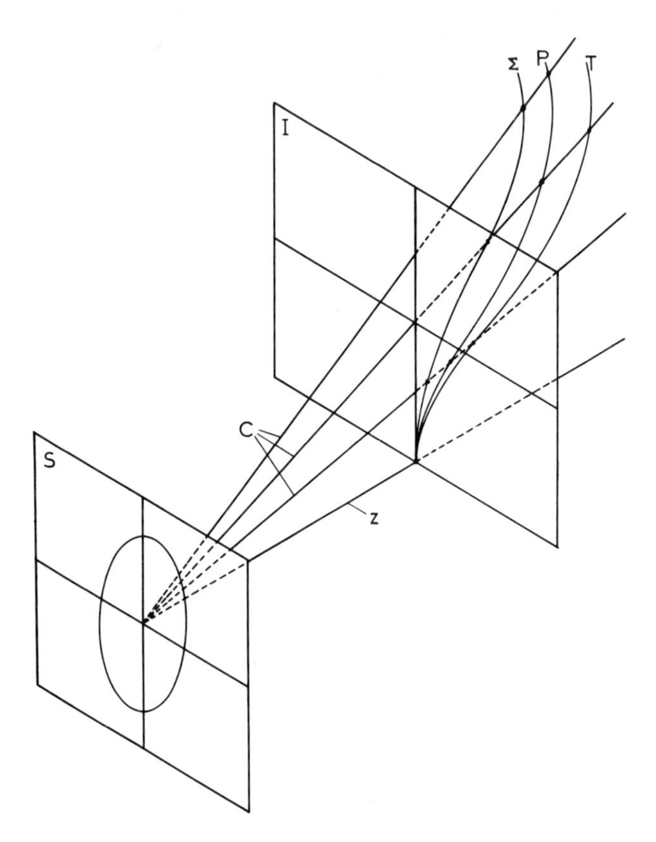

FIG. XII-7. Field curvature.

Principal rays C from different object points all pass through the axial point of S, the exit pupil. The rays from each object point intersect the principal ray at the sagittal and meridian foci. As the object point moves over the object plane, these foci generate the *sagittal surface* and the *tangential surface* T. The *Petzval surface* P is a compromise. Its curvature is the *field curvature.*

is no astigmatism. We can then regard $\mathfrak{C}$ as the aberration coefficient associated with astigmatism.

On the other hand, the distances from the center of the exit pupil to the sagittal and tangential foci are proportional, respectively, to $\mathfrak{D}$ and $\mathfrak{D} + 2\mathfrak{C}$. Therefore, as the object point is allowed to vary on the object plane, the principal ray will no longer be fixed and the sagittal and tangential foci will generate surfaces whose radii of curvature are, respectively, $\mathfrak{D}$ and $\mathfrak{D} + 2\mathfrak{C}$. These two quantities are referred to as the *sagittal curvature* and the *tangential curvature* while their arithmetic mean is referred to as simply the *field curvature*. Figure XII-7 illustrates field curvature, the fourth Seidel aberration.

The final Seidel aberration is *distortion* associated with the coefficient $\mathfrak{E}$:

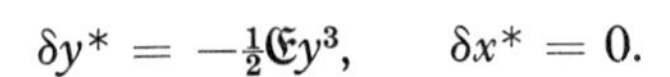

$$\delta y^* = -\tfrac{1}{2}\mathfrak{E}y^3, \qquad \delta x^* = 0.$$

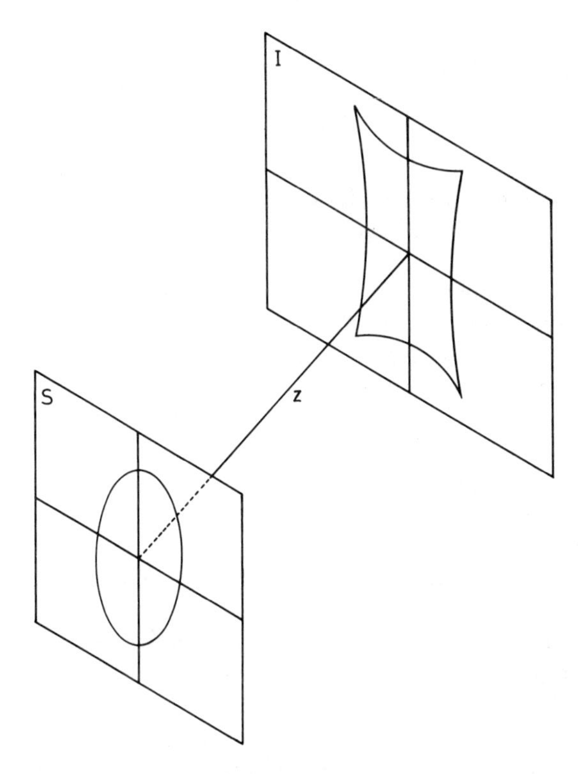

FIG. XII-8. Distortion.

Principal rays from different objects points pass through the axial point of S, the exit pupil. Since *distortion* can be interpreted as a magnification error, the image of a square will look like either a pincushion (as illustrated) or a barrel.

This, like field curvature, is a field aberration depending on object point position alone. Thus a radial line through the center of the object plane will be imaged as a radial line through the center of the image plane. However, it will be stretched or shrunk depending on the sign and magnitude of the distortion aberration coefficient $\mathfrak{E}$ and in proportion to the cube of the height y of the object point.

Another way to look at it is through Eqs. (XII-46) and (XII-47):

$$y^* = y^{\#} + \delta y^* = my - \tfrac{1}{2}\mathfrak{E}y^3 = y(m - \tfrac{1}{2}\mathfrak{E}y^2).$$

Here distortion appears as an error in magnification proportional to the square of the object height. Distortion is shown in Fig. XII-8.

So ends our rather brief discussion of the classical third-order aberrations. It has been short and qualitative rather than quantitative, and a number of important points have been omitted. For full details on these, many fine references are available; among these are Buchdahl (1968, 1970), Born and Wolf (1970, Chapter 5), and Brouwer (1964).

HERZBERGER'S DIAPOINT THEORY

The theory of aberrations we have been looking at is concerned with measuring the departure of an orthotomic system of rays from its ideal image point. The measuring is done on some plane perpendicular to the axis of symmetry of the optical system. An aggregate of *real* rays (as opposed to *paraxial* rays) traced through a lens will intercept this plane to form an array of points in the general region of the ideal image point. Such an array is called a spot diagram. We have been describing a method of estimating the distribution of the points in a spot diagram.

The principal advantage of an aberration theory that is an analog to the spot diagram is that it can be compared with the simplest method of evaluating the overall quality of a lens. The spot diagram is after all the geometrical analog of the star image test in which the image of a distant point object is viewed through the tested lens. The image of the point can then be recorded on photographic film or viewed directly through a microscope. There one sees the point distributions that we have already identified as the Seidel aberrations. Thus, this aberration theory is true to life. Its results correspond faithfully to what one actually sees, excluding diffraction effects, on the image plane of a lens.

Herzberger's diapoints (Herzberger, 1948) are similar to spot diagrams in that they are formed by the interaction of the rays of an orthotomic system on a plane. But there the similarity ends. Rather than forming a distribution of points on an image plane, like a spot diagram, diapoints lie on the meridian plane. Strictly speaking, diapoints are defined only for skew rays. Diapoints for meridian rays must be described by a limiting process. An image error theory and a third-order aberration theory are available. These we will look at next (Herzberger, 1944).

Assume a rotationally symmetric optical system. This time we will consider a fixed object point at infinity, determined by the three constant reduced direction cosines, ξ_0, η_0, ζ_0, connected by the relationship $\xi_0^2 + \eta_0^2 + \zeta_0^2 = n^2$, where n is the refractive index of object space.

The meridian plane is defined as the plane containing both the object point and the axis of symmetry of the optical system. (We have tacitly assumed that this axis coincides with the z coordinate axis.) The equation of this plane must then be

$$x\eta_0 - y\xi_0 = 0. \tag{XII-64}$$

Each ray from the object point will intersect the Gaussian focal plane in image space at a point whose coordinates we will call x^* and y^*. The reduced direction cosines satisfy $\xi^{*2} + \eta^{*2} + \zeta^{*2} = n^{*2}$, where n^* is the refractive index of image space.

The coordinates $\bar{x}^*$, $\bar{y}^*$, and $\bar{z}^*$ of any point on a ray in image space are given by

$$\bar{x}^* = x^* + \lambda\xi^*, \qquad \bar{y}^* = y^* + \lambda\eta^*, \qquad \bar{z}^* = z^* + \lambda\zeta^*,$$

where the parameter λ represents the distance along the ray between the Gaussian focal plane and the point in question. Eliminating λ from these equations results in

$$\bar{x}^* = x^* + \xi^*\bar{z}^*/\zeta^*, \qquad \bar{y}^* = y^* + \eta^*\bar{z}^*/\zeta^*. \tag{XII-65}$$

In spite of the obvious resemblance to the transfer equations, bear in mind that $\bar{z}^*$ represents a coordinate of a point on a ray and not a fixed thickness or separation.

We need to find the diapoint associated with each skew ray. If the point $(\bar{x}^*, \bar{y}^*, \bar{z}^*)$ on the ray is indeed a diapoint, then it lies on the meridian plane and its coordinates must satisfy its equation (XII-64). The substitution of Eqs. (XII-65) into Eqs. (XII-64) results in the equation for the diapoint

$$(\eta_0 x^* - \xi_0 y^*) + (\eta_0\xi^* - \xi_0\eta^*)\, z^*/\zeta^* = 0.$$

Recall Eqs. (XII-17), obtained by differentiating the angle eikonal. We are particularly concerned with the second pair of these,

$$x^* = -(W_s\xi_0 + W_{w^*}\xi^*), \qquad y^* = -(W_s\eta_0 + W_{w^*}\eta^*), \quad \text{(XII-66)}$$

which when substituted into the above give us

$$(-W_{w^*} + \bar{z}^*/\zeta^*)(\eta_0\xi^* - \xi_0\eta^*) = 0,$$

from which we obtain

$$\bar{z}_d^* = W_{w^*}\zeta^*. \quad \text{(XII-67a)}$$

Combining this with Eqs. (XII-65) and (XII-66), we obtain

$$\bar{x}_d^* = -W_s\xi, \qquad \bar{y}_d^* = -W_s\eta, \quad \text{(XII-67b, c)}$$

the coordinates of the diapoint associated with this skew ray.

The height of this diapoint, the perpendicular distance between it and the optical axis, is clearly

$$h_d^* = -2W_s w_0 \,, \quad \text{(XII-67d)}$$

while Eq. (XII-67a) gives its distance to the Gaussian image plane.

This describes the general situation in which the diapoints cover a region on the meridian plane. Suppose, however, that all the diapoints lie on a curve in the meridian plane. Then a functional relationship between h_d^* and z_d^* must exist. It follows that the Jacobian must vanish,

$$\begin{vmatrix} \partial h_d^*/\partial w^* & \partial h_d^*/\partial s \\ \partial z_d^*/\partial w^* & \partial z_d^*/\partial s \end{vmatrix} = 0.$$

Recalling that $\zeta^{*2} = n^{*2} - 2w^*$, we calculate this and obtain the partial differential equation

$$(W_{sw^*}^2 - W_{ss}W_{w^*w^*})\,\zeta^* + W_{w^*}W_{ss}/\zeta^* = 0.$$

The trick we use to solve this equation in any just universe would probably be regarded as unfair. Write this equation in the form

$$\zeta^* W_{sw^*}(\partial W_s/\partial w^*) - (W_{w^*w^*}\zeta^* - W_{w^*}/\zeta^*)(\partial W_s/\partial s) = 0.$$

It is clearly a first-order homogeneous partial differential equation where the unknown is W_s. Using the method of characteristics (see Chapter VII), we obtain the characteristic equation

$$dW^*/\zeta^* W_{sw^*} = ds/(W_{w^*w^*}\zeta^* - W_{w^*}/\zeta^*),$$

which boils down to $d(\zeta^* W_{w^*}) = 0$. It follows that the characteristic is $\zeta^* W_{w^*}$ and that the solution is in the form

$$W_s = F(\zeta^* W_{w^*}, w_0). \tag{XII-68}$$

Here we throw in the fixed quantity w_0 to show that there is indeed a dependence on the object point. Here F is an arbitrary function.

If the distribution of diapoints does indeed degenerate into a curve on the meridian plane, then the image is said to be, after Herzberger, free from *deformation errors*.

It may happen that the curve is actually a straight line. Then Eq. (XII-68) becomes

$$W_s = A(w)\, W_{w^*}\zeta^* + B(w), \tag{XII-69}$$

where A and B depend on the object point alone. Incidentally, comparing Eqs. (XII-68) and (XII-69) with Eq. (XII-67d) makes it clear that these equations represent, respectively, a curve and a straight line on the meridian plane.

If the diapoints all lie on a straight line, then the image is said to be free of *asymmetry errors*. This term is also due to Herzberger.

Finally, the image is said to be *sharp* if the line degenerates to a point. This occurs if

$$h_d^* = (2u_0)^{1/2}\, W_s = \alpha(u_0), \qquad z_d^* = \zeta^* W_{w^*} = \beta(u_0). \tag{XII-70}$$

We have described a hierarchy of aberrations. Beginning with the most general situation and proceeding to the perfect image, we have

deformation error,
asymmetry error (or half-symmetric error),
symmetric error,
sharp image.

This list includes only defects in the sharpness of the image of a point. The field errors, distortion and field curvature, are contained in the functions $\alpha(u_0)$ and $\beta(u_0)$ in Eqs. (XII-70).

To see what these aberrations look like we need to prove a short lemma. Consider an orthotomic system of rays in a homogeneous medium that has the property that each ray of the system passes through a fixed space curve. What we will prove is that all rays passing through a point on the curve will make the same angle with the tangent to the curve at that point. The proof, which follows, is a delightful thing that illustrates perfectly the advantages of the notation and the point of view that we have been using.

Let $\mathbf{W}(u, v)$, a vector function of the parameters u and v, represent a wavefront associated with the given orthotomic system of rays. Since the medium is homogeneous, all the rays are straight lines perpendicular to the surface $\mathbf{W}$. The ray directions therefore are given by $\mathbf{N}(u, v)$, the unit normal vector to $\mathbf{W}$. Suppose that each ray of this system passes through a space curve $\mathbf{P}(s)$, a vector function of the arc length parameters.

Each pair of values for the parameters u and v determines a point on $\mathbf{W}(u, v)$ and a ray direction $\mathbf{N}(u, v)$ through that point. This ray then intersects the curve $\mathbf{P}(s)$. Let $\lambda(u, v)$ be the distance along a ray between the wavefront $\mathbf{W}(u, v)$ and the curve $\mathbf{P}(s)$. The value of s must also be determined by u and v. We express this in the following equations:

$$\mathbf{P}(s) = \mathbf{W}(u, v) + \lambda(u, v)\,\mathbf{N}(u, v), \qquad s = f(u, v).$$

However, the choice of the parameters u and v is arbitrary. Let us choose two new parameters s and t to replace u and v, where s is the arc length parameter for the curve $\mathbf{P}(s)$. Making this transformation of the parameters, we obtain

$$\mathbf{P}(s) = \mathbf{W}(s, t) + \lambda(s, t)\,\mathbf{N}(s, t).$$

Taking the derivatives of this expression with respect to s and t gives us

$$\mathbf{P}' = \mathbf{W}_s + \lambda_s \mathbf{N} + \lambda \mathbf{N}_s, \qquad \mathbf{0} = \mathbf{W}_t + \lambda_t \mathbf{N} + \lambda \mathbf{N}_t.$$

The scalar product of each of these expressions with $\mathbf{N}$ results in

$$\mathbf{P}' \cdot \mathbf{N} = \lambda_s \qquad \text{and} \qquad 0 = \lambda_t.$$

It follows that λ, and therefore λ_s, is independent of t and depends on s alone. The same is true of $\mathbf{P}(s)$ and $\mathbf{P}'(s)$. But $\mathbf{N}(s, t)$ depends on both s and t. The equation

$$\mathbf{P}'(s) \cdot \mathbf{N}(s, t) = \lambda_s(s)$$

must be satisfied for all s and t. This proves the lemma. For s fixed, all rays through the point $\mathbf{P}(s)$ make a constant angle with the tangent vector $\mathbf{P}'(s)$ to the curve at that point. Thus, all the rays through a point on the curve $\mathbf{P}(s)$ form a hollow circular cone with its vertex at that point and with its axis tangent to the curve at that point.

The point of all this is to show what asymmetry error and symmetry error look like. Figure XII-9 is an attempt to do this. Clearly, asymmetry

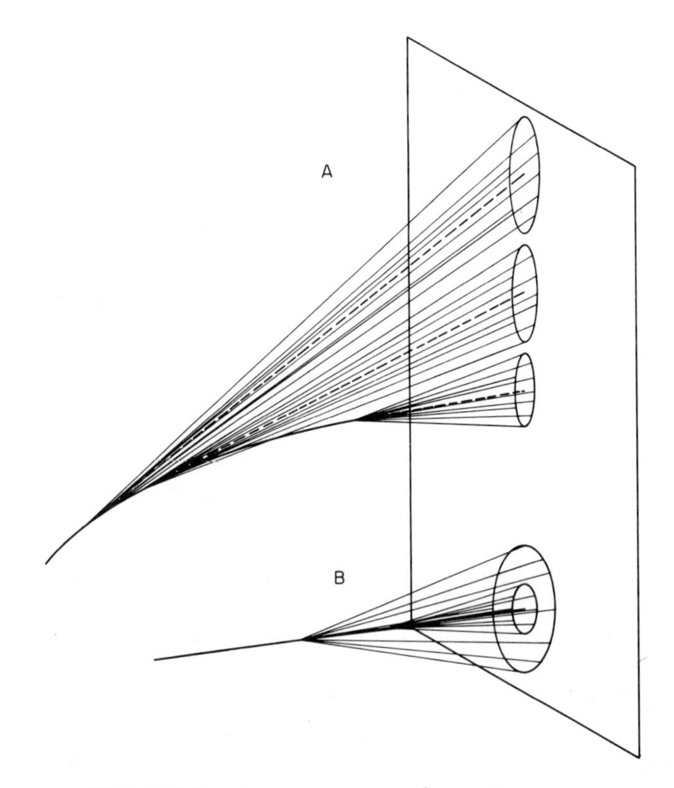

FIG. XII-9. Asymmetry error and symmetry error.

Part (A) shows asymmetry error in which all rays from an object point pass through a plane curve. All rays through a single point on this curve make a constant angle with the curve and thus generate the surface of a cone. These rays intersect an image plane as a conic section. All of these form a coma-like aberration figure. (Compare with Fig. XII-5.) Part (B) shows symmetry error. The above curve is now a straight line. Rays intersect an image plane perpendicular to the straight line in a system of concentric circles. Symmetry error is then a form of spherical aberration. (Compare with Fig. XII-4.)

error is a coma-like aberration and symmetry error is a spherical-aberration-like image error.

Approximations to these aberrations are easy to come by. Use Eqs. (XII-20) and (XII-22) in Eqs. (XII-70) to get

$$\begin{aligned} h_d{}^* &= (2w_0)^{1/2}[J_s + J_{ws}w + J_{ss}s + J_{sw^*}w^*], \\ z_d{}^* &= \zeta^*[J_{w^*} + J_{ww^*}w + J_{sw^*}s + J_{w^*w^*}w^*]. \end{aligned} \qquad \text{(XII-71)}$$

Dropping all terms higher than the first, we have the first-order approximation

$$h_d^{\#} = (2w_0)^{1/2} J_s, \qquad z_d^{\#} = n^* J_{w^*}.$$

Now the first-order terms represent the ideal image position, and therefore $z_d^{\#}$, since it is measured from the Gaussian image plane, must be zero. It follows that $J_{w^*} = 0$. The other quantity, h_d^*, represents the ideal image height, which is related to the focal length of the lens and the angle the infinite object point makes with the optical axis. Thus

$$h_d^{\#} = f \tan \beta = f \cdot 2u_0/n$$

to the first order. It follows that $J_s = f/n$. The equations for the diapoint, Eqs. (XII-71), now become

$$\begin{aligned} h_d^* &= (2w_0)^{1/2}[f/n + J_{ws}w_0 + J_{ss}s + J_{sw^*}w^*], \\ z_d^* &= \zeta^*[J_{ww^*}w_0 + J_{sw^*}s + J_{w^*w^*}w^*]. \end{aligned} \tag{XII-72}$$

Now let us assume that the image is free from deformation errors. Then Eq. (XII-69) must be satisfied, and as a consequence

$$J_{sw^*}^2 - J_{ss}J_{w^*w^*} = 0. \tag{XII-73}$$

Using this to eliminate s and w^* in Eqs. (XII-72), we get

$$[h_d^* - (2w_0)^{1/2}f/n]\, J_{sw^*} = (2w_0)^{1/2}[(J_{ws}J_{sw^*} - J_{ww^*}J_{ss})\, W_0 + J_{ss}z_d^*/\zeta^*].$$

This curve is rather difficult to visualize. However, for rays so close to the axis that $\zeta^* = n^*$, this equation become the equation of a straight line.

There are several final points to be made. The diapoint theory does not refer to any particular reference plane. Therefore, the terms representing the aberrations do not contain expressions to account for a shift in the image plane. These aberrations are also less dependent on the stop position. Rather than looking for symmetry around a principal ray that depends on the position of the stop, one tries to find a curve that, in some sense, is a best fit to the diapoints. A stop shift in a certain sense changes this curve only in that it changes the distribution of the diapoints to which the curve is fitted. Finally, note that these aberration functions depend more directly on the coefficients of the power series expansion in Eq. (XII-20).

We have examined two approaches to aberration theory: the classical interpretation of the Seidel aberrations and Herzberger's diapoint theory. There is yet a third approach. This is a theory based on the measures of the errors in the wavefronts emerging from a lens. This is the approach that is generally used today, and it is covered well and in great detail by some excellent expositions such as Buchdahl (1968, 1970). There is therefore no need for us to cover it here.

THE CONVERGENCE PROBLEM

The work of this chapter has been based almost entirely on the power series expansion, Eq. (XII-20). We have differentiated it, rearranged its terms, truncated it, and otherwise abused it as though it were everywhere absolutely and uniformly convergent. The fact of the matter is that the convergence properties of this series are not at all understood and many of the steps that we have performed are questionable.

That the series is not everywhere convergent is illustrated by this short example, which I owe to Donald P. Feder of the Eastman Kodak Laboratories.

Consider the optical system consisting only of a plane interface between air ($n = 1$) and a glass ($n^* > 1$). An object point O is placed on the air side a distance d from the surface. Regard the perpendicular dropped from the object point to the plane refracting surface as the axis of rotational symmetry. The object point then becomes an axial object point (see Fig. XII-10). Its Gaussian image must also lie on this axis.

Take as an image space reference plane a plane perpendicular to the axis a distance d^* from the refracting surface. In Fig. XII-10 this is shown as being to the left of the refracting surface. Choose a point $\bar{P}$ on the refracting surface. Then a ray from O to P will be refracted at P and will intersect the image reference plane at O^*. Let $\bar{y}$ be the height of the point P on the refracting surface and let y^* be the height of the point P on the refracting surface and let y^* be the height of the intersection of the ray with the O^* image reference plane. From Fig. XII-10 we see that

$$\sin i = \bar{y}/(d^2 + \bar{y}^2)^{1/2}$$

while

$$\tan i^* = (\bar{y} - y^*)/d^*.$$

Here i and i^* represent the angle of incidence and the angle of refraction,

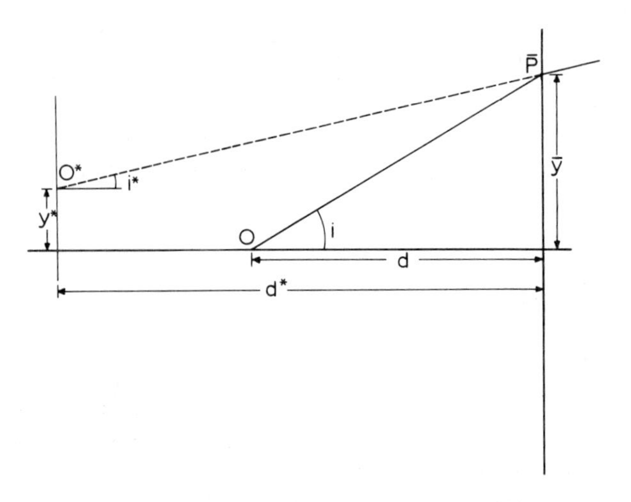

FIG. XII-10. Illustrating the convergence problem.
The refracting surface is the vertical line on the right side of the illustation. A ray from object point O intercepts the refracting surface at $\bar{P}$ and is refracted. The image of O at O^* is virtual and is obtained by extending the refracted ray backward (dotted line) to the image plane. Segments d and d^* are the distances of the object and image points, respectively, from the refracting surface. Lines $\bar{y}$ and y^* are the heights of the ray on the refracting surface and the image planes, respectively.

respectively. To these two expressions is applied the scalar version of Snell's law, Eq. (V-20),

$$n \sin i = n^* \sin i^*,$$

which leads to

$$\begin{aligned}(\bar{y} - y^*)/d^* &= \frac{(n/n^*)\bar{y}}{(d^2 + \bar{y}^2)^{1/2}} \left\{1 - \frac{(n/n^*)^2\bar{y}^2}{(d^2 + \bar{y}^2)}\right\}^{-1/2} \\ &= n\bar{y}/[n^{*2}(d^2 + \bar{y}^2) - n^2\bar{y}^2]^{1/2}.\end{aligned}$$

This leads at once to

$$y^* = \bar{y}\{1 - nd^*/[n^{*2}d^2 + \bar{y}^2(n^{*2} - n^2)]^{1/2}\}. \qquad \text{(XII-74)}$$

Now we need to find the Gaussian image plane. At this plane O^*, the image of the axial object point O, must also lie on the axis. We need first to calculate the limit

$$\lim_{\bar{y}\to 0}(y^*/\bar{y}) = 1 - nd^*/n^*d.$$

From this we see that the paraxial approximation is given by

$$y^* = \bar{y}(1 - nd^*/n^*d).$$

For y^* to vanish, the coefficient of $\bar{y}$ must also be zero. This happens when $d^* = n^*d/n$. This value of d^* must therefore correspond to the distance from the Gaussian image plane to the refracting surface.

Substituting this into Eq. (XII-74) gives

$$\delta y^* = \bar{y}\{1 - n^*d/[n^{*2}d^2 + \bar{y}^2(n^{*2} - n^2)]^{1/2}\}. \qquad \text{(XII-75)}$$

We use δy^* here because this is now an aberration function, representing the deviation of the ray's point of intersection from the ideal image point.

As long as $n^* > n$, δy^* is a real function defined for all values of $\bar{y}$. Its power series expansion begins with a term of degree *three*, exactly as advertised,

$$\delta y^* = \tfrac{1}{2}\bar{y}^3(n^{*2} - n^2)/n^{*2}d^2 - \cdots .$$

But what of the region of convergence? If we consider $\bar{y}$ as a variable on the complex plane, then its circle of convergence will extend out to the nearest complex singularity. This is located where the denominator in Eq. (XII-75) vanishes, at the point

$$\bar{y} = \pm i n^*d/(n^{*2} - n^2)^{1/2}.$$

The radius of the circle of convergence is therefore

$$r_c = n^*d/(n^{*2} - n^2)^{1/2}.$$

This is the point that needs to be made: For real values of $\bar{y}$ exceeding r_c the power series expansion for the aberration function δy^* will diverge even though the function is well defined for all real values of $\bar{y}$.

REFERENCES

Born, M., and Wolf, E. (1970). "Principles of Optics," 4th ed. Pergamon, Oxford.

Brouwer, W. (1964). "Matrix Methods in Optical Instrument Design." Benjamin, New York.

Bruns, H. (1895). Das Eikonal. *Leipzig, Abh. Sachs. Ges. Wiss., Math. Phys. Kl.* **21**, 325–436.

Buchdahl, H. A. (1968). "Optical Aberration Coefficients." Dover, New York.

Buchdahl, H. A. (1970). "An Introduction to Hamilton's Optics." Cambridge Univ. Press, London and New York.

Conway, A. W., and Synge, J. L. (1931). "The Mathematical Papers of Sir William Rowan Hamilton," Vol. I: Geometrical Optics. Cambridge Univ. Press, London and New York.

Conway, A. W., and Synge, J. L. (1940). "The Mathematical Papers of Sir William Rowan Hamilton," Vol. II: Dynamics. Cambridge Univ. Press, London and New York.

Herzberger, M. (1936). On the characteristic function of Hamilton, the eikonal of Bruns and their use in optics. *J. Opt. Soc. Amer.* **26**, 177.

Herzberger, M. (1937). Hamilton's characteristic function and Bruns' eikonal. *J. Opt. Soc. Amer.* **27**, 133.

Herzberger, M. (1944). Symmetry and Asymmetry in Optical Images. Supplementary Note No. IV in Luneburg (1966).

Herzberger, M. (1948). Image error theory for finite aperture and field. *J. Opt. Soc. Amer.* **38**, 736.

Luneburg, R. K. (1966). "Mathematical Theory of Optics." Univ. of California Press, Berkeley, California.

Steward, G. C. (1928). "The Symmetrical Optical System." Cambridge Univ. Press, London and New York.

Synge, J. L. (1937a). Hamilton's method in geometrical optics. *J. Opt. Soc. Amer.* **27**, 75.

Synge, J. L. (1937b). Hamilton's characteristic function and Bruns' eikonal. *J. Opt. Soc. Amer.* **27**, 138.

Synge, J. L. (1937c). "Geometrical Optics: An Introduction to Hamilton's Method." Cambridge Univ. Press, London and New York.

XIII

The Fundamental Optical Invariant

The usual route, the traditional route to formal techniques for analyzing optical systems, has been by way of Hamilton's method and the eikonal function. There is yet another approach, based on exactly the same premises but making use of an entirely different point of view. This approach makes use of the *fundamental optical invariant*, which appears to have been discovered by Herzberger (1931; 1958, Chapter 20), who had studied it intensively. As an approach to the problem of optical design it looks promising; as a practical tool it has never lived up to this promise. Nevertheless, as a point of view it is interesting and attractive. And in our quest for new techniques we cannot afford to overlook the possibility that something useful has been missed.

The usual derivation of the fundamental optical invariant (Herzberger, 1944) begins with the setting up of an optical system consisting of m refracting surfaces that separate $m + 1$ media of constant refractive index. Through this lens we trace a ray using forms of the equations of transfer and refraction that we have derived earlier. This ray has a *starting point* $\mathbf{R}_0$ in the first medium, which we will call *object space*, and a *terminal point* $\mathbf{R}_{m+1}$ in the last medium, called *image space*. Sometimes each of the regions between refracting surfaces within the lens is called an *in-between space*, a rather awkward expression in English but nice in German (*Zwischenraum*). We will represent the point of intersection of the ray with the ith surface by the vector $\mathbf{R}_i$. The unit normal vector to the ith surface at $\mathbf{R}_i$ will be denoted by $\bar{\mathbf{N}}_i$. The distance along a ray segment between $\mathbf{R}_i$ and $\mathbf{R}_{i+1}$ will be denoted by λ_i; the direction vector of this ray segment will be $\mathbf{N}_i$, and the refractive index of the medium in which it lies will be n_i. All this is shown in Fig. XIII-1.

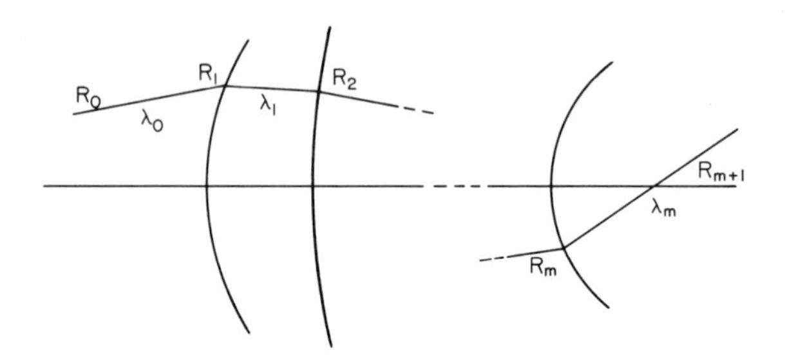

FIG. XIII-1. A ray path through a lens.

Points R_1, R_2,..., R_m represent points of intersection of the ray at each refracting surface. R_0 and R_{m+1} represent points in object space and image space, respectively. Lines λ_0, λ_1,..., λ_m are distances along the ray between each surface.

From Eq. (IV-9) the transfer equation from the ith to the $(i+1)$th surface is

$$\mathbf{R}_{i+1} = \mathbf{R}_i + t_i\mathbf{A}_i + \lambda_i\mathbf{N}_i ,$$

where $\mathbf{A}_i$ is the unit vector along a coordinate axis in the ith medium and t_i is the distance between the surfaces along $\mathbf{A}_i$. We also use the refraction equation from Eq. (IV-1),

$$\mathbf{N}_i = \mu_i\mathbf{N}_{i-1} + \gamma_i\bar{\mathbf{N}}_i ,$$

where $\mu_i = n_{i-1}/n_i$ and where γ_i is given by Eq. (IV-2). These equations lead to a sequence of formulas describing the path of a ray through the lens:

$$\begin{aligned}
\mathbf{R}_1 &= \mathbf{R}_0 + t_0\mathbf{A}_0 + \lambda_0\mathbf{N}_0 , & \mathbf{N}_1 &= \mu_1\mathbf{N}_0 + \gamma_1\bar{\mathbf{N}}_1 , \\
\mathbf{R}_2 &= \mathbf{R}_1 + t_1\mathbf{A}_1 + \lambda_1\mathbf{N}_1 , & \mathbf{N}_2 &= \mu_2\mathbf{N}_0 + \gamma_2\bar{\mathbf{N}}_2 , \\
&\vdots & &\vdots \\
\mathbf{R}_{m+1} &= \mathbf{R}_m + t_m\mathbf{A}_m + \lambda_m\mathbf{N}_m , & \mathbf{N}_m &= \mu_m\mathbf{N}_{m-1} + \gamma_m\bar{\mathbf{N}}_m .
\end{aligned} \tag{XIII-1}$$

Now let us assume that we are dealing with a two-parameter family of rays rather than a single ray. Suppose that the starting point $\mathbf{P}_0$ and the initial ray direction $\mathbf{N}_0$ depend on the two parameters u and v. Then, at each surface, $\mathbf{P}_i$ and $\mathbf{N}_i$ will also depend on these two parameters as well as on the coefficients λ_i and γ_i occurring in the ray tracing equations. Of course, t_i, $\mathbf{A}_i$, and μ_i are all constants. We assume that every quantity possesses continuous second-order derivatives with respect to u and v.

Differentiating each of the transfer equations in Eqs. (XIII-1) with respect to u, we get

$$\frac{\partial \mathbf{R}_1}{\partial u} = \frac{\partial \mathbf{R}_0}{\partial u} + \lambda_0 \frac{\partial \mathbf{N}_0}{\partial u} + \frac{\partial \lambda_0}{\partial u} \mathbf{N}_0 ,$$

$$\frac{\partial \mathbf{R}_2}{\partial u} = \frac{\partial \mathbf{R}_1}{\partial u} + \lambda_1 \frac{\partial \mathbf{N}_1}{\partial u} + \frac{\partial \lambda_1}{\partial u} \mathbf{N}_1 ,$$

$$\vdots$$

$$\frac{\partial \mathbf{R}_{m+1}}{\partial u} = \frac{\partial \mathbf{R}_m}{\partial u} + \lambda_m \frac{\partial \mathbf{N}_m}{\partial u} + \frac{\partial \lambda_m}{\partial u} \mathbf{N}_m .$$

Next we multiply the first of these by $n_0\mathbf{N}_0$, the second by $n_1\mathbf{N}_1$, etc. Recall that $\mathbf{N}_i^2 = 1$ so that $\mathbf{N}_i \cdot \partial \mathbf{N}_i/\partial u = 0$:

$$n_0\mathbf{N}_0 \cdot \partial \mathbf{R}_1/\partial u = n_0\mathbf{N}_0 \cdot \partial \mathbf{R}_0/\partial u + n_0(\partial \lambda_0/\partial u) ,$$

$$n_1\mathbf{N}_1 \cdot \partial \mathbf{R}_2/\partial u = n_1\mathbf{N}_1 \cdot \partial \mathbf{R}_1/\partial u + n_1(\partial \lambda_1/\partial u) ,$$

$$\vdots$$

$$n_m\mathbf{N}_m \cdot \partial \mathbf{R}_{m+1}/\partial u = n_m\mathbf{N}_m \cdot \partial \mathbf{R}_m/\partial u + n_m(\partial \lambda_m/\partial u).$$

Adding all of these gives us

$$n_m\mathbf{N}_m \cdot \frac{\partial \mathbf{R}_{m+1}}{\partial u} - (n_m\mathbf{N}_m - n_{m-1}\mathbf{N}_{m-1}) \cdot \frac{\partial \mathbf{R}_m}{\partial u} - \cdots - (n_1\mathbf{N}_1 - n_0\mathbf{N}_0) \cdot \frac{\partial \mathbf{R}_1}{\partial u}$$

$$= n_0\mathbf{N}_0 \cdot \frac{\partial \mathbf{R}_0}{\partial u} + \sum_{i=1}^{m} n_i \frac{\partial \lambda_i}{\partial u}. \tag{XIII-2}$$

The expressions in parentheses, by the refraction equations in Eqs. (XIII-1), are all proportional to the surface normals at the point of incidence:

$$n_i\mathbf{N}_i - n_{i-1}\mathbf{N}_{i-1} = \gamma_i \bar{\mathbf{N}}_i .$$

Moreover, $\partial \mathbf{R}_i/\partial u$, being the derivative of a vector representing a point on the ith surface, must be a vector tangent to that surface. Therefore it and the surface normal are perpendicular and their scalar product must vanish. Therefore, at each surface,

$$(n_i\mathbf{N}_i - n_{i-1}\mathbf{N}_{i-1}) \cdot \partial\mathbf{R}_i/\partial u = 0$$

and Eq. (XIII-2) reduces to

$$n_m\mathbf{N}_m \cdot \mathbf{R}_{m+1}/\partial u - n_0\mathbf{N}_0 \cdot \partial\mathbf{R}_0/\partial u = \partial L/\partial u,$$

where $L = \sum_{i=1}^{m} n_i\lambda_i$ and can be recognized as the optical path length along the ray from its starting point to its end point. In fact, L is nothing more than the point eikonal defined in Eq. (XII-1).

We can show in exactly the same way that an identical expression holds involving the second parameter v,

$$n_m\mathbf{N}_m \cdot (\partial\mathbf{R}_{m+1}/\partial v) - n_0\mathbf{N}_0 \cdot (\partial\mathbf{R}_0/\partial v) = \partial L/\partial v.$$

Since we have assumed that everything has continuous second derivatives, $\partial^2 L/\partial u\,\partial v = \partial^2 L/\partial v\,\partial u$ and therefore

$$n_m\frac{\partial\mathbf{N}_m}{\partial u}\cdot\frac{\partial\mathbf{R}_{m+1}}{\partial v} + n_m\mathbf{N}_m\cdot\frac{\partial^2\mathbf{R}_{m+1}}{\partial u\partial v} - n_0\frac{\partial\mathbf{N}_0}{\partial u}\cdot\frac{\partial\mathbf{R}_0}{\partial v} - n_0\mathbf{N}_0\cdot\frac{\partial^2\mathbf{R}_0}{\partial u\partial v}$$

$$= n_m\frac{\partial\mathbf{N}_m}{\partial v}\cdot\frac{\partial\mathbf{R}_{m+1}}{\partial u} + n_m\mathbf{N}_m\cdot\frac{\partial^2\mathbf{R}_{m+1}}{\partial v\partial u} - n_0\frac{\partial\mathbf{N}_0}{\partial v}\cdot\frac{\partial\mathbf{R}_0}{\partial u} - n_0\mathbf{N}_0\cdot\frac{\partial^2\mathbf{R}_0}{\partial v\partial u}.$$

This in turn reduces to

$$n_m\left(\frac{\partial\mathbf{R}_{m+1}}{\partial u}\cdot\frac{\partial\mathbf{N}_m}{\partial v} - \frac{\partial\mathbf{R}_{m+1}}{\partial v}\cdot\frac{\partial\mathbf{N}_m}{\partial u}\right) = n_0\left(\frac{\partial\mathbf{R}_0}{\partial u}\cdot\frac{\partial\mathbf{N}_0}{\partial v} - \frac{\partial\mathbf{R}_0}{\partial v}\cdot\frac{\partial\mathbf{N}_0}{\partial u}\right). \tag{XIII-3}$$

Of course in this last step use is made of the fact that $\mathbf{R}_{m+1}$ and $\mathbf{R}_0$ have continuous second derivatives.

The expressions in each member of the above equation are identical except for the subscripts. Therefore, the quantity

$$n\left[\frac{\partial\mathbf{R}}{\partial u}\cdot\frac{\partial\mathbf{N}}{\partial v} - \frac{\partial\mathbf{R}}{\partial v}\cdot\frac{\partial\mathbf{N}}{\partial u}\right] \tag{XIII-4}$$

has the same value whether it be calculated in object space, image space or, for that matter, any of the "in-between" spaces. This quantity is called the *fundamental optical invariant*.

As it stands, the fundamental optical invariant refers to the points of intersection $\mathbf{R}_i$ of the ray with a refracting surface. It turns out that the invariant is valid for any point on the ray. In particular, it is convenient to refer to the intersection of the ray and an arbitrary plane in object space, in image space, and in each medium of the lens. If we call such a point $\tilde{\mathbf{R}}$, then, by the transfer equations $\tilde{\mathbf{R}} = \mathbf{R} + \lambda\mathbf{N}$, where $\tilde{\mathbf{R}}$ and λ, as well as $\mathbf{R}$ and $\mathbf{N}$, are functions of u and v. Then differentiating with respect to u and v and substituting these into the fundamental optical invariant, Eq. (XIII-4), we obtain

$$\frac{\partial\mathbf{N}}{\partial u}\cdot\frac{\partial\mathbf{R}}{\partial v} - \frac{\partial\mathbf{N}}{\partial v}\cdot\frac{\partial\mathbf{R}}{\partial u} = \frac{\partial\mathbf{N}}{\partial u}\cdot\frac{\partial\mathbf{R}}{\partial v} - \frac{\partial\mathbf{N}}{\partial v}\cdot\frac{\partial\mathbf{R}}{\partial u}.$$

Once again we use the fact that $\mathbf{N}\cdot(\partial\mathbf{N}/\partial u) = \mathbf{N}\cdot(\partial\mathbf{N}/\partial v) = 0$. All this tells us is that the transfer operation satisfies the invariant.

Thus, if a ray is specified by its point of intersection with an arbitrary plane, the fundamental optical invariant still holds. From this point on, we will assume that all rays are determined relative to some such reference plane. We will choose that plane as the y, z plane of some coordinate system in the medium so that the point $\mathbf{R}$ will have the form $\mathbf{R} = (x, y, 0)$. In addition, rather than specifying a ray direction by the unit vector $\mathbf{N}$, we will use the reduced direction vector Ξ defined by

$$\Xi = (\xi, \eta, \zeta) = n\mathbf{N}$$

so that

$$\xi^2 + \eta^2 + \zeta^2 = n^2.$$

Finally, we drop the subscripts and denote the vectors identifying a ray in object space by $\mathbf{R}$ and Ξ. Those denoting a ray in image space will be $\mathbf{R}^*$ and Ξ^*. The fundamental optical invariant then becomes

$$\frac{\partial\Xi^*}{\partial u}\cdot\frac{\partial\mathbf{R}^*}{\partial v} - \frac{\partial\Xi^*}{\partial v}\cdot\frac{\partial\mathbf{R}^*}{\partial u} = \frac{\partial\Xi}{\partial u}\cdot\frac{\partial\mathbf{R}}{\partial v} - \frac{\partial\Xi}{\partial v}\cdot\frac{\partial\mathbf{R}}{\partial u}. \tag{XIII-5}$$

Now suppose that the two-parameter family of rays emanates from a single fixed object point. Then $\partial\mathbf{R}/\partial u = \partial\mathbf{R}/\partial v = 0$ and the invariant vanishes in object space. It therefore must also vanish in image space and in every medium in the interior of the lens. It follows that, if a two-parameter family of rays is an orthotomic system, then the fundamental optical invariant vanishes. The converse is also true. If the fundamental optical invariant vanishes, then the system of rays associated with it is an orthotomic system.

THE INHOMOGENEOUS MEDIUM

Compared with what we have just gone through, the derivation of the fundamental optical invariant for the inhomogeneous medium is absurdly simple. As in the case already studied, the key is the fact that the optical path length λ has continuous partial derivatives so that $\partial^2\lambda/\partial u\,\partial v = \partial^2\lambda/\partial v\,\partial u$.

For an inhomogeneous medium we write the optical path length in the form $\lambda = \int n\,ds$, the familiar variational integral by means of which Fermat's principle is expressed. We have already seen that, along any ray path for which λ is an extremum, given by the vector function $\mathbf{P}(s)$,

$$d\mathbf{P} = \mathbf{P}'\,ds,$$

from which we obtained, in Eq. (XII-2),

$$\mathbf{P}' \cdot d\mathbf{P} = ds.$$

Substituting this into the integral for λ converts it into a line integral as in Eq. (XII-3),

$$\lambda = \int_{\mathbf{P}_0}^{\mathbf{P}^*} n\mathbf{P}' \cdot d\mathbf{P}, \tag{XIII-6}$$

where $\mathbf{P}_0$ and $\mathbf{P}^*$ represent some starting point and end point of the ray path along which λ is measured.

As before, we assume that $\mathbf{P}_0$ and $\mathbf{P}_0'$, the latter being the ray's initial direction vector, are functions of the two parameters u and v. Each point on the ray path as well as the tangent vector at each point must therefore also depend on these two parameters. In particular the terminal point $\mathbf{P}^*$ and the final ray direction $\mathbf{P}^{*\prime}$ must be functions of u and v.

It follows that λ is now a function of u and v. We differentiate λ with respect to each of these two variables using Eq. (XIII-6),

$$\partial\lambda/\partial u = n^*\mathbf{P}^{*\prime} \cdot (\partial\mathbf{P}^*/\partial u) - n_0\mathbf{P}_0' \cdot (\partial\mathbf{P}_0/\partial u),$$
$$\partial\lambda/\partial v = n^*\mathbf{P}^{*\prime} \cdot (\partial\mathbf{P}^*/\partial v) - n_0\mathbf{P}_0' \cdot (\partial\mathbf{P}_0/\partial v).$$

Using the notation of the previous section, this can be written in the form

$$\partial\lambda/\partial u = \Xi^* \cdot (\partial\mathbf{R}^*/\partial u) - \Xi \cdot (\partial\mathbf{R}/\partial u),$$
$$\partial\lambda/\partial v = \Xi^* \cdot (\partial\mathbf{R}^*/\partial v) - \Xi \cdot (\partial\mathbf{R}/\partial v).$$

When the second partial derivatives are taken and equated, we get the fundamental optical invariant as given in Eq. (XIII-5).

THE LENS EQUATION

The next step in this development is to write the fundamental optical invariant in its scalar form. The rays in any two-parameter family are designated by the reduced direction vector

$$\Xi = (\xi, \eta, \zeta), \qquad \text{where} \quad \zeta = (n^2 - \xi^2 - \eta^2)^{1/2},$$

and by the intersection of the ray with the plane $z = 0$, $\mathbf{R} = (x, y, 0)$. Both of these vectors are presumed to be functions of the two parameters u and v. Substituting these into the vector form of the fundamental optical invariant, Eq. (XIII-5), we obtain

$$x_u{}^*\xi_v{}^* + y_u{}^*\eta_v{}^* - x_v{}^*\xi_u{}^* - y_v{}^*\eta_u{}^* = x_u\xi_v + y_u\eta_v - x_v\xi_u - y_v\eta_u\,. \quad \text{(XIII-7)}$$

Here the subscripts denote partial differentiation. (See Herzberger, 1943.)

The totality of all rays comprises a four-parameter system of rays. These four parameters may be chosen at will. We will select as parameters the four numbers that determine a ray in object space, x, y, ξ, η, any pair of which determines a two-parameter family of rays for which the fundamental optical invariant is valid.

We make all possible substitutions. We replace u and v by every pair of the four symbols x, y, ξ, and η. Ringing all the changes, we obtain a set of exactly six distinct equations,

$$\begin{aligned}
x_x{}^*\xi_y{}^* + y_x{}^*\eta_y{}^* - x_y{}^*\xi_x{}^* - y_y{}^*\eta_x{}^* &= 0,\\
x_x{}^*\xi_\xi{}^* + y_x{}^*\eta_\xi{}^* - x_\xi{}^*\xi_x{}^* - y_\xi{}^*\eta_x{}^* &= 1,\\
x_x{}^*\xi_\eta{}^* + y_x{}^*\eta_\eta{}^* - x_\eta{}^*\xi_x{}^* - y_\eta{}^*\eta_x{}^* &= 0,\\
x_y{}^*\xi_\xi{}^* + y_y{}^*\eta_\xi{}^* - x_\xi{}^*\xi_y{}^* - y_\xi{}^*\eta_y{}^* &= 0,\\
x_y{}^*\xi_\eta{}^* + y_y{}^*\eta_\eta{}^* - x_\eta{}^*\xi_y{}^* - y_\eta{}^*\eta_y{}^* &= 1,\\
x_\xi{}^*\xi_\eta{}^* + y_\xi{}^*\eta_\eta{}^* - x_\eta{}^*\xi_\xi{}^* - y_\eta{}^*\eta_\xi{}^* &= 0.
\end{aligned} \qquad \text{(XIII-8)}$$

Although this array is rather formidable, it can be expressed in a much more compact form using a matrix notation. Let

$$\mathsf{X} = \begin{pmatrix} x_x{}^* & x_y{}^* & x_\xi{}^* & x_\eta{}^* \\ y_x{}^* & y_y{}^* & y_\xi{}^* & y_\eta{}^* \\ \xi_x{}^* & \xi_y{}^* & \xi_\xi{}^* & \xi_\eta{}^* \\ \eta_x{}^* & \eta_y{}^* & \eta_\xi{}^* & \eta_\eta{}^* \end{pmatrix}, \qquad \text{(XIII-9)}$$

which is recognizable as a Jacobian matrix. In addition, let

$$\mathsf{J} = \begin{pmatrix} 0 & 0 & 1 & 0 \\ 0 & 0 & 0 & 1 \\ -1 & 0 & 0 & 0 \\ 0 & -1 & 0 & 0 \end{pmatrix}, \tag{XIII-10}$$

a constant matrix. Finally denote by X^{T} the transpose of X, the matrix obtained from X by interchanging rows and columns. Thus,

$$\mathsf{X}^{\mathrm{T}} = \begin{pmatrix} x_x{}^* & y_x{}^* & \xi_x{}^* & \eta_x{}^* \\ x_y{}^* & y_y{}^* & \xi_y{}^* & \eta_y{}^* \\ x_\xi{}^* & y_\xi{}^* & \xi_\xi{}^* & \eta_\xi{}^* \\ x_\eta{}^* & y_\eta{}^* & \xi_\eta{}^* & \eta_\eta{}^* \end{pmatrix}.$$

Then the set of six equations, Eqs. (XIII-8), can be expressed as a single matrix equation

$$\mathsf{X}^{\mathrm{T}}\mathsf{J}\,\mathsf{X} = \mathsf{J}. \tag{XIII-11}$$

This will be referred to as the *lens equation.*

From the lens equation we can deduce some of the elementary properties of the X matrix. Each member of the lens equation is a square matrix, and therefore its determinant can be calculated. Since the rules for multiplying determinants are identical to those for matrices, the determinant of a product of matrices must equate the product of the determinants of those matrices. Taking the determinant of both sides of the lens equation, we obtain

$$|\,\mathsf{X}^{\mathrm{T}}\mathsf{J}\,\mathsf{X}\,| = |\,\mathsf{X}^{\mathrm{T}}\,|\,|\,\mathsf{J}\,|\,|\,\mathsf{X}\,| = |\,\mathsf{J}\,|.$$

Since $|\,\mathsf{X}^{\mathrm{T}}\,| = |\,\mathsf{X}\,|$ and since $|\,\mathsf{J}\,| = 1$, this expression reduces to

$$|\,\mathsf{X}\,|^2 = 1 \qquad \text{or} \qquad |\,\mathsf{X}\,| = \pm 1.$$

It will turn out that $|\,\mathsf{X}\,| = +1$. If the determinant is expanded by cofactors and if the six scalar equations are substituted, after a finite but large amount of labor one can show that the value of the determinant is indeed unity.

But by being a little clever we can avoid this long and tedious calculation. Consider all lens systems that can be completely described by a certain number m of constants. These constants could consist of the refractive indices of the glasses used, the thicknesses and curvatures of each element, and the separation of each element. Should a system require fewer than m constants, we can always adjoin a sufficient number of dummy parameters to bring the number of constants up to m. We can regard the set of lenses as depending continuously on the m parameters. Associated with each lens is a four-parameter family of rays. These we may represent by the four functions

$$x^* = x^*(x, y, \xi, \eta; a_1, a_2, ..., a_m),$$
$$y^* = y^*(x, y, \xi, \eta; a_1, a_2, ..., a_m),$$
$$\xi^* = \xi^*(x, y, \xi, \eta; a_1, a_2, ..., a_m),$$
$$\eta^* = \eta^*(x, y, \xi, \eta; a_1, a_2, ..., a_m),$$

which depend continuously on the m lens parameters. The Jacobian matrix X, or more precisely each element of X, must also be continuous functions of the m parameters. It follows that the determinant $|\mathsf{X}|$ must also be continuous in $a_1, a_2, ..., a_m$. But $|\mathsf{X}| = \pm 1$. Therefore the value of $|\mathsf{X}|$ must be either $+1$ or -1 over the entire set of lenses.

We can include in our set of lenses the trivial lens whose rays are given by

$$x^* = x, \qquad y^* = y, \qquad \xi^* = \xi, \qquad \eta^* = \eta,$$

which can be realized by, for example, a perfect copy lens with unit magnification. For this case $|\mathsf{X}| = +1$. Therefore, for all lenses

$$|\mathsf{X}| = 1. \qquad \text{(XIII-12)}$$

We have established the fact that X is nonsingular; X therefore possesses an inverse X^{-1} so that

$$\mathsf{X}\,\mathsf{X}^{-1} = \mathsf{X}^{-1}\,\mathsf{X} = \mathsf{I},$$

where I is the 4×4 identity matrix. We may therefore rewrite the lens equation, Eq. (XIII-11), as

$$\mathsf{X}^{\mathrm{T}} = \mathsf{J}\,\mathsf{X}^{-1}\,\mathsf{J}^{-1}.$$

A simple calculation will show that

$$\mathsf{J}^{-1} = \mathsf{J}^{\mathrm{T}} = -\mathsf{J}$$

so that

$$\mathsf{X}^{\mathrm{T}} = -\mathsf{J}\,\mathsf{X}^{-1}\,\mathsf{J}.$$

Next,

$$\mathsf{X}\,\mathsf{J}\,\mathsf{X}^{\mathrm{T}} = \mathsf{X}\,\mathsf{J}(-\mathsf{J}\,\mathsf{X}^{-1}\,\mathsf{J}) = \mathsf{X}(-\mathsf{J}^2)\,\mathsf{X}^{-1}\,\mathsf{J}.$$

Another simple calculation will show that $\mathsf{J}^2 = -\mathsf{I}$, where I is the 4×4 identity matrix, and the above reduces to

$$\mathsf{X}\,\mathsf{J}\,\mathsf{X}^{\mathrm{T}} = \mathsf{J}, \tag{XIII-13}$$

which we shall call the *equivalent lens equation.* In scalar form it is

$$\begin{aligned}
F_1 &= x_x{}^* y_\xi{}^* + x_y{}^* y_\eta{}^* - x_\xi{}^* y_x{}^* - x_\eta{}^* y_y{}^* = 0,\\
F_2 &= x_x{}^* \xi_\xi{}^* + x_y{}^* \xi_\eta{}^* - x_\xi{}^* \xi_x{}^* - x_\eta{}^* \xi_y{}^* = 1,\\
F_3 &= x_x{}^* \eta_\xi{}^* + x_y{}^* \eta_\eta{}^* - x_\xi{}^* \eta_x{}^* - x_\eta{}^* \eta_y{}^* = 0,\\
F_4 &= y_x{}^* \xi_\xi{}^* + y_y{}^* \xi_\eta{}^* - y_\xi{}^* \xi_x{}^* - y_\eta{}^* \xi_y{}^* = 0,\\
F_5 &= y_x{}^* \eta_\xi{}^* + y_y{}^* \eta_\eta{}^* - y_\xi{}^* \eta_x{}^* - y_\eta{}^* \eta_y{}^* = 1,\\
F_6 &= \xi_x{}^* \eta_\xi{}^* + \xi_y{}^* \eta_\eta{}^* - \xi_\xi{}^* \eta_x{}^* - \xi_\eta{}^* \eta_y{}^* = 0.
\end{aligned} \tag{XIII-14}$$

This is a system of nonlinear first-order partial differential equations. We need to know under what conditions they have a solution in common. Examine the first two and regard them as a simultaneous pair of equations in the unknown x^* and assume for the moment that y^* and ξ^* are known. We have seen in Chapter VIII that for F_1 and F_2 to have a solution in common their bracket, defined in Eqs. (VIII-6)–(VIII-9), must vanish. Using Eq. (VIII-8), we go through the following calculations

$$\begin{aligned}
[F_1, F_2] = {} & \frac{\partial F_1}{\partial x}\frac{\partial F_2}{\partial x_x{}^*} - \frac{\partial F_1}{\partial x_x{}^*}\frac{\partial F_2}{\partial x} + \frac{\partial F_1}{\partial y}\frac{\partial F_2}{\partial x_y{}^*} - \frac{\partial F_1}{\partial x_y{}^*}\frac{\partial F_2}{\partial y}\\
& + \frac{\partial F_1}{\partial \xi}\frac{\partial F_2}{\partial x_\xi{}^*} - \frac{\partial F_1}{\partial x_\xi{}^*}\frac{\partial F_2}{\partial \xi} + \frac{\partial F_1}{\partial \eta}\frac{\partial F_2}{\partial x_\eta{}^*} - \frac{\partial F_1}{\partial x_\eta{}^*}\frac{\partial F_2}{\partial \eta}.
\end{aligned}$$

The rule to follow here is to calculate the partial derivatives with respect to the quantities that appear *explicitly* in the expressions for F_1 and F_2. At first glance it would seem that x, y, ξ, and η do not appear explicitly in these equations. However, we have assumed that y^* and ξ^* were known and depend on the four variables x, y, ξ, and η. It is correct to assume that these appear explicitly in F_1 and F_2. The calculation proceeds as follows:

$$\begin{aligned}
[F_1, F_2] = {} & (x_x{}^* y_{x\xi}^* + x_y{}^* y_{x\eta}^* - x_\xi{}^* y_{xx}^* - x_\eta{}^* y_{xy}^*)\, \xi_\xi{}^*\\
& - (x_x{}^* \xi_{x\xi}^* + x_y{}^* \xi_{x\eta}^* - x_\xi{}^* \xi_{xx}^* - x_\eta{}^* \xi_{xy}^*)\, y_\xi{}^*\\
& + (x_x{}^* y_{y\xi}^* + x_y{}^* y_{y\eta}^* - x_\xi{}^* y_{xy}^* - x_\eta{}^* y_{yy}^*)\, \xi_\eta{}^*
\end{aligned}$$

Equation continues

$$
\begin{aligned}
&- (x_x{}^*\xi^*_{y\xi} + x_y{}^*\xi^*_{y\eta} - x_\xi{}^*\xi^*_{xy} - x_\eta{}^*\xi^*_{yy})\, y_\eta{}^* \\
&+ (x_x{}^*y^*_{\xi\xi} + x_y{}^*y^*_{\xi\eta} - x_\xi{}^*y^*_{x\xi} - x_\eta{}^*y^*_{y\xi})(-\xi_x{}^*) \\
&- (x_x{}^*\xi^*_{\xi\xi} + x_y{}^*\xi^*_{\xi\eta} - x_\xi{}^*\xi^*_{x\xi} - x_\eta{}^*\xi^*_{y\xi})(-y_x{}^*) \\
&+ (x_x{}^*y^*_{\xi\eta} + x_y{}^*y^*_{\eta\eta} - x_\xi{}^*y^*_{x\eta} - x_\eta{}^*y^*_{y\eta})(-\xi_y{}^*) \\
&- (x_x{}^*\xi^*_{\xi\eta} + x_y{}^*\xi^*_{\eta\eta} - x_\xi{}^*\xi^*_{x\eta} - x_\eta{}^*\xi^*_{y\eta})(-y_y{}^*) \\
= {}& x_x{}^*(y^*_{x\xi}\xi_\xi{}^* + y_x{}^*\xi^*_{\xi\xi} + y^*_{y\xi}\xi_\eta{}^* + y_y{}^*\xi^*_{\xi\eta} \\
&- y^*_{\xi\xi}\xi_x{}^* - y_\xi{}^*\xi^*_{x\xi} - y^*_{\xi\eta}\xi_y{}^* - y_\eta{}^*\xi^*_{y\xi}) \\
&+ x_y{}^*(y^*_{x\eta}\xi_\xi{}^* + y_x{}^*\xi^*_{\xi\eta} + y^*_{y\eta}\xi_\eta{}^* + y_y{}^*\xi^*_{\eta\eta} \\
&- y^*_{\xi\eta}\xi_x{}^* - y_\xi{}^*\xi^*_{x\eta} - y^*_{\eta\eta}\xi_y{}^* - y_\eta{}^*\xi^*_{y\eta}) \\
&- x_\xi{}^*(y^*_{xx}\xi_\xi{}^* + y_x{}^*\xi^*_{x\xi} + y^*_{xy}\xi_\eta{}^* + y_y{}^*\xi^*_{x\eta} \\
&- y^*_{x\xi}\xi_x{}^* - y_\xi{}^*\xi^*_{xx} - y^*_{x\eta}\xi_y{}^* - y_\eta{}^*\xi^*_{xy}) \\
&- x_\eta{}^*(y^*_{xy}\xi_\xi{}^* + y_x{}^*\xi^*_{y\xi} + y^*_{yy}\xi_\eta{}^* + y_y{}^*\xi^*_{y\eta} \\
&- y^*_{y\xi}\xi_x{}^* - y_\xi{}^*\xi^*_{xy} - y^*_{y\eta}\xi_y{}^* - y_\eta{}^*\xi^*_{yy}).
\end{aligned}
$$

From this it follows that

$$[F_1, F_2] = x_x{}^*(dF_4/d\xi) + x_y{}^*(dF_4/d\eta) - x_\xi{}^*(dF_4/dx) - x_\eta{}^*(dF_4/dy),$$

where F_4 is the fourth equation in Eqs. (XIII-14). Now if y^* and ξ^* are not only assumed to be known but are also assumed to satisfy the fourth equation of Eqs. (XIII-13), $F_4 = 0$, then the right-hand member of the above equation vanishes and we are left with $[F_1, F_2] = 0$. We conclude that F_1 and F_2, considered as a pair of differential equations in x^*, have a common solution if and only if y^* and ξ^* satisfy $F_4 = 0$.

We can go through this procedure with any two of the six equations, and find a third equation in the set that must be satisfied before the first two have a solution in common. It is evident that the existence of two functions that satisfy any one of the equations ensures the existence of a set of solutions for the entire simultaneous system.

THE ROTATIONALLY SYMMETRIC SYSTEM

In Chapter XII the rotationally symmetric optical system was encountered. Recall that such a lens consists of a sequence of rotationally symmetric refracting surfaces so arranged that their axes of symmetry all lie on the same straight line. This line coincides with the z axis of any coordinate system we may use. The skewness was defined, Eq. (XII-17), as

$$p = x\eta - y\xi \tag{XIII-15}$$

and we proved, Eq. (XII-16), that the skewness was an invariant in a rotationally symmetric lens

$$x^*\eta^* - y^*\xi^* = x\eta - y\xi. \tag{XIII-16}$$

The skewness vanishes for meridian rays, that is, rays lying entirely on a plane containing the axis. For skew rays, rays not lying on the meridian plane, skewness is a measure of the angle the ray makes with that plane.

Differentiating the skewness invariant with respect to x, y, ξ, and η produces four equations

$$\begin{aligned}
x_x{}^*\eta^* - y_x{}^*\xi^* - \xi_x{}^*y^* + \eta_x{}^*x^* &= \eta,\\
x_y{}^*\eta^* - y_y{}^*\xi^* - \xi_y{}^*y^* + \eta_y{}^*x^* &= -\xi,\\
x_\xi{}^*\eta^* - y_\xi{}^*\xi^* - \xi_\xi{}^*y^* + \eta_\xi{}^*x^* &= -y,\\
x_\eta{}^*\eta^* - y_\eta{}^*\xi^* - \xi_\eta{}^*y^* + \eta_\eta{}^*x^* &= x.
\end{aligned} \tag{XIII-17}$$

This we cast in matrix form

$$\begin{pmatrix} x_x{}^* & y_x{}^* & \xi_x{}^* & \eta_x{}^* \\ x_y{}^* & y_y{}^* & \xi_y{}^* & \eta_y{}^* \\ x_\xi{}^* & y_\xi{}^* & \xi_\xi{}^* & \eta_\xi{}^* \\ x_\eta{}^* & y_\eta{}^* & \xi_\eta{}^* & \eta_\eta{}^* \end{pmatrix} \begin{pmatrix} \eta^* \\ -\xi^* \\ -y^* \\ x^* \end{pmatrix} = \begin{pmatrix} \eta \\ -\xi \\ -y \\ x \end{pmatrix},$$

which, using the notation introduced previously, can be written as

$$\mathsf{X}^{\mathsf{T}}\mathsf{W}^* = \mathsf{W}, \tag{XIII-18}$$

where

$$\mathsf{W}^* = \begin{pmatrix} \eta^* \\ -\xi^* \\ -y^* \\ x^* \end{pmatrix}, \qquad \mathsf{W} = \begin{pmatrix} \eta \\ -\xi \\ -y \\ x \end{pmatrix}.$$

Now multiplying the equivalent lens equation, Eq. (XIII-13), on the right by W^* gives us

$$\mathsf{X}\,\mathsf{J}\,\mathsf{X}^{\mathsf{T}}\mathsf{W}^* = \mathsf{J}\,\mathsf{W}^*,$$

which, using Eq. (XIII-18), reduces to

$$\mathsf{X}\,\mathsf{J}\,\mathsf{W} = \mathsf{J}\,\mathsf{W}^*. \tag{XIII-19}$$

In scalar form this is

$$\begin{aligned} yx_x{}^* - xx_y{}^* + \eta x_\xi{}^* - \xi x_\eta{}^* &= y^*, \\ yy_x{}^* - xy_y{}^* + \eta y_\xi{}^* - \xi y_\eta{}^* &= -x^*, \\ y\xi_x{}^* - x\xi_y{}^* + \eta\xi_\xi{}^* - \xi\xi_\eta{}^* &= \eta^*, \\ y\eta_x{}^* - x\eta_y{}^* + \eta\eta_\xi{}^* - \xi\eta_\eta{}^* &= -\xi^*, \end{aligned} \tag{XIII-20}$$

a system of linear, inhomogeneous, first-order partial differential equations.

A GENERAL SOLUTION

By multiplying the first equation of Eqs. (XIII-20) by x^*, the second by y^*, and adding, we obtain

$$yu_x{}^* - xu_y{}^* + \eta u_\xi{}^* - \xi u_\eta{}^* = 0, \tag{XIII-21}$$

where $u^* = \frac{1}{2}(x^{*2} + y^{*2})$. Similarly, multiplying the third by ξ^* and the fourth by η^* and adding, we obtain

$$yw_x{}^* - xw_y{}^* + \eta w_\xi{}^* - \xi w_\eta{}^* = 0, \tag{XIII-22}$$

where

$$w^* = \tfrac{1}{2}(\xi^{*2} + \eta^{*2}) = \tfrac{1}{2}(n^{*2} - \zeta^{*2}).$$

Finally, by multiplying the first by η^*, the second by $-\xi^*$, the third by $-y^*$, and the fourth by x^*, and adding, we obtain

$$yp_x{}^* - xp_y{}^* + \eta p_\xi{}^* - \xi p_\eta{}^* = 0. \tag{XIII-23}$$

Here, $p^* = x^*\eta^* - y^*\xi^*$ is the skewness defined in Eq. (XIII-15).

Each of the three equations, (XIII-21), (XIII-22) and (XIII-23), is a linear first-order partial differential equation with exactly the same coefficients. When we apply the method of Lagrange, described in Chapter VIII, we find that these equations have exactly the same set of characteristic equations (see Eq. VIII-1):

$$dx/y = -dy/x = d\xi/\eta = -d\eta/\xi.$$

Clearly, we may choose as characteristics the quantities

$$u = \tfrac{1}{2}(x^2 + y^2),$$
$$w = \tfrac{1}{2}(\xi^2 + \eta^2) = \tfrac{1}{2}(n^2 - \zeta^2),$$
$$p = x\eta - y\xi.$$

Then the solutions to the three homogeneous equations are

$$u^* = u^*(u, w, p), \qquad w^* = w^*(u, w, p), \qquad p^* = p, \tag{XIII-24}$$

where u^* and w^* are arbitrary functions of u, v, and p. We know the last relation from the skewness invariant itself, Eq. (XIII-16).

These constitute general solutions of Eqs. (XIII-21), (XIII-22), and (XIII-23). With these behind us we turn now to the four equations in Eqs. (XIII-20). Define

$$x = (2u)^{1/2}\cos\phi, \qquad \xi = (2w)^{1/2}\cos\psi,$$
$$y = (2u)^{1/2}\sin\phi, \qquad \eta = (2w)^{1/2}\sin\psi.$$

Then $p = (4uw)^{1/2}\sin(\psi - \phi)$. Then define $v = x\xi + y\eta$. Then it is a simple matter to show that $v^2 + p^2 = 4uw$ and that $v = (4uw)^{1/2}\cos(\psi - \phi)$. We solve these two equations for $\sin\psi$ and $\cos\psi$ and obtain

$$\begin{aligned} x &= (2u)^{1/2}\cos\phi, \qquad \xi = (2u)^{-1/2}(v\cos\phi - p\sin\phi), \\ y &= (2u)^{1/2}\sin\phi, \qquad \eta = (2u)^{-1/2}(v\sin\phi + p\cos\phi), \end{aligned} \tag{XIII-25}$$

where

$$v = (4uw - p^2)^{1/2}. \tag{XIII-26}$$

In like manner

$$\begin{aligned} x^* &= (2u^*)^{1/2}\cos\phi^*, \qquad \xi^* = (2u^*)^{-1/2}(v^*\cos\phi^* - p\sin\phi^*), \\ y^* &= (2u^*)^{1/2}\sin\phi^*, \qquad \eta^* = (2u^*)^{-1/2}(v^*\sin\phi^* + p\cos\phi^*), \end{aligned} \tag{XIII-27}$$

where

$$v^* = (4u^*w^* - p^2)^{1/2}. \tag{XIII-28}$$

Note that in Eqs. (XIII-27) and (XIII-28) we make use of the skewness invariant, Eq. (XIII-16).

From Eq. (XIII-27) it follows that $\tan\phi^* = y^*/x^*$. Multiplying the first equation of Eqs. (XIII-20) by y^* and the second by $-x^*$, and adding, we obtain after a few steps the linear, nonhomogeneous partial differential equation for ϕ^*,

$$y\phi_x^* - x\phi_y^* + \eta\phi_\xi^* - \xi\phi_\eta^* = -1. \tag{XIII-29}$$

Again using the methods of Chapter VIII we get the characteristic equations

$$dx/y = -dy/x = d\xi/\eta = -d\eta/\xi = -d\phi^*.$$

We may use the three characteristics already chosen, u, w, and p. Now using the first two equations of Eqs. (XIII-25), remembering that u is characteristic, we obtain the fact that

$$d \arctan(y/x) = d\phi = d\phi^*.$$

The final characteristic is obtained from the equation $d(\phi^* - \phi) = 0$. The general solution for Eq. (XIII-29) is therefore given by

$$A(u, w, p, \phi^* - \phi) = \text{const.}$$

or equivalently by

$$\phi^* = \phi + \alpha(u, w, p), \tag{XIII-30}$$

where A and α are arbitrary functions.

This completes the task. The general solution of the set of simultaneous equations in Eqs. (XIII-20) can be taken from Eqs. (XIII-24), (XIII-27), and (XIII-30):

$$\begin{aligned} x^* &= (2u^*)^{1/2} \cos(\phi + \alpha), \\ y^* &= (2u^*)^{1/2} \sin(\phi + \alpha), \\ \xi^* &= (2u^*)^{-1/2}[v^* \cos(\phi + \alpha) - p \sin(\phi + \alpha)], \\ \eta^* &= (2u^*)^{-1/2}[v^* \sin(\phi + \alpha) + p \cos(\phi + \alpha)]. \end{aligned} \tag{XIII-31}$$

The parameter ϕ can be eliminated between these and Eq. (XIII-25), resulting in a rather convenient form of the solution:

$$\begin{aligned} x^* &= (u^*/u)^{1/2}(x \cos\alpha - y \sin\alpha), \\ y^* &= (u^*/u)^{1/2}(x \sin\alpha + y \cos\alpha), \\ \xi^* &= (u/u^*)^{1/2}(\xi \cos\alpha - \eta \sin\alpha) + [(v^* - v)/2(uu^*)^{1/2}](x \cos\alpha - y \sin\alpha), \\ \eta^* &= (u/u^*)^{1/2}(\xi \sin\alpha + \eta \cos\alpha) + [(v^* - v)/2(uu^*)^{1/2}](x \sin\alpha + y \cos\alpha), \end{aligned} \tag{XIII-32}$$

where, of course, u^*, w^*, and α are arbitrary functions of u, w, and p and where $v = (4uw - p^2)^{1/2}$ and $v^* = (4u^*w^* - p^2)^{1/2}$.

These represent a general solution to Eqs. (XIII-20). They are not solutions to the lens equation, however.

THE RESIDUAL EQUATIONS

The solutions just obtained, given in Eq. (XIII-31) or (XIII-32), constitute a general solution of the linear equations contained in Eqs. (XIII-20). They are not solutions of the lens equation in any of its forms, Eqs. (XIII-8), (XIII-11), (XIII-13), or (XIII-14). When the solutions, Eqs. (XIII-31), are substituted into the lens equation, we obtain three first order partial differential equations in the functions u^*, w^*, and α, which up to this point have been regarded as arbitrary functions of u, w, and p. These differential equations we shall call the *residual equations*.

Their derivation is a rather formidable task. The easiest way is to first calculate the X matrix, Eq. (XIII-9), making use of the fact that it is a Jacobian matrix, and substituting it into either the lens equation, Eq. (XIII-11), or the equivalent lens equation, Eq. (XIII-13).

First the X matrix is decomposed into three factors,

$$\mathsf{X} = \frac{\partial(x^*y^*\xi^*\eta^*)}{\partial(xy\xi\eta)} = \frac{\partial(x^*y^*\xi^*\eta^*)}{\partial(p^*\phi^*u^*w^*)}\,\frac{\partial(p^*\phi^*u^*w^*)}{\partial(p\phi uw)}\,\frac{\partial(p\phi uw)}{\partial(xy\xi\eta)}.$$

Using Eq. (XIII-27), the first of these factors is calculated. It turns out to be

$$\frac{\partial(x^*y^*\xi^*\eta^*)}{\partial(p^*\phi^*u^*w^*)} = \begin{pmatrix} 0 & -y^* & x^*/2u^* & 0 \\ 0 & x^* & y^*/2u^* & 0 \\ -\eta^*/v^* & -\eta^* & p\eta^*/2u^*v^* & x^*/v^* \\ \xi^*/v^* & \xi^* & -p\xi^*/2u^*v^* & y^*/v^* \end{pmatrix}.$$

In doing this calculation, we make repeated use of Eq. (XIII-28). Next the third of the three factors is calculated:

$$\frac{\partial(p\phi uw)}{\partial(xy\xi\eta)} = \begin{pmatrix} \eta & -\xi & -y & x \\ -y/2u & x/2u & 0 & 0 \\ x & y & 0 & 0 \\ 0 & 0 & \xi & \eta \end{pmatrix}.$$

Notice that, if the asterisk is dropped from the first of these matrices, then it and the third will be inverses. This allows us to denote these factors as

$$\mathsf{U} = \begin{pmatrix} \eta & -\xi & -y & x \\ -y/2u & x/2u & 0 & 0 \\ x & y & 0 & 0 \\ 0 & 0 & \xi & \eta \end{pmatrix}, \tag{XIII-33}$$

$$\mathsf{U}^{*-1} = \begin{pmatrix} 0 & -y^* & x^*/2u^* & 0 \\ 0 & x^* & y^*/2u^* & 0 \\ -\eta^*/v^* & -\eta^* & p\eta^*/2u^*v^* & x^*/v^* \\ \xi^*/v^* & \xi^* & -p\xi^*/2u^*v^* & y^*/v^* \end{pmatrix}. \tag{XIII-34}$$

The middle matrix we will denote by M:

$$\mathsf{M} = \frac{\partial(p^*\phi^*u^*w^*)}{\partial(p\phi uw)} = \begin{pmatrix} 1 & 0 & 0 & 0 \\ \alpha_p & 1 & \alpha_u & \alpha_w \\ u_p{}^* & 0 & u_u{}^* & u_w{}^* \\ w_p{}^* & 0 & w_u{}^* & w_w{}^* \end{pmatrix}. \tag{XIII-35}$$

Now the X matrix may be written as

$$\mathsf{X} = \mathsf{U}^{*-1}\,\mathsf{M}\,\mathsf{U}, \tag{XIII-36}$$

which is then substituted into the equivalent lens equation, Eq. (XIII-13), resulting in

$$(\mathsf{U}^{*-1}\,\mathsf{M}\,\mathsf{U})\,\mathsf{J}\,(\mathsf{U}^{*-1}\,\mathsf{M}\,\mathsf{U})^{\mathsf{T}} = \mathsf{J},$$

which reduces easily to

$$\mathsf{M}(\mathsf{U}\,\mathsf{J}\,\mathsf{U}^{\mathsf{T}})\,\mathsf{M}^{\mathsf{T}} = \mathsf{U}^*\,\mathsf{J}\,\mathsf{U}^{*\mathsf{T}}. \tag{XIII-37}$$

The next calculation is involved but straightforward:

$$\mathsf{U}\,\mathsf{J}\,\mathsf{U}^{\mathsf{T}} = \begin{pmatrix} 0 & -1 & 0 & 0 \\ 1 & 0 & 0 & p/2u \\ 0 & 0 & 0 & v \\ 0 & -p/2u & -v & 0 \end{pmatrix}.$$

It follows at once that

$$\mathsf{U}^*\,\mathsf{J}\,\mathsf{U}^{*\mathsf{T}} = \begin{pmatrix} 0 & -1 & 0 & 0 \\ 1 & 0 & 0 & p/2u^* \\ 0 & 0 & 0 & v^* \\ 0 & -p/2u^* & -v^* & 0 \end{pmatrix}.$$

Finally, we calculate $\mathsf{M}(\mathsf{U}\,\mathsf{J}\,\mathsf{U}^{\mathsf{T}})\,\mathsf{M}^{\mathsf{T}}$, which when substituted into Eq. (XIII-37), yields the three *residual equations*

$$\begin{aligned} v(\alpha_u u_w{}^* - \alpha_w u_u{}^*) + u_p{}^* + (p/2u)\, u_w{}^* &= 0, \\ v(\alpha_u w_w{}^* - \alpha_w w_u{}^*) + w_p{}^* + (p/2u)\, w_w{}^* &= p/2u^*, \\ v(u_u{}^* w_w{}^* - u_w{}^* w_u{}^*) &= v^*. \end{aligned} \tag{XIII-38}$$

The first two of these can be solved for α_u and α_v with the aid of the third:

$$\begin{aligned} \alpha_u &= (p/2u^*v^*)\, u_u{}^* - (1/v^*)(u_u{}^* w_p{}^* - u_p{}^* w_u{}^*) - p/2uv, \\ \alpha_w &= (p/2u^*v^*)\, u_w{}^* - (1/v^*)(u_w{}^* w_p{}^* - u_p{}^* w_w{}^*). \end{aligned} \tag{XIII-39}$$

A CHANGE OF VARIABLES

The three variables we are currently using were three characteristics obtained from the characteristic equations for the first-order partial differential equations, Eqs. (XIII-21), (XIII-22), and (XIII-23). Of all possible characteristics we could have chosen, these, at the time, seemed reasonable and appropriate. However, we can well ask whether another choice could have been made that would result in a more convenient set of variables than u, p, and w. It turns out that the following change of variables results in some important simplifications. Let

$$\beta = (2u)^{1/2}, \qquad \delta = v(2u)^{-1/2}, \tag{XIII-40}$$

or equivalently $2u = \beta^2$, $v = \beta\delta$, where v is defined in Eq. (XIII-26). It follows that

$$2w = \delta^2 + p^2/\beta^2. \tag{XIII-41}$$

The residual equations, Eqs. (XIII-38), now become

$$\begin{aligned} \beta_\beta{}^*\alpha_\delta - \beta_\delta{}^*\alpha_\beta &= \beta_p{}^*, \\ \delta_\beta{}^*\alpha_\delta - \delta_\delta{}^*\alpha_\beta &= \delta_p{}^*, \\ \beta_\beta{}^*\delta_\delta{}^* - \beta_\delta{}^*\delta_\beta{}^* &= 1, \end{aligned} \tag{XIII-42}$$

while the equations in Eqs. (XIII-39) are

$$\alpha_\beta = \beta_p{}^*\delta_\beta{}^* - \delta_p{}^*\beta_\beta{}^*, \qquad \alpha_\delta = \beta_p{}^*\delta_\delta{}^* - \delta_p{}^*\beta_\beta{}^*. \tag{XIII-43}$$

These last equations lead us to some rather startling conclusions. Calculate the total differential of α:

$$
\begin{aligned}
d\alpha &= \alpha_\beta\, d\beta + \alpha_p\, dp + \alpha_\delta\, d\delta \\
&= (\beta_p{}^*\delta_\beta{}^* - \delta_p{}^*\beta_\beta{}^*)\, d\beta + \alpha_p\, dp + (\beta_p{}^*\delta_\delta{}^* - \delta_p{}^*\beta_\delta{}^*)\, d\delta \\
&= \beta_p{}^*(\delta_\beta{}^*\, d\beta + \delta_\delta{}^*\, d\delta) + \alpha_p\, dp - \delta_p{}^*(\beta_\beta{}^*\, d\beta + \beta_\delta{}^*\, d\delta) \\
&= \beta_p{}^*(\delta_\beta{}^*\, d\beta + \delta_\delta{}^*\, d\delta + \delta_p{}^*\, dp) + \alpha_p\, dp - \delta_p{}^*(\beta_\beta{}^*\, d\beta + \beta_\delta{}^*\, d\delta + \beta_p{}^*\, dp) \\
&= \beta_p{}^*\, d\delta^* + \alpha_p\, dp - \delta_p{}^*\, d\beta^*.
\end{aligned}
$$

From this we may conclude that we may regard α as a function of β^*, δ^*, and p instead of β, δ, and p. Then

$$
\partial\alpha/\partial\delta^* = \partial\beta^*/\partial p, \qquad \partial\alpha/\partial\beta^* = -\partial\delta^*/\partial p. \tag{XIII-44}
$$

Note the similarity between these and the canonical equations in the Hamilton–Jacobi theory of the calculus of variations. We will return to this point later.

The matrix U, defined in Eq. (XIII-33), now becomes

$$
\mathsf{U} = \begin{pmatrix} \eta & -\xi & -y & x \\ -y/\beta^2 & x/\beta^2 & 0 & 0 \\ x & y & 0 & 0 \\ 0 & 0 & \xi & \eta \end{pmatrix} \tag{XIII-45}
$$

and from Eq. (XIII-34),

$$
\mathsf{U}^{*-1} = \begin{pmatrix} 0 & -y^* & x^*/\beta^{*2} & 0 \\ 0 & x^* & y^*/\beta^{*2} & 0 \\ -n^*/\beta^*\delta^* & -\eta^* & p\eta^*/\beta^{*3}\delta^* & x^*/\beta^*\delta^* \\ \xi^*/\beta^*\delta^* & \xi^* & -p\xi^*/\beta^{*3}\delta^* & y^*/\beta^*\delta^* \end{pmatrix}. \tag{XIII-46}
$$

The M matrix in Eq. (XIII-35) becomes

$$
\begin{aligned}
\mathsf{M} &= \begin{pmatrix} 1 & 0 & 0 & 0 \\ \alpha_p & 1 & \alpha_u & \alpha_w \\ u_p{}^* & 0 & u_u{}^* & u_w{}^* \\ w_p{}^* & 0 & w_u{}^* & w_w{}^* \end{pmatrix} \\
&= \begin{pmatrix} 0 & 0 & 0 & 1 \\ 0 & 1 & 0 & 0 \\ \beta^* & 0 & 0 & 0 \\ -p^2/\beta^{*3} & 0 & \delta^* & p/\beta^{*2} \end{pmatrix} \begin{pmatrix} \beta_\beta{}^* & 0 & \beta_\delta{}^* & \beta_p{}^* \\ \alpha_\beta & 1 & \alpha_\delta & \alpha_p \\ \delta_\beta{}^* & 0 & \delta_\delta{}^* & \delta_p{}^* \\ 0 & 0 & 0 & 1 \end{pmatrix} \\
&\quad \times \begin{pmatrix} 0 & 0 & 1/\beta & 0 \\ 0 & 1 & 0 & 0 \\ -p/\beta^2\delta & 0 & p^2/\beta^4\delta & 1/\delta \\ 1 & 0 & 0 & 0 \end{pmatrix}.
\end{aligned} \tag{XIII-47}
$$

Substituting these expressions for U, U^{*-1}, and M into the expression for the X matrix, Eq. (XIII-36), we get

$$\mathsf{X} = \mathsf{V}^{*-1}\mathsf{N}\mathsf{V}, \tag{XIII-48}$$

where

$$\mathsf{V} = \begin{pmatrix} 0 & 0 & 1/\beta & 0 \\ 0 & 1 & 0 & 0 \\ -p/\beta^2\delta & 0 & p^2/\beta^4\delta & 1/\delta \\ 1 & 0 & 0 & 0 \end{pmatrix} \mathsf{U},$$

(XIII-49)

$$\mathsf{U} = \begin{pmatrix} x/\beta & y/\beta & 0 & 0 \\ -y/\beta^2 & x/\beta^2 & 0 & 0 \\ -py/\beta^3 & px/\beta^3 & x/\beta & y/\beta \\ \eta & -\xi & -y & x \end{pmatrix},$$

where

$$\mathsf{V}^{*-1} = \mathsf{U}^{*-1}\begin{pmatrix} 0 & 0 & 0 & 1 \\ 0 & 1 & 0 & 0 \\ \beta^* & 0 & 0 & 0 \\ -p^2/\beta^{*3} & 0 & \delta^* & p/\beta^{*2} \end{pmatrix}$$

$$= \begin{pmatrix} x^*/\beta^* & -y^* & 0 & 0 \\ y^*/\beta^* & x^* & 0 & 0 \\ py^*/\beta^{*3} & -\eta^* & x^*/\beta^* & -y^*/\beta^{*2} \\ -px^*/\beta^{*3} & \xi^* & y^*/\beta^* & x^*/\beta^{*2} \end{pmatrix}, \tag{XIII-50}$$

and where

$$\mathsf{N} = \begin{pmatrix} \beta_\beta{}^* & 0 & \beta_\delta{}^* & \beta_p{}^* \\ \alpha_\beta & 1 & \alpha_\delta & \alpha_p \\ \delta_\beta{}^* & 0 & \delta_\delta{}^* & \delta_p{}^* \\ 0 & 0 & 0 & 1 \end{pmatrix}. \tag{XIII-51}$$

We have chosen our variables and arranged the elements in the matrices in such a way that $\mathsf{V}\,\mathsf{J}\,\mathsf{V}^{\mathrm{T}} = \mathsf{J}$. Thus, since from Eq. (XIII-13), $\mathsf{X}\,\mathsf{J}\,\mathsf{X}^{\mathrm{T}} = \mathsf{J}$, it follows that

$$\mathsf{N}\,\mathsf{J}\,\mathsf{N}^{\mathrm{T}} = \mathsf{J}. \tag{XIII-52}$$

When written out in scalar form, this last equation can be seen to be identical with the residual equations, Eqs. (XIII-42). By using these new forms of the residual equations, it is possible to factor the N matrix in two distinct ways. Substituting the first two equations in Eqs. (XIII-42) into Eq. (XIII-51), we get

$$\mathsf{N} = \begin{pmatrix} \beta_\beta^* & 0 & \beta_\delta^* & \beta_\beta^*\alpha_\delta - \beta_\delta^*\alpha_\beta \\ \alpha_\beta & 1 & \alpha_\delta & \alpha_p \\ \delta_\beta^* & 0 & \delta_\delta^* & \delta_\beta^*\alpha_\delta - \delta_\delta^*\alpha_\beta \\ 0 & 0 & 0 & 1 \end{pmatrix}$$

$$= \begin{pmatrix} \beta_\beta^* & 0 & \beta_\delta^* & 0 \\ 0 & 1 & 0 & 0 \\ \delta_\beta^* & 0 & \delta_\delta^* & 0 \\ 0 & 0 & 0 & 1 \end{pmatrix} \begin{pmatrix} 1 & 0 & 0 & \alpha_\delta \\ \alpha_\beta & 1 & \alpha_\delta & \alpha_p \\ 0 & 0 & 1 & -\alpha_\beta \\ 0 & 0 & 0 & 1 \end{pmatrix}. \quad \text{(XIII-53)}$$

The alternative factorization is obtained by substituting the two equations in Eqs. (XIII-43) into Eq. (XIII-51):

$$\mathsf{N} = \begin{pmatrix} \beta_\beta^* & 0 & \beta_\delta^* & \beta_p^* \\ \beta_p^*\delta_\beta^* - \delta_p^*\beta_\beta^* & 1 & \beta_p^*\delta_\delta^* - \delta_p^*\beta_\delta^* & \alpha_p \\ \delta_\beta^* & 0 & \delta_\delta^* & \delta_p^* \\ 0 & 0 & 0 & 1 \end{pmatrix}$$

$$= \begin{pmatrix} 1 & 0 & 0 & \alpha_\delta \\ \alpha_\beta & 1 & \alpha_\delta & \alpha_p \\ 0 & 0 & 1 & -\alpha_\beta \\ 0 & 0 & 0 & 1 \end{pmatrix} \begin{pmatrix} \beta_\beta^* & 0 & \beta_\delta^* & 0 \\ 0 & 1 & 0 & 0 \\ \delta_\beta^* & 0 & \delta_\delta^* & 0 \\ 0 & 0 & 0 & 1 \end{pmatrix}. \quad \text{(XIII-54)}$$

In this last step use was made of Eq. (XIII-44).

THE FINAL FACTORIZATION

As a final touch we will obtain a partial solution to the residual equations, Eqs. (XIII-42), reducing their number from *three* to *two.* It has already been noted that, Eqs. (XIII-44),

$$\partial\alpha/\partial\delta^* = \partial\beta^*/\partial p, \qquad \partial\alpha/\partial\beta^* = -\partial\delta^*/\partial p,$$

in which α is regarded as a function of β^*, δ^*, and p, resemble the canonical equations of the Hamilton–Jacobi theory, Eqs. (VII-26),

$$\partial H/\partial u = dy/dx, \qquad \partial H/\partial y = -du/dx.$$

These latter equations arise from a two-dimensional variational problem

$$I = \int f(x, y, y')\, dx$$

and the associated Euler equation, Eq. (II-4),

$$(d/dx)(\partial f/\partial y') = \partial f/\partial y.$$

The canonical variable is defined by Eq. (VII-23), $u = \partial f/\partial y'$, and the Hamiltonian by Eq. (VII-25),

$$H(x, y, u) = y'(\partial f/\partial y') - f.$$

By identifying α, δ^*, β^*, and p with H, u, y, and x, we can regard them as variables associated with a variational problem

$$I = \int f(p, \beta^*, \beta_p{}^*)\, dp$$

whose Euler equation is

$$(d/dp)(\partial f/\partial \beta_p{}^*) = \partial f/\partial \beta^*,$$

with the canonical variable defined as $\delta^* = \partial f/\partial \beta_p{}^*$ and with α as the Hamiltonian

$$\alpha = \beta_p{}^* \delta^* - f(p, \beta^*, \beta_p{}^*).$$

For convenience let $\partial f/\partial p = f_1$, $\partial f/\partial \beta^* = f_2$, and $\partial f/\partial \beta_p{}^* = f_3$. Then

$$\delta^* = f_3, \qquad \alpha = \beta_p{}^* \delta^* - f, \tag{XIII-55}$$

where f is taken to be an arbitrary function of p, β^*, and $\beta_p{}^*$. This can be regarded as a solution to Eq. (XIII-44) provided that the condition equivalent to the Euler equation is satisfied:

$$(d/dp)\, f_3 = f_2\,.$$

This we expand to obtain

$$f_{31} + f_{32}\beta_p{}^* + f_{33}\beta_{pp}^* = f_2\,. \tag{XIII-56}$$

Finally we substitute Eqs. (XIII-55) into the last of the remaining residual equations, Eqs. (XIII-42). We obtain

$$f_{33}(\beta_{p\beta}^* \beta_\delta{}^* - \beta_\beta{}^* \beta_{p\delta}^*) = 1. \tag{XIII-57}$$

The solution of these two equations is beyond our reach.

There is a certain amount of symmetry in all of this. We could just as easily have written $H = -\alpha$, $u = \beta^*$, $y = \delta^*$, and $x = p$. In this case the variational problem would have led us to a slightly different solution:

$$\beta^* = f_3, \qquad \alpha = -\delta_p{}^*\beta^* + f(p, \delta^*, \delta_p{}^*).$$

This concludes our attempt to describe lenses from a purely abstract point of view. In the next chapter we shall develop an *ad hoc* solution for a rotationally symmetric lens.

REFERENCES

Herzberger, M. (1931). "Strahlenoptik." Springer-Verlag, Berlin and New York.

Herzberger, M. (1943). Direct methods in geometrical optics. *Trans. Amer. Math. Soc.* **53**, 218–229.

Herzberger, M. (1944). Mathematics and Geometrical Optics, Supplementary Note No. III. See Luneburg (1964).

Herzberger, M. (1958). "Modern Geometrical Optics." Wiley (Interscience), New York.

Luneburg, R. K. (1964). "Mathematical Theory of Optics." Univ. of California Press, Berkeley, California.

XIV The Lens Equation

It is clear that general solutions to the lens equation, at least in terms of the methods we have been applying, are unobtainable. In our frustration we turn now to a method for obtaining particular solutions, solutions applicable to a particular optical design. This will be the concern of this chapter.

To fix ideas, suppose that we begin by considering two lenses. Associated with each of these will be an object reference plane and an image reference plane. It is not implied that these planes are conjugates; they may be selected arbitrarily. Their only purpose is to provide a pair of coordinate axes to which we can reference rays.

The set of all rays in the first lens will be described by the four functions

$$x_1 = x_1(x, y, \xi, \eta), \qquad \xi_1 = \xi_1(x, y, \xi, \eta),$$
$$y_1 = y_1(x, y, \xi, \eta), \qquad \eta_1 = \eta_1(x, y, \xi, \eta),$$

with which is associated a Jacobian matrix X_1 that satisfies the lens equation (XIII-11).

The second lens is described in the same way; four functions

$$x^* = x^*(x, y, \xi, \eta), \qquad \xi^* = \xi^*(x, y, \xi, \eta),$$
$$y^* = y^*(x, y, \xi, \eta), \qquad \eta^* = \eta^*(x, y, \xi, \eta),$$

with which is associated a second Jacobian matrix X_2 that also satisfies the lens equation.

With these two lenses we form a third by simply placing them next to each other so that the object reference plane of the second coincides with the image reference plane of the first. The functional relationship of the rays associated with the new lens is described completely by replacing x, y, ξ, and η in the second set of functions by x_1, y_1, ξ_1, and η_1, the latter being the dependent variables of the first. Thus

$$x^* = x^*(x_1, y_1, \xi_1, \eta_1), \qquad x_1 = x_1(x, y, \xi, \eta),$$
$$y^* = y^*(x_1, y_1, \xi_1, \eta_1), \qquad y_1 = y_1(x, y, \xi, \eta),$$
$$\xi^* = \xi^*(x_1, y_1, \xi_1, \eta_1), \qquad \xi_1 = \xi_1(x, y, \xi, \eta),$$
$$\eta^* = \eta^*(x_1, y_1, \xi_1, \eta_1), \qquad \eta_1 = \eta_1(x, y, \xi, \eta).$$

Associated with the new lens is a Jacobian matrix that we shall call X. To determine the relationship between X and X_1 and X_2, we calculate the partial derivatives of the four functions x^*, y^*, ξ^*, η^* with respect to the four variables x, y, ξ, η. For example,

$$\partial x^*/\partial x = (\partial x^*/\partial x_1)(\partial x_1/\partial x) + (\partial x^*/\partial y_1)(\partial y_1/\partial x)$$
$$+ (\partial x^*/\partial \xi_1)(\partial \xi_1/\partial x) + (\partial x^*/\partial \eta_1)(\partial \eta_1/\partial x),$$

the right-hand member of which can be written as a matrix product

$$\begin{pmatrix} \dfrac{\partial x^*}{\partial x_1} & \dfrac{\partial x^*}{\partial y_1} & \dfrac{\partial x^*}{\partial \xi_1} & \dfrac{\partial x^*}{\partial \eta_1} \end{pmatrix} \begin{pmatrix} \partial x_1/\partial x \\ \partial y_1/\partial x \\ \partial \xi_1/\partial x \\ \partial \eta_1/\partial x \end{pmatrix}.$$

From this it is clear that the Jacobian matrix X of the new lens is equal to the product of the Jacobian matrices of the second and first lenses, *in that order*: $\mathsf{X} = \mathsf{X}_2\mathsf{X}_1$. Since both X_1 and X_2 are assumed to satisfy the lens equation, their product does also:

$$\mathsf{X}^{\mathsf{T}}\,\mathsf{J}\,\mathsf{X} = (\mathsf{X}_2\mathsf{X}_1)^{\mathsf{T}}\,\mathsf{J}\,(\mathsf{X}_2\mathsf{X}_1) = \mathsf{X}_1^{\mathsf{T}}(\mathsf{X}_2^{\mathsf{T}}\,\mathsf{J}\,\mathsf{X}_2)\,\mathsf{X}_1 = \mathsf{X}_1^{\mathsf{T}}\,\mathsf{J}\,\mathsf{X}_1 = \mathsf{J}\,.$$

Details of the mechanics of matrix algebra, which will be used extensively in this chapter and the next, can be found in any standard text such as MacDuffee (1943).

All this says is that, if an optical system can be decomposed into component parts, each of which is represented by a Jacobian matrix satisfying the lens equation, then the Jacobian matrix of the entire system is equal to the product of the Jacobians of the component parts.

We shall see that every lens can be thought of as consisting of a sequence of operations that are applied to the rays traced through it. These operations are *transfer*, the tracing of a ray from one refracting surface to the next; and *refraction*, the tracing of a ray across a refracting surface. This concept has already been encountered in earlier chapters. These operations we will regard as the ultimate components of an optical system. We shall see that each of these operations, slightly modified, can be represented by a Jacobian matrix satisfying the lens equation (XIII-11). The Jacobian matrix for a composite lens then may be represented by a product of the Jacobian matrices for refraction and transfer.

THE TRANSFER MATRIX

The first item on the agenda is a redefinition of what is meant by transfer and refraction. We have described transfer as the operation of tracing a ray from one refracting surface to the next and refraction as the process of tracing a ray through a refracting surface. However, the ray coordinates that we have used to define the X matrices are referred to planes, which we have called the object reference plane and the image reference plane. Some modifications are clearly necessary.

We will use as a reference plane for both refraction and transfer a plane tangent to the refracting surface. The point of contact of plane and surface will be the origin of the coordinate system to which the rays will be referenced, the z axis coinciding with the normal to the surface at this point. The distances between these reference planes along the z axis will correspond exactly to what we have called t, the vertex separation of successive refracting surfaces.

Transfer will henceforth be regarded as the tracing of a ray from one reference plane to the next. Refraction will be the operation of tracing a ray from a reference plane to a refracting surface, refraction across that surface, then a transfer back to the reference plane. These steps appear unnecessarily complicated. Nevertheless, these changes do result in some definite advantages. The X matrix for transfer will depend only on the vertex separation t, and the X matrix for refraction will depend only on the geometrical properties of the refracting surface itself.

Modifying the transfer equations (VI-5) to conform to these changes, we have

$$x^* = x + t\xi/\zeta, \qquad y^* = y + t\eta/\zeta, \tag{XIV-1}$$

to which we adjoin the trivial equations

$$\xi^* = \xi, \qquad \eta^* = \eta.$$

Here t is the vertex separation, the distance from one refracting surface to the next along the z axis. Also, $\zeta = (n^2 - \xi^2 - \eta^2)^{1/2}$, where n is the refractive index of the medium through which the ray is being traced. The elements of the X matrix for transfer are simply

$$\mathsf{X}_t = \begin{pmatrix} 1 & 0 & t(n^2 - \eta^2)/\zeta^3 & t\xi\eta/\zeta^3 \\ 0 & 1 & t\xi\eta/\zeta^3 & t(n^2 - \xi^2)/\zeta^3 \\ 0 & 0 & 1 & 0 \\ 0 & 0 & 0 & 1 \end{pmatrix}. \tag{XIV-2}$$

Note that, when $t = 0$, X_t becomes the unity matrix. Note also that

$$\mathsf{X}_{t_1}\mathsf{X}_{t_2} = \mathsf{X}_{t_1+t_2}$$

so that

$$\mathsf{X}_t\mathsf{X}_{-t} = \mathsf{X}_0 = \mathsf{I},$$

whence

$$\mathsf{X}_{-t} = \mathsf{X}_t^{-1}.$$

Thus the X matrix for transfer, X_t, behaves like an exponential function of t. So, we define

$$\mathsf{T} = \begin{pmatrix} 1 & 0 & (n^2 - \eta^2)/\zeta^3 & \xi\eta/\zeta^3 \\ 0 & 1 & \xi\eta/\zeta^3 & (n^2 - \xi^2)/\zeta^3 \\ 0 & 0 & 1 & 0 \\ 0 & 0 & 0 & 1 \end{pmatrix} \tag{XIV-3}$$

and write $\mathsf{X}_t = \mathsf{T}^t$. By this we mean raising the matrix T to the power t.

THE REFRACTION MATRIX

We are concerned only with rotationally symmetric surfaces. These can be represented by a function $\bar{z}$ of a symmetric function $\bar{x}^2 + \bar{y}^2$. The "barred" coordinates represent a point on the refracting surface.

In Eq. (XIII-40) we defined certain symmetric functions, which will be used here in the form

$$\bar{\beta}^2 = \bar{x}^2 + \bar{y}^2, \qquad \bar{\delta} = (\bar{x}\xi + \bar{y}\eta)/\bar{\beta}, \qquad \delta^* = (\bar{x}\xi^* + \bar{y}\eta^*)/\bar{\beta}. \quad \text{(XIV-4)}$$

The equation for the refracting surface then can be written as

$$\bar{z} = \bar{z}(\bar{\beta}). \quad \text{(XIV-5)}$$

To make things a little simpler, we place the coordinate axes so that the z axis is along the axis of symmetry of the refracting surface and the x, y plane coincides with the reference plane. Under these conditions $\bar{z} = 0$ whenever $\bar{\beta} = 0$.

The refraction operation will consist of a transfer from the reference plane to the refracting surface, i.e.,

$$\bar{x} = x + \bar{z}\xi/\zeta, \qquad \bar{y} = y + \bar{z}\eta/\zeta, \quad \text{(XIV-6)}$$

followed by refraction across that surface [see Eq. (VI-1)], i.e.,

$$\xi^* = \xi + n^*\gamma\bar{\xi}, \qquad \eta^* = \eta + n^*\gamma\bar{\eta}, \quad \text{(XIV-7)}$$

followed by a transfer back to the reference plane, i.e.,

$$x^* = \bar{x} - \bar{z}\xi^*/\zeta^*, \qquad y^* = \bar{y} - \bar{z}\eta^*/\zeta^*. \quad \text{(XIV-8)}$$

In Eq. (XIV-7), $\bar{\xi}$ and $\bar{\eta}$ are two of the direction cosines of the normal to the refracting surface at the point of intersection.

These equations are considerably more complicated than they appear. In the first transfer equations (XIV-6) the intersection of the ray with the refracting surface, given by the coordinates $\bar{x}$, $\bar{y}$, and $\bar{z}$, is given implicitly. If we square and add the two equations of Eq. (XIV-7), we obtain, making use of Eq. (XIII-41),

$$\begin{aligned} \bar{\beta}^2 &= \beta^2 + 2\beta\delta\bar{z}/\zeta + (\delta^2 + p^2/\beta^2)\,\bar{z}^2/\zeta^2 \\ &= (\beta + \delta\bar{z}/\zeta)^2 + (p\bar{z}/\beta\zeta)^2. \end{aligned} \quad \text{(XIV-9)}$$

This must be solved for $\bar{\beta}$ before $\bar{z}$ can be calculated.

The direction cosines of the surface normal are obtained by differentiating the equation for the surface, Eq. (XIV-5),

$$\bar{\xi} = (\partial\bar{z}/\partial\bar{x})/k, \qquad \bar{\zeta} = -1/k,$$

$$\bar{\eta} = (\partial\bar{z}/\partial\bar{y})/k, \qquad k^2 = 1 + (\partial\bar{z}/\partial\bar{x})^2 + (\partial\bar{z}/\partial\bar{y})^2.$$

Applying Eqs. (XIV-4) and (XIV-5) to these expressions, we get

$$\bar{\xi} = \bar{z}'\bar{x}/\bar{\beta}k, \qquad \bar{\zeta} = -1/k,$$
$$\bar{\eta} = \bar{z}'\bar{y}/\bar{\beta}k, \qquad k^2 = 1 + \bar{z}'^2. \tag{XIV-10}$$

Our purpose is to calculate the X matrix for the refraction operation. It is easy to see that this will be the product of three matrices, $\mathsf{X}_3\mathsf{X}_2\mathsf{X}_1$, where X_1 is the Jacobian of the transformation representing the first transfer, Eq. (XIV-6), and X_2 is the Jacobian of the transformation representing refraction, etc.

To calculate X_1 we need to differentiate Eq. (XIV-6) with respect to x and y using Eq. (XIV-5) and again Eq. (XIV-6). This gives us

$$\bar{x}_x = 1 + \bar{z}'(\overline{xx}_x + \overline{yy}_x)\,\xi/\bar{\beta}\zeta, \qquad \bar{x}_y = \bar{z}'(\overline{xx}_y + \overline{yy}_y)\,\xi/\bar{\beta}\zeta,$$

$$\bar{y}_x = \bar{z}'(\overline{xx}_x + \overline{yy}_x)\,\eta/\bar{\beta}\zeta, \qquad \bar{y}_y = 1 + \bar{z}'(\overline{xx}_y + \overline{yy}_y)\,\eta/\bar{\beta}\zeta.$$

Solving these for the derivatives of $\bar{x}$ and $\bar{y}$, we get

$$\bar{x}_x = 1 + \bar{z}'\bar{x}\xi/\bar{\beta}H, \qquad \bar{x}_y = \bar{z}'\bar{y}\xi/\bar{\beta}H,$$

$$\bar{y}_x = \bar{z}'\bar{x}\eta/\bar{\beta}H, \qquad \bar{y}_y = 1 + \bar{z}'\bar{y}\eta/\bar{\beta}H,$$

where

$$H = \zeta - \overline{z\delta}'. \tag{XIV-11}$$

Note that the cosine of the angle of incidence equals $-H/Kn$.

The derivatives of x and y with respect to ξ and η proceed in exactly the same way. Omitting the details of the calculations, we obtain

$$\bar{x}_\xi = \bar{z}[\bar{\beta}(\zeta^2 + \xi^2) - \bar{z}'\bar{y}\eta\zeta]/\bar{\beta}H\zeta^2, \qquad \bar{x}_\eta = \bar{z}(\bar{\beta}\xi\eta + \bar{z}'\bar{y}\xi\zeta)/\bar{\beta}H\zeta^2,$$

$$\bar{y}_\xi = \bar{z}(\bar{\beta}\xi\eta + \bar{z}'\bar{x}\eta\zeta)/\bar{\beta}H\zeta^2, \qquad \bar{y}_\eta = \bar{z}[\bar{\beta}(\zeta^2 + \eta^2) - \bar{z}'\bar{x}\xi\zeta]/\bar{\beta}H\zeta^2.$$

The X matrix for the first transfer is thus

$$\mathsf{X}_1 = \begin{pmatrix} 1 + \bar{z}'\bar{x}\xi/\bar{\beta}H & \bar{z}'\bar{y}\xi/\bar{\beta}H & \bar{z}(\zeta^2+\xi^2)/H\zeta^2 - \bar{z}\bar{z}'\bar{y}\eta/\bar{\beta}H\zeta & \bar{z}\xi\eta/H\zeta^2 + \bar{z}\bar{z}'\bar{y}\xi/\bar{\beta}H\zeta \\ \bar{z}'\bar{x}\eta/\bar{\beta}H & 1 + \bar{z}'\bar{y}\eta/\bar{\beta}H & \bar{z}\xi\eta/H\zeta^2 + \bar{z}\bar{z}'\bar{y}\eta/\bar{\beta}H\zeta & \bar{z}(\zeta^2+\eta^2)/H\zeta^2 - \bar{z}\bar{z}'\bar{x}\xi/\bar{\beta}H\zeta \\ 0 & 0 & 1 & 0 \\ 0 & 0 & 0 & 1 \end{pmatrix}. \quad \text{(XIV-12)}$$

To do the X matrix for the second transfer is much simpler. Differentiating the two equations in Eqs. (XIV-8) with respect to $\bar{x}$, $\bar{y}$, ξ^*, and η^* is straightforward and leads at once to

$$\mathsf{X}_3 = \begin{pmatrix} 1 - \bar{z}'\bar{x}\xi^*/\bar{\beta}\zeta^* & -\bar{z}'\bar{y}\xi^*/\bar{\beta}\zeta^* & -\bar{z}(n^{*2}-\eta^{*2})/\zeta^{*3} & -\bar{z}\xi^*\eta^*/\zeta^{*3} \\ -\bar{z}'\bar{x}\eta^*/\bar{\beta}\zeta^* & 1 - \bar{z}'\bar{y}\eta^*/\bar{\beta}\zeta^* & -\bar{z}\xi^*\eta^*/\zeta^{*3} & -\bar{z}(n^{*2}-\xi^{*2})/\zeta^{*3} \\ 0 & 0 & 1 & 0 \\ 0 & 0 & 0 & 1 \end{pmatrix}.$$

Now X_3 can be factored into a product of two matrices:

$$\mathsf{X}_3 = \begin{pmatrix} 1 & 0 & -\bar{z}(n^{*2}-\eta^{*2})/\zeta^{*3} & -\bar{z}\xi^*\eta^*/\zeta^{*3} \\ 0 & 1 & -\bar{z}\xi^*\eta^*/\zeta^{*3} & -\bar{z}(n^{*2}-\xi^{*2})/\zeta^{*3} \\ 0 & 0 & 1 & 0 \\ 0 & 0 & 0 & 1 \end{pmatrix} \times \begin{pmatrix} 1 - \bar{z}'\bar{x}\xi^*/\bar{\beta}\zeta^* & -\bar{z}'\bar{y}\xi^*/\bar{\beta}\zeta^* & 0 & 0 \\ -\bar{z}'\bar{x}\eta^*/\bar{\beta}\zeta^* & 1 - \bar{z}'\bar{y}\eta^*/\bar{\beta}\zeta^* & 0 & 0 \\ 0 & 0 & 1 & 0 \\ 0 & 0 & 0 & 1 \end{pmatrix}.$$

The first of these products can be written, using the notation introduced in Eq. (XIV-3), as $\mathsf{T}^{*-\bar{z}}$. If the second factor is denoted by S^*, then the above may be written in the form

$$\mathsf{X}_3 = \mathsf{T}^{*-\bar{z}}\mathsf{S}^*. \quad \text{(XIV-13)}$$

Using the same sort of notation, define

$$\mathsf{S} = \begin{pmatrix} 1 - \bar{z}'\bar{x}\xi/\bar{\beta}\zeta & -\bar{z}'\bar{y}\xi/\bar{\beta}\zeta & 0 & 0 \\ -\bar{z}'\bar{x}\eta/\bar{\beta}\zeta & 1 - \bar{z}'\bar{y}\eta/\bar{\beta}\zeta & 0 & 0 \\ 0 & 0 & 1 & 0 \\ 0 & 0 & 0 & 1 \end{pmatrix}, \tag{XIV-14}$$

and note that the determinant of this matrix is, from Eq. (XIV-11), $|\,\mathsf{S}\,| = H/\zeta$. Moreover, one can easily find the inverse of S

$$\mathsf{S}^{-1} = \begin{pmatrix} (\bar{\beta}\zeta - \bar{z}'\bar{y}\eta)/\bar{\beta}H & \bar{z}'\bar{y}\xi/\bar{\beta}H & 0 & 0 \\ \bar{z}'\bar{x}\eta/\bar{\beta}H & (\bar{\beta}\zeta - \bar{z}'\bar{x}\xi)/\bar{\beta}H & 0 & 0 \\ 0 & 0 & 1 & 0 \\ 0 & 0 & 0 & 1 \end{pmatrix},$$

or, using Eq. (XIV-11), this becomes

$$\mathsf{S}^{-1} = \begin{pmatrix} 1 + \bar{z}'\bar{x}\xi/\bar{\beta}H & \bar{z}'\bar{y}\xi/\bar{\beta}H & 0 & 0 \\ \bar{z}'\bar{x}\eta/\bar{\beta}H & 1 + \bar{z}'\bar{y}\eta/\bar{\beta}H & 0 & 0 \\ 0 & 0 & 1 & 0 \\ 0 & 0 & 0 & 1 \end{pmatrix}. \tag{XIV-15}$$

Now, going back to the matrix representation of X_1, Eq. (XIV-12), we find that it, too, can be written as a matrix product,

$$\mathsf{X}_1 = \mathsf{S}^{-1}\mathsf{T}^{\bar{z}}. \tag{XIV-16}$$

To show this, a number of rather long calculations are involved. These are omitted here.

What this really means is that X_1 is the inverse of X_3 when the asterisks are left out. This is obvious. Suppose $n^* = n$. Then no refraction occurs and $\xi^* = \xi$ and $\eta^* = \eta$; therefore the matrix X_2 representing refraction degenerates into the unity matrix. Then the product $\mathsf{X}_3\mathsf{X}_1$ must also equal the unity matrix. It follows that in this case X_1 and X_3 are reciprocals.

THE REFRACTION CALCULATIONS

We begin this rather difficult sequence of calculations by combining Eqs. (XIV-7) and (XIV-10) to get

$$\xi^* = \xi + n^*\gamma\bar{z}'\bar{x}/\bar{\beta}k, \qquad \eta^* = \eta + n^*\gamma\bar{z}'\bar{y}/\bar{\beta}k,$$

where γ is the positive branch of the solution of a certain quadratic that we have encountered before and that led up to Eq. (VI-2). Using the notation introduced in this section, it becomes

$$F = n^{*2}\gamma^2 - 2n^*H\gamma/k - (n^{*2} - n^2) = 0.$$

At this point it will be most convenient to make a substitution. Let $\gamma = \bar{\beta}k\psi/n^*\bar{z}'$ and substitute it into the three preceding equations. Then the refraction equations become

$$\xi^* = \xi + \bar{x}\psi, \qquad \eta^* = \eta + \bar{y}\psi, \tag{XIV-17}$$

where ψ is the positive branch of the solution to the quadratic

$$F = \bar{\beta}^2k^2\psi^2 - 2\overline{\beta z'}H\psi - \bar{z}'^2(n^{*2} - n^2) = 0. \tag{XIV-18}$$

Into the vector form of Snell's law, Eq. (V-19),

$$(\xi^*, \eta^*, \zeta^*) \times (\bar{\xi}, \bar{\eta}, \bar{\zeta}) = (\xi, \eta, \zeta) \times (\bar{\xi}, \bar{\eta}, \bar{\zeta}),$$

we substitute the value of the unit normal vector to the refracting surface calculated in Eq. (XIV-10), and obtain

$$\xi^* + \zeta^*\overline{xz'}/\bar{\beta} = \xi + \zeta\overline{xz'}/\bar{\beta},$$

$$\eta^* + \zeta^*\overline{yz'}/\bar{\beta} = \eta + \zeta\overline{yz'}/\bar{\beta},$$

$$\bar{x}\eta^* - \bar{y}\xi^* = \bar{x}\eta - \bar{y}\xi.$$

The third of these is nothing more than a restatement of the skewness invariant given in Eq. (XII-12) or (XIII-16). Substituting Eq. (XIV-17) into the other two, we get

$$\psi + \zeta^*\bar{z}'/\bar{\beta} = \zeta\bar{z}'/\bar{\beta},$$

a more useful form of which is

$$\zeta^* = \zeta - \bar{\beta}\psi/\bar{z}'. \tag{XIV-19}$$

From the definition of $\bar{\delta}^*$ given in Eq. (XIV-4), to which we apply Eq. (XIV-17), we get

$$\bar{\delta}^* = \bar{\delta} + \bar{\beta}\psi. \tag{XIV-20}$$

Next, we define H^*:

$$H^* = \zeta^* - \bar{\delta}^*\bar{z}'. \tag{XIV-21}$$

To this we apply Eqs. (XIV-19) and (XIV-20) and obtain

$$H^* = H - \bar{\beta}k^2\psi/\bar{z}'. \tag{XIV-22}$$

Note that the cosine of the angle of refraction becomes $-H^*/kn^*$. We might as well at this point write the solution of the quadratic, Eq. (XIV-18),

$$\psi = \bar{z}'\{H + [H^2 + k^2(n^{*2} - n^2)]^{1/2}\}/\bar{\beta}k^2, \tag{XIV-23}$$

which, when combined with Eq. (XIV-22), shows us that

$$H^* = -[H^2 + k^2(n^{*2} - n^2)]^{1/2}. \tag{XIV-24}$$

Now we need to calculate the derivatives of ψ. To do this we calculate the derivatives of F in Eq. (XIV-18) with respect to $\bar{\beta}$, δ, ζ, and ψ. In doing these calculations we note that, from Eq. (XIV-4),

$$\bar{\delta}_{\bar{x}} = -p\bar{y}/\bar{\beta}^3, \qquad \bar{\delta}_{\bar{y}} = p\bar{x}/\bar{\beta}^3,$$
$$\bar{\delta}_{\xi} = \bar{x}/\bar{\beta}, \qquad \bar{\delta}_{\eta} = \bar{y}/\bar{\beta}.$$

The result gives us the refraction matrix

$$\mathsf{X}_2 = \begin{pmatrix} 1 & 0 & 0 & 0 \\ 0 & 1 & 0 & 0 \\ \psi + \bar{x}(\bar{x}\bar{\beta}^2\psi_{\bar{\beta}} - \bar{y}p\psi_{\delta})/\bar{\beta}^3 & \bar{x}(\bar{y}\bar{\beta}^2\psi_{\bar{\beta}} + \bar{x}p\psi_{\delta})/\bar{\beta}^3 & 1 + \bar{x}(\bar{x}\zeta\psi_{\delta} - \xi\bar{\beta}\psi_{\zeta})/\bar{\beta}\zeta & \bar{x}(\bar{y}\zeta\psi_{\delta} - \eta\bar{\beta}\psi_{\zeta})/\bar{\beta}\zeta \\ \bar{y}(\bar{x}\bar{\beta}^2\psi_{\bar{\beta}} - \bar{y}p\psi_{\delta})/\bar{\beta}^3 & \psi + \bar{y}(\bar{y}\bar{\beta}^2\psi_{\bar{\beta}} + \bar{x}p\psi_{\delta})/\bar{\beta}^3 & \bar{y}(\bar{x}\zeta\psi_{\delta} - \xi\bar{\beta}\psi_{\zeta})/\bar{\beta}\zeta & 1 + \bar{y}(\bar{y}\zeta\psi_{\delta} - \eta\bar{\beta}\psi_{\zeta})/\bar{\beta}\zeta \end{pmatrix}. \tag{XIV-25}$$

Our next concern is the simplification of X_2. First, note that

$$\xi = \bar{x}\bar{\delta}/\bar{\beta} - \bar{y}p/\bar{\beta}^2, \qquad \eta = \bar{y}\bar{\delta}/\bar{\beta} + \bar{x}p/\bar{\beta}^2.$$

Here we have used Eq. (XIV-4) and the definition of p, the skewness, given in Eq. (XIII-15). When these are substituted into Eq. (XIV-14), we obtain

$$\mathsf{S} = \begin{pmatrix} 1 - \bar{z}'\bar{x}(\bar{x}\bar{\beta}\bar{\delta} - \bar{y}p)/\bar{\beta}^3\zeta & -\bar{z}'\bar{y}(\bar{x}\bar{\beta}\bar{\delta} - \bar{y}p)/\bar{\beta}^3\zeta & 0 & 0 \\ -\bar{z}'\bar{x}(\bar{y}\bar{\beta}\bar{\delta} + \bar{x}p)/\bar{\beta}^3\zeta & 1 - \bar{z}'\bar{y}(\bar{y}\bar{\beta}\bar{\delta} + \bar{x}p)/\bar{\beta}^3\zeta & 0 & 0 \\ 0 & 0 & 1 & 0 \\ 0 & 0 & 0 & 1 \end{pmatrix}.$$

This factors into $\mathsf{S} = \mathsf{W}^{-1}\mathsf{V}\mathsf{W}$, where

$$\mathsf{V} = \begin{pmatrix} H/\zeta & 0 & 0 & 0 \\ -\bar{z}'p/\bar{\beta}\zeta & 1 & 0 & 0 \\ 0 & 0 & 1 & 0 \\ 0 & 0 & 0 & 1 \end{pmatrix} \tag{XIV-26}$$

and where

$$\mathsf{W} = \begin{pmatrix} \bar{x}/\bar{\beta} & \bar{y}/\bar{\beta} & 0 & 0 \\ -\bar{y}/\bar{\beta} & \bar{x}/\bar{\beta} & 0 & 0 \\ 0 & 0 & \bar{x}/\beta & \bar{y}/\beta \\ 0 & 0 & -\bar{y}/\beta & \bar{x}/\beta \end{pmatrix}$$

and

$$\mathsf{W}^{-1} = \begin{pmatrix} \bar{x}/\bar{\beta} & -\bar{y}/\bar{\beta} & 0 & 0 \\ \bar{y}/\bar{\beta} & \bar{x}/\bar{\beta} & 0 & 0 \\ 0 & 0 & \bar{x}/\bar{\beta} & -\bar{y}/\bar{\beta} \\ 0 & 0 & \bar{y}/\bar{\beta} & \bar{x}/\bar{\beta} \end{pmatrix}. \tag{XIV-27}$$

The refraction matrix is known to be in the form $\mathsf{X}_R = \mathsf{X}_3\mathsf{X}_2\mathsf{X}_1$, which, from Eqs. (XIV-13) and (XIV-16), can be written as

$$\mathsf{X}_R = \mathsf{T}^{*-\bar{z}}\mathsf{S}^*\mathsf{X}_2\mathsf{S}^{-1}\mathsf{T}^{\bar{z}}.$$

Using the factorization of S, this becomes

$$\mathsf{X}_R = \mathsf{T}^{*-\bar{z}}\mathsf{W}^{-1}\mathsf{V}^*\mathsf{W}\mathsf{X}_2\mathsf{W}^{-1}\mathsf{V}^{-1}\mathsf{W}\mathsf{T}^{\bar{z}}. \tag{XIV-28}$$

Next is calculated the product $\mathsf{W}\mathsf{X}_2\mathsf{W}^{-1}$ in which the factorization of S is used:

$$\mathsf{W}\mathsf{X}_2\mathsf{W}^{-1} = \begin{pmatrix} 1 & 0 & 0 & 0 \\ 0 & 1 & 0 & 0 \\ \psi + \bar{\beta}\psi_{\bar{\beta}} & p\psi_{\bar{\delta}}/\bar{\beta} & 1 + \bar{\beta}(\zeta\psi_{\bar{\delta}} - \bar{\delta}\psi_{\zeta})/\zeta & -p\psi_{\zeta}/\zeta \\ 0 & \psi & 0 & 1 \end{pmatrix}. \tag{XIV-29}$$

The derivatives of ψ are now required. First we calculate the derivatives of F in Eq. (XIV-18) with respect to $\bar{\beta}$, $\bar{\delta}$, ζ, and ψ. First,

$$F_{\psi} = 2\bar{\beta}(\bar{\beta}k^2\psi - \bar{z}'H) = -2\bar{\beta}\bar{z}'H^*,$$

where we used Eq. (XIV-22). Next

$$F_{\bar{\beta}} = 2[\bar{\beta}(k^2 + \bar{\beta}\bar{z}'\bar{z}'')\psi^2 - (\bar{z}'H + \bar{\beta}\zeta\bar{z}'' - 2\bar{\beta}\bar{\delta}\bar{z}'\bar{z}'')\psi - \bar{z}'\bar{z}''(n^{*2} - n^2)].$$

From this expression we eliminate the last term using Eq. (XIV-18), obtaining

$$\begin{aligned}\tfrac{1}{2}\bar{z}'F_{\bar{\beta}} &= \bar{\beta}(k^2 + \bar{\beta}\bar{z}'\bar{z}'')\,\psi^2 - (\bar{z}'H + \bar{\beta}\zeta\bar{z}'' - 2\bar{\beta}\bar{\delta}\bar{z}'\bar{z}'')\psi - \bar{z}''(\bar{\beta}^2k^2\psi^2 - 2\bar{\beta}\bar{z}'H\zeta\psi)\\ &= \psi[\bar{\beta}\bar{z}'\bar{z}''(\zeta - \bar{\beta}\psi/\bar{z}') - \bar{z}'^2(H - \bar{\beta}k^2\psi/\bar{z}')].\end{aligned}$$

Using Eqs. (XIV-19) and (XIV-22), this is seen to reduce to

$$\tfrac{1}{2}F_{\bar{\beta}} = \psi(\bar{\beta}\bar{z}''\zeta^* - \bar{z}'H^*).$$

The remaining two derivatives are trivial:

$$\tfrac{1}{2}F_{\bar{\delta}} = \bar{\beta}\bar{z}'^2\psi, \qquad \tfrac{1}{2}F_{\delta} = -\bar{\beta}\bar{z}'\psi.$$

From these the derivatives of ψ are obtained:

$$\begin{aligned}\psi_{\bar{\beta}} &= -F_{\bar{\beta}}/F_{\psi} = \psi(\bar{\beta}\bar{z}''\zeta^* - \bar{z}'H^*)/\bar{\beta}\bar{z}'H^*,\\ \psi_{\bar{\delta}} &= -F_{\bar{\delta}}/F_{\psi} = \bar{z}'\psi/H^*,\\ \psi_{\delta} &= -F_{\bar{\delta}}/F_{\psi} = -\psi/H^*.\end{aligned}$$

Now we calculate the matrix elements in Eq. (XIV-29):

$$\begin{aligned}\psi + \bar{\beta}\psi_{\bar{\beta}} &= \psi\bar{\beta}\bar{z}''\zeta^*/\bar{z}'H^*,\\ 1 + \bar{\beta}(\zeta\psi_{\bar{\delta}} - \bar{\delta}\psi_{\zeta}) &= H\zeta^*/\zeta H^*.\end{aligned}$$

Substituting these into Eq. (XIV-29) results in

$$\mathsf{WX_2W^{-1}} = \begin{pmatrix} 1 & 0 & 0 & 0\\ 0 & 1 & 0 & 0\\ \bar{\beta}\bar{z}''\zeta^*\psi/\bar{z}'H^* & p\bar{z}'\psi/\bar{\beta}H^* & H\zeta^*/\zeta H^* & p\psi/\zeta H^*\\ 0 & 0 & 0 & 1\end{pmatrix}$$

$$= \begin{pmatrix} 1 & 0 & 0 & 0\\ 0 & 1 & 0 & 0\\ 0 & 0 & \zeta^*/H^* & 0\\ 0 & -\bar{z}'\zeta^*/\bar{\beta} & 0 & 1\end{pmatrix}\begin{pmatrix} 1 & 0 & 0 & 0\\ 0 & 1 & 0 & 0\\ \bar{\beta}\bar{z}''\psi/\bar{z}' & 0 & 1 & p\psi/\zeta\zeta^*\\ 0 & 0 & 0 & 1\end{pmatrix}$$

$$\times \begin{pmatrix} 1 & 0 & 0 & 0\\ 0 & 1 & 0 & 0\\ 0 & 0 & H/\zeta & 0\\ 0 & \bar{z}'\zeta/\bar{\beta} & 0 & 1\end{pmatrix}.$$

Note that the first and last matrices in this product are inverses except for the asterisks. From Eq. (XIV-19) we get

$$\psi = \bar{z}'(\zeta - \zeta^*)/\bar{\beta},$$

which, when substituted into the preceding, yields

$$\mathsf{WX_2W^{-1}} = \begin{pmatrix} 1 & 0 & 0 & 0 \\ 0 & 1 & 0 & 0 \\ 0 & 0 & \zeta^*/H^* & 0 \\ 0 & -\bar{z}'\zeta^*/\bar{\beta} & 0 & 1 \end{pmatrix} \begin{pmatrix} 1 & 0 & 0 & 0 \\ 0 & 1 & 0 & 0 \\ \bar{z}''(\zeta - \zeta^*) & 0 & 1 & \dfrac{p\bar{z}'}{\bar{\beta}}\left(\dfrac{1}{\zeta^*} - \dfrac{1}{\zeta}\right) \\ 0 & 0 & 0 & 1 \end{pmatrix}$$

$$\times \begin{pmatrix} 1 & 0 & 0 & 0 \\ 0 & 1 & 0 & 0 \\ 0 & 0 & H/\zeta & 0 \\ 0 & \bar{z}'\zeta/\bar{\beta} & 0 & 1 \end{pmatrix}.$$

This factors. We are left with

$$\mathsf{WX_2W^{-1}} = \mathsf{U^*U^{-1}}, \tag{XIV-30}$$

where

$$\mathsf{U^*} = \begin{pmatrix} 1 & 0 & 0 & 0 \\ 0 & 1 & 0 & 0 \\ -\bar{z}''\zeta^{*2}/H^* & 0 & \zeta^*/H^* & p\bar{z}'/\bar{\beta}H^* \\ 0 & -\bar{z}'\zeta^*/\bar{\beta} & 0 & 1 \end{pmatrix} \tag{XIV-31}$$

and where

$$\mathsf{U^{-1}} = \begin{pmatrix} 1 & 0 & 0 & 0 \\ 0 & 1 & 0 & 0 \\ \bar{z}''\zeta & -p\bar{z}'^2/\bar{\beta}^2 & H/\zeta & -p\bar{z}'/\bar{\beta}\zeta \\ 0 & \bar{z}'\zeta/\bar{\beta} & 0 & 1 \end{pmatrix}. \tag{XIV-32}$$

Thus, from Eqs. (XIV-28) and (XIV-30), we obtain

$$\mathsf{X_R} = \mathsf{T^{*-\bar{z}}W^{-1}V^*U^*U^{-1}V^{-1}WT^{\bar{z}}}.$$

From Eqs. (XIV-26) and (XIV-31) we calculate

$$\mathsf{L^*} = \mathsf{V^*U^*} = \begin{pmatrix} H^*/\zeta^* & 0 & 0 & 0 \\ -\bar{z}'p/\bar{\beta}\zeta^* & 1 & 0 & 0 \\ -\bar{z}''\zeta^{*2}/H^* & 0 & \zeta^*/H^* & p\bar{z}'/\bar{\beta}H^* \\ 0 & -\bar{z}'\zeta^*/\bar{\beta} & 0 & 1 \end{pmatrix} \tag{XIV-33}$$

and

$$\mathsf{L}^{-1} = \mathsf{U}^{-1}\mathsf{V}^{-1} = \begin{pmatrix} \zeta/H & 0 & 0 & 0 \\ \bar{z}'p/\bar{\beta}H & 1 & 0 & 0 \\ \bar{z}''\zeta^2/H - p^2\bar{z}'^3/\bar{\beta}^3H & -p\bar{z}'^2/\bar{\beta}^2 & H/\zeta & -p\bar{z}'/\bar{\beta}\zeta \\ \bar{z}'^2p/\bar{\beta}^2H & \bar{z}'\zeta/\bar{\beta} & 0 & 1 \end{pmatrix}. \quad \text{(XIV-34)}$$

This is about as far as we can go. The refraction matrix can then be written in the form

$$\mathsf{X}_R = \mathsf{T}^{*-\bar{z}}\mathsf{W}^{-1}\mathsf{L}^*\mathsf{L}^{-1}\mathsf{W}\mathsf{T}^{\bar{z}}. \quad \text{(XIV-35)}$$

Now we are in a position to construct a solution to the lens equation, Eq.(XIII-8) or (XIII-11), or to the equivalent lens equation, Eq.(XIII-13) or (XIII-14). Such a solution is not a general solution, of course, but an *ad hoc* solution constructed for a specific lens design. In principle, the construction of such a solution is simple. Suppose the lens in question consists of a sequence of refracting surfaces separating media of constant refractive index. The X matrix solution would then consist of a product of refraction matrices and transfer matrices that correspond to the media and to the surfaces, respectively. The overall characteristics of the lens design then are contained in the X matrix. The extraction of this information is a process that goes beyond the scope of this book and the capabilities of its author.

A few further observations on the structure of these matrices need to be made. It is convenient at this point to introduce the notion of a *partitioned matrix* (Korn and Korn, 1968, p. 408). This involves the splitting of a matrix into an array of smaller submatrices. For example, the 4×4 matrices we are considering can be partitioned into an array of four 2×2 matrices. These can then be regarded as a 2×2 matrix of 2×2 matrices. Products of matrices partitioned in this way have as elements the products of the 2×2 submatrices. From Eq. (XIV-3) the transfer matrix can be partitioned as

$$\mathsf{T} = \begin{pmatrix} \mathsf{I} & \mathsf{A} \\ (0) & \mathsf{I} \end{pmatrix}, \quad \text{(XIV-36a)}$$

where I is the 2×2 identity matrix, (0) is the 2×2 zero matrix, and

$$\mathsf{A} = \begin{pmatrix} (n^2 - \eta^2)/\zeta^3 & \xi\eta/\zeta^3 \\ \xi\eta/\zeta^3 & (n^2 - \xi^2)/\zeta^3 \end{pmatrix}, \quad \text{(XIV-36b)}$$

The matrices that form the refraction matrix can be partitioned in the same way. From Eq. (XIV-27),

$$\mathsf{W} = \begin{pmatrix} \mathsf{B} & (0) \\ (0) & \mathsf{B} \end{pmatrix}, \quad \text{(XIV-37a)}$$

where

$$\mathsf{B} = \begin{pmatrix} \bar{x}/\bar{\beta} & \bar{y}/\bar{\beta} \\ -\bar{y}/\bar{\beta} & \bar{x}/\bar{\beta} \end{pmatrix}. \tag{XIV-37b}$$

Also from Eq. (XIV-33),

$$\mathsf{L} = \begin{pmatrix} \mathsf{C} & (0) \\ \mathsf{D} & \mathsf{E} \end{pmatrix}, \tag{XIV-37c}$$

where

$$\mathsf{C} = \begin{pmatrix} H/\zeta & 0 \\ -\bar{z}'p/\bar{\beta}\zeta & 1 \end{pmatrix}, \tag{XIV-37d}$$

$$\mathsf{D} = -\begin{pmatrix} \bar{z}''\zeta^2/H & 0 \\ 0 & \bar{z}'\zeta/\bar{\beta} \end{pmatrix}, \tag{XIV-37e}$$

and

$$\mathsf{E} = \begin{pmatrix} \zeta/H & \bar{z}'p/\bar{\beta}H \\ 0 & 1 \end{pmatrix}. \tag{XIV-37f}$$

The thing to note here is that the matrices A and D are symmetric. Moreover, E is the inverse of the transpose of C, $\mathsf{E} = \mathsf{C}^{\mathrm{T}-1}$. This last property applies also to the matrix W since $\mathsf{B} = \mathsf{B}^{\mathrm{T}-1}$. The significance of these properties will be discussed in the next chapter.

THE PERFECT LENS

We have reached the point where we can say a little about the process of designing a lens. We can regard lens design as the construction of a product of transfer and refraction matrices to form an X matrix representing the finished design. This X matrix needs to possess certain properties; it needs to satisfy certain prescribed requirements that the finished lens must have. There must be a standard X matrix that embodies the requirements to be satisfied by the lens design and that can be compared with the X matrix of the design which has been constructed.

The problem we consider now is the problem of the construction of the X matrix for an ideal lens. We consider first the specifications for a copy lens. Such a lens will have associated with it an object plane and an image plane with the property that every point on the object plane is imaged perfectly on the image plane with a magnification m. That is,

$$x^* = mx, \qquad y^* = my.$$

This is shown in Fig. XIV-1. From this the derivatives are calculated:

$$x_x{}^* = y_y{}^* = m,$$

$$x_y{}^* = y_x{}^* = x_\xi{}^* = x_\eta{}^* = y_\xi{}^* = y_\eta{}^* = 0.,$$

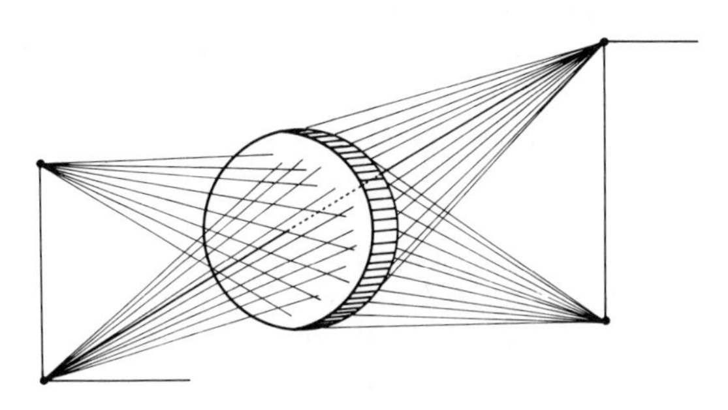

FIG. XIV-1. The perfect copy lens.
Every point on the object plane is imaged perfectly on the image plane.

These are then substituted into the equivalent lens equation (XIII-14). The first of these is ephemeral. The next four give

$$\begin{aligned} \xi_\xi{}^* &= 1/m, \qquad & \eta_\xi{}^* &= 0, \\ \xi_\eta{}^* &= 0, \qquad & \eta_\eta{}^* &= 1/m, \end{aligned} \tag{XIV-38}$$

which we integrate to obtain

$$\xi^* = \xi/m + a(x, y), \qquad \eta^* = \eta/m + b(x, y), \tag{XIV-39}$$

where a and b are arbitrary functions of x and y. Substituting these into the skewness invariant, Eq. (XII-1b) or Eq. (XIII-16), we obtain $xb - ya = 0$, so that $b = ya/x$. Differentiating Eq. (XIV-39) with respect to x and y, we obtain

$$\begin{aligned} \xi_x{}^* &= a_x\,, \qquad & \eta_x{}^* &= -ya/x^2 + ya_x/x, \\ \xi_y{}^* &= a_y\,, \qquad & \eta_y{}^* &= a/x + ya_y/x. \end{aligned}$$

When this is substituted into the sixth equation of the equivalent lens equation, Eq. (XIII-14), we obtain a partial differential equation in a,

$$ya_x - xa_y = ya/x. \tag{XIV-40}$$

We obtain a general solution for this equation by using the method of Lagrange discussed in Chapter VIII. The characteristic equations are formed, after Eq. (VIII-1),

$$dx/y = -dy/x = x\,da/ya,$$

from which we extract two characteristics. From the first two parts of the above we can see that $\beta = (x^2 + y^2)^{1/2}$ is a characteristic and from the first and third members we find another, a/x. The general solution is then obtained by solving for a the equation

$$A(\beta, a/x) = 0,$$

where A is an arbitrary function. This gives us $a = x\alpha(\beta)$ as a general solution. The function α is an arbitrary function of the single variable β. Substituting this back into Eq. (XIV-39), we get the general solution

$$\xi^* = \xi/m + x\alpha(\beta), \qquad \eta^* = \eta/m + y\alpha(\beta). \tag{XIV-41}$$

The derivatives of Eq. (XIV-38) with respect to x, y, ξ, and η provide the elements of the X matrix for the perfect copy lens. This is

$$\mathsf{X}_c = \begin{pmatrix} m & 0 & 0 & 0 \\ 0 & m & 0 & 0 \\ \alpha + x^2\alpha'/\beta & xy\alpha'/\beta & 1/m & 0 \\ xy\alpha'/\beta & \alpha + y^2\alpha'/\beta & 0 & 1/m \end{pmatrix}. \tag{XIV-42}$$

This matrix can also be partitioned, and we can obtain quite easily the fact that

$$\mathsf{X}_c = \begin{pmatrix} m\mathsf{I} & (0) \\ \mathsf{K} & (1/m)\,\mathsf{I} \end{pmatrix}, \tag{XIV-43a}$$

where

$$\mathsf{K} = \begin{pmatrix} \alpha + x^2\alpha'/\beta & xy\alpha'/\beta \\ xy\alpha'/\beta & \alpha + y^2\alpha'/\beta \end{pmatrix}. \tag{XIV-43b}$$

Now let us translate the object plane by an amount t and the image plane by an amount t^*. The X matrix for the system thus obtained is given by multiplying X_c by the two transfer matrices representing these displacements:

$$\bar{\mathsf{X}}_c = \mathsf{T}^{*-t^*}\mathsf{X}_c\mathsf{T}^t.$$

Using Eqs. (XIV-36) and (XIV-43), this becomes

$$\bar{\mathsf{X}}_c = \begin{pmatrix} \mathsf{I} & -t^*\mathsf{A}^* \\ (0) & \mathsf{I} \end{pmatrix} \begin{pmatrix} m\mathsf{I} & (0) \\ \mathsf{K} & (1/m)\mathsf{I} \end{pmatrix} \begin{pmatrix} \mathsf{I} & t\mathsf{A} \\ (0) & \mathsf{I} \end{pmatrix}$$

$$= \begin{pmatrix} m\mathsf{I} - t^*\mathsf{A}^*\mathsf{K} & mt\mathsf{A} - (1/m)\,t^*\mathsf{A}^* - tt^*a^*\mathsf{K}\mathsf{A} \\ \mathsf{K} & (1/m)\mathsf{I} + t\mathsf{K}\mathsf{A} \end{pmatrix}. \tag{XIV-44}$$

Now let us require that the lens be a perfect copy lens with respect to these two new planes. Let the new magnification be $\bar{m}$. In this case the X matrix, $\bar{\mathsf{X}}_c$, would have exactly the same form as Eq. (XIV-42) or (XIV-43). The equality in Eq. (XIV-44) would then yield four equations in the 2×2 matrices whose solutions presumably would give appropriate values of t and t^*. These equations are

$$\begin{aligned} \bar{m}\,\mathsf{I} &= m\,\mathsf{I} - t^*\mathsf{A}^*\mathsf{K}, \\ (0) &= mt\mathsf{A} - (1/m)\,t^*\mathsf{A}^* - tt^*\mathsf{A}^*\mathsf{K}\mathsf{A}, \\ \bar{\mathsf{K}} &= \mathsf{K}, \\ (1/\bar{m})\,\mathsf{I} &= (1/m)\,\mathsf{I} + t\mathsf{K}\mathsf{A}. \end{aligned}$$

The second of these is obtainable from the first and third, is therefore redundant, and may be discarded. From the third of these the matrix K must be a constant. A consequence of this and Eq. (XIV-43b) is that α is a constant. Therefore $\mathsf{K} = \alpha\mathsf{I}$.

The remaining three equations now become

$$\bar{m}\,\mathsf{I} = m\,\mathsf{I} - t^*\alpha\mathsf{A}^*, \qquad (1/\bar{m})\,\mathsf{I} = (1/m)\,\mathsf{I} + t\alpha\mathsf{A}.$$

Since A and A^* are not constant matrices, these two equations cannot be solved for t and t^*. We must conclude that, even though a lens may be perfect in relation to one pair of conjugate planes, its performance decreases to something less than perfection in relation to a different pair.

However, things look a little different when we consider rays arbitrarily near the axis. We consider, in other words, the paraxial approximation in which all powers of the variables x, y, ξ, and η greater than the first are neglected. Then from Eq. (XIV-36b) we get

$$\mathsf{A} = (1/n)\,\mathsf{I}, \qquad \mathsf{A}^* = (1/n^*)\,\mathsf{I},$$

and the above matrix equations reduce to

$$\begin{aligned} \bar{m}\,\mathsf{I} &= m\mathsf{I} - t^*\alpha(1/n^*)\,\mathsf{I}, \\ (1/\bar{m})\,\mathsf{I} &= (1/m)\,\mathsf{I} + t\alpha(1/n)\,\mathsf{I}, \end{aligned}$$

which are equivalent to the scalar equations

$$\begin{aligned} \bar{m} &= m - t^*\alpha/n^*, \\ 1/\bar{m} &= 1/m + t\alpha/n. \end{aligned} \qquad \text{(XIV-45)}$$

These we compare with Eq. (XII-36a) and find that (apart from a discrepancy in sign)

$$n^*/\alpha = f^*, \qquad n/\alpha = f,$$

where f^* and f are the rear and front focal lengths of the lens.

This discussion could be continued by exploring this line of reasoning further and in greater detail as well as by considering different types of lenses that lead to different kinds of X matrices. Although this would be desirable, lack of space and time prohibits it. Let the work we have done here serve as an introduction to the subject and pass on to other things.

REFERENCES

MacDuffee, C. C. (1943). "Vectors and Matrices." Math. Assoc. Amer., Buffalo, New York. (Distributed by Open Court Publ., LaSalle, Illinois.)

Korn, G. A., and Korn, T. M. (1968). "Mathematical Handbook for Scientists and Engineers," 2nd ed. McGraw-Hill, New York.

XV

The Lens Group

Algebra, of which group theory is a part, has been called one of the "richest and boldest intellectual achievements of the human race" (Levi, 1961). Our culture, however, scorns endeavors that are not productive in the physical or economic sense. One suspects that algebraists, especially in earlier days, had to develop particularly tough hides as a defense against the thrusts of the demand for "practicality." Since the advent of the Born–Heisenberg quantum mechanics and its subsequent developments, algebraists should feel no need for protective armor on this account. On the contrary, the ease and completeness with which group theory has been absorbed into the most esoteric fields of modern physics is the highest kind of compliment that can be paid to a sister science and its practitioners. Such compliments, although nice to have, are unnecessary. Algebra provides its own motivation and its own justification.

This discussion is about *groups*. A group can be thought of as consisting of two entities, a set of objects and a *law of composition* for those objects. One bends over backward not to say more than absolutely necessary in defining these terms. One must eschew preconceived notions that might inhibit the range of the imagination.

A group G then will consist of a collection of objects, each of which will be denoted by a lowercase letter, e.g., a, b, c,... . These we will call the group *elements*. The law of composition will be denoted by the asterisk $*$. To constitute a group, the objects and their law of composition must obey four postulates. These are (and we follow van der Waerden, 1949, Chapter II):

1. *Combination Rule.* For every a and b in G, $a * b$ is also an element of G.
2. *Associative Law.* If a, b, and c are elements of G, then $a * (b * c)$ and $(a * b) * c$ represent the same element of G. In other words,

$$a * (b * c) = (a * b) * c$$

   is an identity.
3. *Unit Element.* There exists an element e of G which has the property that, for every a in G, a and $e * a$ represent the same element of G. That is,

$$e * a = a$$

   is true for every a. Such an element e is called a *unit element.*
4. *Inverse.* For every a in G there exists an element u in G with the the property that $a * u$ is the unit element. The element u is called an inverse of a.

These four postulates are the bare bones; their consequences provide the organic material that properly comprises group theory. Notice that nothing has been said about *commutativity*. In general it is *not* the case that $a * b$ and $b * a$ represent the same group element. Indeed, when all group elements commute, the group is said to be *Abelian*. Such groups, however, will not concern us; our interest lies only in noncommutative groups.

Immediate consequences of the postulates are the following statements. The unit element is unique; each group possesses exactly one. Moreover it commutes with every other group element; e.g.,

$$e * a = a * e = a.$$

The inverse is also unique. For every element a of a group there exists one and only one inverse. This inverse element of a, being unique, can be designated by a special symbol a^{-1}. Thus

$$a^{-1} * a = a * a^{-1} = e.$$

There can be subgroups. Suppose there is a subset S of the elements of a group G, which, under the same law of composition, is also a group. Then the group S is called a subgroup of G.

Suppose there are two groups G_1 and G_2 with the property that the elements of G_1 can be put into a one-to-one correspondence with the elements of G_2. That is to say, corresponding to every a_1 belonging to G_1, there exists a unique element a_2 of G_2. And *vice versa*, to every element a_2 of G_2, there corresponds one and only one element a_1 of G_1. This correspondence is often represented symbolically by

$$a_1 \leftrightarrow a_2 .$$

If, moreover, the group element $a_1 * b_1$ corresponds to $a_2 * b_2$ whenever a_1 and b_1 correspond to a_2 and b_2, respectively, then the two groups are said to be *isomorphic*. Every element and operation pertaining to one group is replicated in the other. An *isomorphism* can be regarded as a functional relationship or a *mapping* of one group onto another.

A *homomorphism* is also a functional relationship or mapping between two groups. It differs from an isomorphism in that the biuniqueness requirement is not present. Suppose to every element of G_1 there corresponds an element of G_2:

$$a_1 \rightarrow a_2 .$$

Suppose H is the set of all elements of G_2 that are images of the elements of G_1 under the homomorphism. In general, H will be a proper subgroup of G_2, in which case the homomorphism is said to be *into*. However, it may happen that H is identical to G_2, that every element of G_2 is also an element of H. In this case the homomorphism is called *onto*.

The *kernel* of a homomorphism is defined as the set of all elements of G_1 that are mapped into the unit element of G_2. The kernel is a proper subgroup of G_1. It turns out that a homomorphism is an isomorphism if and only if it is *onto* and if the kernel contains only the unit element of G_1.

The definition of *group* is certainly abstract, abstract almost to the point of being abstruse. Yet groups are ubiquitous. The set of all possible rotations of a solid object in space, the collection of all possible permutations of n objects, the positive rational numbers under multiplication, the positive and negative integers under addition—these and many others are all groups.

The point to be made here is that lenses form a group. Let L denote the collection of all possible lenses. The law of composition will be *conjuction*, the placing of one lens next to the other. Two lenses, placed so that light exiting from one enters the other, form, from our point of view, a legitimate new lens.

As we have defined it, conjunction is a legal law of composition. Since it operates for any pair of lenses whatsoever, the first of the group postulates is satisfied. The associative law is also satisfied. The sequence of operations by which, say, three lens elements are combined has no effect on the final lens. Lenses are on the other hand noncommutative. The order and the orientation in which a pair of lenses is put together are important.

The question of the existence of the unit element is a little more troublesome to satisfy. Here we can use an abstract "no lens at all" for the unit element. Or, if we must have a concrete example, we can consider as the unit lens that optical system that is used to make a contact print. The unit lens at any rate must be an object that maps a plane exactly onto itself.

The final postulate concerning the existence of inverses is another difficult point, which will be discussed later.

THE FULL LINEAR GROUP

The aggregate of all $n \times n$ nonsingular matrices, with matrix multiplication as a law of composition, constitutes a group. That the product of any pair of $n \times n$ matrices is an $n \times n$ matrix and that matrix multiplication is associative enable us to dispose of the first two group postulates. The unit element is the $n \times n$ identity matrix I. The existence of inverses is assured by the restriction to nonsingular matrices. This group is called the *full linear group* (Weyl, 1946, Chapter 4; Chevalley, 1946, Chapter I) and is designated by $GL(n)$.

Let M be some $n \times n$ matrix belonging to $GL(n)$ and let V be an n-component column vector. The product MV is also a column vector, which we might designate by V^*. The matrix M then can be considered as an operator on a vector space. The group of all such matrices $GL(n)$ can be regarded as a collection of linear transformations of a vector space into itself. Such a collection of transformations or operators also comprises a group, the law of composition being the act of taking successive linear transformations.

We thus have two groups: one, the group of nonsingular, $n \times n$ matrices; the other the group of linear transformations on a vector space. To each element of one group there corresponds an element of the other.

The two are obviously isomorphic to one another. The group of linear transformations of a vector space of n-element column vectors is said to be a *realization* of the matrix group $GL(n)$, and the matrix group $GL(n)$ is called a *representation* of the transformation group (Boerner, 1963, Chapter III).

THE LIE GROUP

We can extend these ideas beyond the strict idea of linear transformation. Any functional relationship between two sets of n variables is a more general kind of transformation. Under certain conditions, a collection of such transformations can also comprise a group. Again, if two transformations, taken in succession, are equivalent to another single transformation in the collection, then the first group postulate is satisfied. Associativity presents no real problem, nor does the existence of a unit element. All that is required is that there be, in the collection of transforms, one that leaves the n variables unchanged. The requirement that there be an inverse is more difficult. Each transformation in the collection must be nonsingular.

Let us be a little more explicit. Let the vector function f of the n variables $\mathsf{V} = (x_1, x_2, \ldots, x_n)$ be an example of such a transform; i.e., $\mathsf{V}* = \mathsf{f}(\mathsf{V})$.

The inverse transform, if it exists, will be $\mathsf{V} = \mathsf{f}^{-1}(\mathsf{V}*)$. The identity transform is $\mathsf{V} = \mathsf{f}(\mathsf{V})$, and the composite of two transforms is

$$\mathsf{V}* = \mathsf{f}_a(\mathsf{V}_1), \qquad \mathsf{V}_1 = \mathsf{f}_b(\mathsf{V}), \qquad \mathsf{V}* = \mathsf{f}_a(\mathsf{f}_b(\mathsf{V})).$$

In the last example, we distinguished two transformations in the group by the subscripts a and b. Suppose all transformations can be parametrized in this way. That is to say, every transformation in the group can be written in the form f_a where a is some real number. Then the composite transformation obtained from taking two transforms in succession must be in the form

$$\mathsf{f}_c(\mathsf{V}) = \mathsf{f}_a(\mathsf{f}_b(\mathsf{V})),$$

where the number c is determined by the two numbers a and b. In other words, c is obtained by some functional relationship from a and b, $c = \phi(a, b)$.

A final supposition is that the transformations f_a be continuous functions of the parameter a. What we mean by this is that two transformations whose parameters differ by only a small amount are almost identical. A more precise way of putting this is as follows. For every $\epsilon > 0$, there exists a $\delta > 0$ such that for every vector V and for every a, the vector

$$\mathsf{V}_1 = \mathsf{f}_{a+\delta}(\mathsf{V})$$

lies within a "sphere" of radius ϵ centered on

$$\mathsf{V} = \mathsf{f}_a(\mathsf{V}).$$

If, in addition, f_a is a continuous function (in the usual sense) of V, then the group is said to be a *continuous transformation group* or a *Lie group.*

Of course, such a group may depend on more than one parameter. There is a minimum number of parameters that can be associated with a continuous transformation group. These are called *essential parameters* (Campbell, 1903, Chapter I; Eisenhart, 1933, Chapter I; Chevalley, 1946, Chapter IV).

If, in addition to being continuous the transformation functions are differentiable as well, then, with each transformation of the group, there is associated a Jacobian matrix. This collection of Jacobian matrices also forms a group that is isomorphic to the continuous group of transformations. The group of Jacobian matrices therefore becomes a representation of the continuous transformation group. This group is again $GL(n)$.

But we have seen that $GL(n)$ can be realized as a group of *linear* transformations. The same is true for a continuous transformation group provided that the vectors being transformed are differentials. To be more explicit, let X_a be the Jacobian matrix obtained from the continuous transform $\mathsf{V}^* = \mathsf{f}_a(\mathsf{V})$. Then $d\mathsf{V}^* = \mathsf{X}_a\, d\mathsf{V}$.

We shall pass over a discussion of these algebraic details just as we omitted a definition of *vector space* earlier. These fine points, of great interest in their own right, are extraneous to our discussion here.

SUBGROUPS

One way that the subgroups of the full linear group can be defined is by expressions that are invariant under the transformations of the subgroup.

Of all the subgroups of $GL(n)$, one of the most familiar is the *orthogonal* group, defined as the transformation group that leaves invariant a nondegenerate, symmetric bilinear form (Weyl, 1946, p. 165). In three dimensions this form is exactly the scalar product of two vectors. In n dimensions it is most conveniently represented as the product of one column vector and the transpose of another

$$\mathsf{V}^{\mathsf{T}}\mathsf{W} = \sum_{i=1}^{n} x_i y_i , \qquad \text{(XV-1)}$$

where

$$\mathsf{V} = (x_1, x_2, \ldots, x_n), \qquad \mathsf{W} = (y_1, y_2, \ldots, y_n).$$

The kind of transformation that leaves this bilinear form invariant is found as follows. Let $\mathsf{V}^* = \mathsf{MV}$, then $\mathsf{V}^{*\mathsf{T}} = \mathsf{V}^{\mathsf{T}}\mathsf{M}^{\mathsf{T}}$. When these are substituted into Eq. (XV-1), we obtain

$$\mathsf{V}^{*\mathsf{T}}\mathsf{W}^* = \mathsf{V}^{\mathsf{T}}\mathsf{M}^{\mathsf{T}}\mathsf{M}\mathsf{W}.$$

Clearly, if $\mathsf{V}^{\mathsf{T}}\mathsf{W}$ is to be an invariant for any choice of V and W, then

$$\mathsf{M}^{\mathsf{T}}\mathsf{M} = \mathsf{I}, \qquad \text{(XV-2)}$$

where I is the $n \times n$ unit matrix. What this says is that the orthogonal group comprises those matrices whose transposes are their inverses. Such matrices, naturally enough, are called orthogonal matrices. These comprise the entire set of transformations that leave the scalar product, Eq. (XV-1), invariant.

It is a simple matter for the reader to convince himself that this collection of matrices does indeed form a group. This group is called the *orthogonal group* and is designated $O(n)$. (See Weyl, 1946, Chapter I.)

The group that really interests us is a different subgroup of $GL(n)$, which leaves invariant a nondegenerate, *skew*-symmetric bilinear form. The requirement of skew symmetry imposes the condition that the dimensionality n of the vector space be even. We will write the skew symmetric form as

$$\mathsf{V}^{\mathsf{T}}\mathsf{K}\mathsf{W}, \qquad \text{(XV-3)}$$

where K is defined in terms of a partitioned matrix

$$\mathsf{K} = \begin{pmatrix} (0) & \mathsf{I} \\ -\mathsf{I} & (0) \end{pmatrix}. \qquad \text{(XV-4)}$$

Here (0) is the $(n/2) \times (n/2)$ zero matrix and I is the $(n/2) \times (n/2)$ unity matrix. When

$$\mathsf{V}^* = \mathsf{MV}, \qquad \mathsf{W}^* = \mathsf{MW}$$

are substituted into the skew symmetric form, Eq. (XV-3), we get

$$\mathsf{V}^{*\mathsf{T}}\mathsf{K}\mathsf{W}^{*} = \mathsf{V}^{\mathsf{T}}\mathsf{M}^{\mathsf{T}}\mathsf{K}\mathsf{M}\mathsf{W},$$

and, as in the case of the orthogonal group, the equation that ensures that Eq. (XV-3) is invariant, for all V and W, is

$$\mathsf{M}^{\mathsf{T}}\mathsf{K}\mathsf{M} = \mathsf{K}. \qquad \text{(XV-5)}$$

Matrices that satisfy Eq. (XV-5) leave Eq. (XV-3) invariant. These matrices, like the orthogonal matrices, constitute what is known as the *symplectic group*, $Sp(n)$ (Chevalley, 1946, p. 22; Weyl, 1946, p. 165). It has also been known as the *Abelian linear group* (Dickson, 1900, Part 2, Chapter II).

The skew-symmetric form, Eq. (XV-3), is frequently written using a bracket notation reminiscent of the definition following Eq. (VIII-8),

$$[\mathsf{V},\mathsf{W}] = \sum_{i=1}^{n/2} (x_i y_{i+n} - x_{i+n} y_i), \qquad \text{(XV-6)}$$

where

$$\mathsf{V} = (x_1, x_2, ..., x_n), \qquad \mathsf{W} = (y_1, y_2, ..., y_n).$$

Of course, n must be even.

THE OPTICAL APPLICATION

Let $n = 4$. Then Eq. (XV-4) becomes identical to the lens equation (XIII-11):

$$\mathsf{X}^{\mathsf{T}}\mathsf{J}\,\mathsf{X} = \mathsf{J}. \qquad \text{(XV-7)}$$

In Chapter XIII, where this equation was derived, it was shown that the X matrix for any lens satisfied the lens equation. The converse is not necessarily true. All we can say is that a transformation satisfying Eq. (XV-6) is *lenslike*; there is no guarantee, at this point, that every lenslike transformation can be realized by an optical system. Moreover, there is no suggestion of uniqueness; it is quite possible that several optical systems will correspond to a single, lenslike transformation.

We can then talk about only the aggregate of lenslike transformations, with the knowledge that all realizable lenses constitute a subset. The Jacobian matrices of a set of lenslike transformations, by Eq. (XV-7), exactly comprise the symplectic group, $Sp(4)$. In other words, lenslike transformations are a realization of the symplectic group. Real, constructible lenses constitute a subset (not necessarily a subgroup) of $Sp(4)$.

Before continuing, there are a few loose ends that we need to ravel up. Recall that the Jacobian matrix of a continuous transformation could be regarded as a linear transformation on a space of differentials of vectors. We also stated that subgroups of the full linear group could be defined by vector functions that were invariant under the subgroup. Further, the skew-symmetric form, the invariant expression by which the symplectic group was defined, could be expressed in terms of a bracket symbol, Eq. (XV-6). Now the fundamental optical invariant, Eq. (XIII-7) can be written using the bracket notation from Eq. (VIII-8). Thus

$$[\mathbf{R}^*, \Xi^*] = [\mathbf{R}, \Xi], \tag{XV-8}$$

where $\mathbf{R} = (x, y, 0)$ and where $\Xi = (\xi, \eta, \zeta)$.

This brings us full cycle back to the fact that the fundamental optical invariant is exactly that—an invariant form that defines the group of lenslike transformations that, as we have seen, comprise the symplectic group.

ESSENTIAL PARAMETERS

Early in this chapter it was mentioned that a continuous group of transformations may be parametrized by means of one or more parameters. The least number of parameters required for a given group constitutes a set of essential parameters for that group. We need to know how many are required for the symplectic group. We use the Cayley parametrization (Weyl, 1946, p. 169).

Let M be an element of the symplectic group so that it satisfies Eq. (XV-5), $\mathsf{M}^T\mathsf{K}\mathsf{M} = \mathsf{K}$. Define the matrix S as

$$\mathsf{S} = (\mathsf{I} - \mathsf{M})(\mathsf{I} + \mathsf{M})^{-1}. \tag{XV-9}$$

In order for this definition to make sense, the matrix $\mathsf{I} + \mathsf{M}$ must be nonsingular. We restrict this definition to those symplectic matrices that are also *nonexceptional*, that is, where

$$|\,\mathsf{I} + \mathsf{M}\,| \neq 0.$$

Then S is also nonexceptional, and

$$\mathsf{M} = (\mathsf{I} - \mathsf{S})(\mathsf{I} + \mathsf{S})^{-1}. \tag{XV-10}$$

Substituting this into the quadratic expression (XV-5), we get the linear matrix equation

$$\mathsf{KS} + \mathsf{S}^{\mathrm{T}}\mathsf{K} = (0). \tag{XV-11}$$

Partition S as

$$\mathsf{S} = \begin{pmatrix} \mathsf{A} & \mathsf{B} \\ \mathsf{C} & \mathsf{D} \end{pmatrix},$$

where A, B, C, and D are $(n/2) \times (n/2)$ matrices. Using Eq. (XV-4), Eq. (XV-11) becomes

$$\mathsf{C} - \mathsf{C}^{\mathrm{T}} = (0), \qquad \mathsf{A} + \mathsf{D}^{\mathrm{T}} = (0),$$

$$\mathsf{D} + \mathsf{A}^{\mathrm{T}} = (0), \qquad -\mathsf{B} + \mathsf{B}^{\mathrm{T}} = (0),$$

from which we obtain the fact that B and C are symmetric matrices and $\mathsf{D} = -\mathsf{A}^{\mathrm{T}}$. Considering these as a set of equations in n^2 variables, the number of degrees of freedom in the solution—and therefore the number of essential parameters in the group—is $\frac{1}{2}n(n + 1)$. For the case where $n = 4$, we may write

$$\mathsf{A} = \begin{pmatrix} a & b \\ c & d \end{pmatrix}, \qquad \mathsf{B} = \begin{pmatrix} e & g \\ g & f \end{pmatrix}, \qquad \mathsf{C} = \begin{pmatrix} h & l \\ l & k \end{pmatrix}.$$

Thus, S (and therefore M) involves exactly ten parameters.

When $n = 2$, we get a much simpler expression

$$\mathsf{S} = \begin{pmatrix} p & q \\ r & -p \end{pmatrix}.$$

Putting this back into Eq. (XV-11), we obtain an expression for the 2×2 symplectic matrix in terms of a set of essential parameters:

$$\mathsf{M} = \begin{pmatrix} 1 - 2p + p^2 + qr & -2q \\ -2r & 1 + 2p + p^2 + qr \end{pmatrix} \Big/ D,$$

where $D = 1 - p^2 - qr$ and $p^2 + qr = 1 - D$:

$$\mathsf{M} = \begin{pmatrix} (2 - 2p - D)/D & -2q/D \\ -2r/D & (2 + 2p - D)/D \end{pmatrix}$$

$$= \begin{pmatrix} -1 + 2(1 - p)/D & -2q/D \\ -2r/D & -1 + 2(1 + p)/D \end{pmatrix}.$$

GENERATORS

It may happen that every element of a group can be constructed from a finite number of group elements by means of repeated operations. Such group elements, if they should exist, are called *generators*. Suppose a group has four generators, a, b, c, and d. Then every element of the group can be written as a *word*

$$b * a * a * c * d * a * b * c.$$

It may happen that the representation of a group element by a word is not unique. Then there will be a collection of words that are all equivalent to the group's unit element. These are referred to as *relations*. The length of a word, i.e., the number of "letters" in the word, may be reduced by the use of relations. When the length of a word is reduced to a minimum, i.e., when no further application of a relation can eliminate additional letters, the word is called a *reduced word*.

An example of a relation is the statement $a * a^{-1} = e$, where e is the unit element. This particular relation is called the *trivial relation*.

If a group has no relations other than the trivial relation, it is termed a *free group*. In this case every group element is uniquely represented by a word (Magnus, Karrass, and Solitar, 1966, Chapter I).

Finally, returning to the idea of generators, a set of generators is said to be *complete* if no member of the set can be represented by a word made up of the other generators.

And, as might have been anticipated, the symplectic group, which is our only concern, has generators. Several systems of generators have been described. The one that best suits our purposes was first described by Dickson (1900, Part 2, Chapter III). Later Hua and Reiner (1949) showed that these generators are independent. [For an alternative set of generators, see Murnaghan (1953) and Stanek (1961, 1963).]

Following Hua and Reiner, the generators of the $n \times n$ symplectic group consist of three types of matrices:

translations

$$\begin{pmatrix} \mathsf{I} & \mathsf{S} \\ (0) & \mathsf{I} \end{pmatrix}, \qquad \text{(XV-12a)}$$

where S is symmetric; i.e., $\mathsf{S} = \mathsf{S}^{\mathrm{T}}$;

rotations

$$\begin{pmatrix} \mathsf{U} & (0) \\ (0) & \mathsf{U}^{\mathrm{T}-1} \end{pmatrix}, \qquad \text{(XV-12b)}$$

where $|\,\mathsf{U}\,| = \pm 1$,

semi-involutions

$$\begin{pmatrix} \mathsf{Q} & \mathsf{I}-\mathsf{Q} \\ -\mathsf{I}+\mathsf{Q} & \mathsf{Q} \end{pmatrix}, \tag{XV-12c}$$

where Q is a diagonal matrix with the diagonal elements consisting of zeros and ones. Moreover, $\mathsf{Q}^2 = \mathsf{Q}$ and $(\mathsf{I} - \mathsf{Q})^2 = \mathsf{I} - \mathsf{Q}$.

Of course, S, U, and Q are all $(n/2)\times(n/2)$ matrices; I is the $(n/2)\times(n/2)$ unit matrix; and (0) is the zero matrix.

Our main interest is in the 4×4 symplectic group. The semi-involutions are treated first. There are four possible Q matrices:

$$\mathsf{Q}_0 = \begin{pmatrix} 0 & 0 \\ 0 & 0 \end{pmatrix}, \qquad \mathsf{Q}_1 = \begin{pmatrix} 1 & 0 \\ 0 & 0 \end{pmatrix},$$

$$\mathsf{Q}_2 = \begin{pmatrix} 0 & 0 \\ 0 & 1 \end{pmatrix}, \qquad \mathsf{Q}_3 = \begin{pmatrix} 1 & 0 \\ 0 & 1 \end{pmatrix},$$

leading to four possible semi-involutions. The first and last lead to nothing more than J and I, respectively. The remaining two are

$$\mathsf{F}_i = \begin{pmatrix} \mathsf{Q}_i & \mathsf{I}-\mathsf{Q}_i \\ -\mathsf{I}+\mathsf{Q}_i & \mathsf{Q}_i \end{pmatrix}, \qquad i = 1, 2,$$

or

$$\mathsf{F}_1 = \begin{pmatrix} 1 & 0 & 0 & 0 \\ 0 & 0 & 0 & 1 \\ 0 & 0 & 1 & 0 \\ 0 & -1 & 0 & 0 \end{pmatrix}, \qquad \mathsf{F}_2 = \begin{pmatrix} 0 & 0 & 1 & 0 \\ 0 & 1 & 0 & 0 \\ -1 & 0 & 0 & 0 \\ 0 & 0 & 0 & 1 \end{pmatrix}. \tag{XV-13}$$

We shall also be interested in $Sp(2)$. In this case, the following are the generators, obtained directly from Eq. (XV-12):

translations

$$\begin{pmatrix} 1 & s \\ 0 & 1 \end{pmatrix}; \tag{XV-14a}$$

rotations

$$\begin{pmatrix} \gamma & 0 \\ 0 & 1/\gamma \end{pmatrix}; \tag{XV-14b}$$

semi-involutions

$$\begin{pmatrix} 1 & 0 \\ 0 & 1 \end{pmatrix}, \qquad \begin{pmatrix} 0 & 1 \\ -1 & 0 \end{pmatrix}. \tag{XV-14c}$$

FURTHER OPTICAL APPLICATIONS

Now we come to the *pièce de résistance.* Compare the translation generator with the transfer matrix, Eqs. (XIV-36). The two are structurally identical, even to the point that the 2×2 A matrix defined in Eq. (XIV-36b) is symmetric. The translation generator therefore becomes identical to the transfer matrix.

The refraction matrix is considerably more complicated. Referring to Eq. (XIV-35),

$$\mathsf{X}_R = \mathsf{T}^{*-z}\mathsf{W}^{-1}\mathsf{L}^{*}\mathsf{L}^{-1}\mathsf{W}\mathsf{T}^{z},$$

we see that T, being a transfer matrix, is one of the generators of the symplectic group. The matrix W, defined by Eqs. (XIV-37a) and (XIV-37b),

$$\mathsf{W} = \begin{pmatrix} \mathsf{B} & (0) \\ (0) & \mathsf{B} \end{pmatrix}, \qquad \mathsf{B} = \begin{pmatrix} \bar{x}/\bar{\beta} & \bar{y}/\bar{\beta} \\ -\bar{y}/\bar{\beta} & \bar{x}/\bar{\beta} \end{pmatrix},$$

clearly belongs to the class of symplectic generators called rotations. The third product is a little more difficult. Equations (XIV-37c) through (XIV-37f), inclusive, lead to

$$\mathsf{L} = \begin{pmatrix} \mathsf{C} & (0) \\ \mathsf{D} & \mathsf{C}^{\mathrm{T}-1} \end{pmatrix}$$

$$\mathsf{C} = \begin{pmatrix} H/\zeta & 0 \\ -\bar{z}'p/\bar{\beta}\zeta & 1 \end{pmatrix}, \qquad \mathsf{D} = \begin{pmatrix} \bar{z}''\zeta^2/H & 0 \\ 0 & \bar{z}'\zeta/\bar{\beta} \end{pmatrix}. \tag{XV-15}$$

Recall that J is a symplectic generator. Then

$$\mathsf{J}\,\mathsf{L}\,\mathsf{J}^{-1} = \begin{pmatrix} \mathsf{C}^{\mathrm{T}-1} & -\mathsf{D} \\ (0) & \mathsf{C} \end{pmatrix} = \begin{pmatrix} \mathsf{C}^{\mathrm{T}-1} & (0) \\ (0) & \mathsf{C} \end{pmatrix}\begin{pmatrix} \mathsf{I} & -\mathsf{C}^{\mathrm{T}}\mathsf{D} \\ (0) & \mathsf{I} \end{pmatrix} \tag{XV-16}$$

so that L may be written as the products

$$\mathsf{L} = \mathsf{J}^{-1}\begin{pmatrix} \mathsf{C}^{\mathrm{T}-1} & (0) \\ (0) & \mathsf{C} \end{pmatrix}\begin{pmatrix} \mathsf{I} & -\mathsf{C}^{\mathrm{T}}\mathsf{D} \\ (0) & \mathsf{I} \end{pmatrix}\mathsf{J}. \tag{XV-17}$$

Thus L, and therefore the whole refraction matrix, can be written as a product of a rotation generator, a translation generator, and J, one of the semi-involution matrices.

Exactly the same can be said for the matrix for the perfect lens, Eq. (XIV-44), consisting of two transfers and a matrix resembling L.

Now we can turn to the question of whether lenses form a group.

Two points must be made. First, every element in a group must be made up of a product of the group's generators *and* their inverses. The translation generator is exactly the transfer matrix; the inverse of the translation generator is therefore the inverse of the transfer matrix. The inverse of the transfer matrix is a transfer in a retrograde direction that, from the point of view of optical design, corresponds to a negative thickness of a lens element.

This brings us back to the starting point of this chapter, where we deferred discussing the existence of the inverse, the last in the list of postulates that must be satisfied by a group. For the aggregate of lenslike transformations to constitute a group, we must admit lenses with negative thicknesses. So, for lenses to be a group, we must include nonconstructible lens elements with negative thicknesses. If we exclude these, we are left with a collection of constructible lenses, which fails to be a group by the failure of the fourth postulate.

The rotation generator, which we know as the refraction matrix, does not have this kind of trouble. The inverse of the refraction matrix is the same matrix as far as the refracting surface is concerned except that the order of the refractive indices is reversed. This is certainly a constructible situation and violates no group postulates.

The second point to be made is that the group of rotationally symmetric lenslike transforms is not the full symplectic group. The group $Sp(4)$ has, as generators, the translation, the rotation, and three semi-involutions, J, F_1, and F_2 [Eq. (XV-13)]. (The identity matrix I also is a generator but can be safely ignored.) The group of lenslike objects is generated only by the translation, the rotation, and one of the semi-involutions J. The group of rotationally symmetric lenslike objects comprises a proper subgroup of $Sp(4)$. The other two generators of the symplectic group may or may not have any optical significance. This interesting aspect ought to be subject to more, as we say in the business, research.

THE PARAXIAL CASE

Paraxial rays are realized by expanding all optical formulas and then discarding all but the linear terms. In this case $\zeta = n$. Referring first to the transfer matrix, Eq. (XIV-36), it becomes

$$\mathsf{T} = \begin{pmatrix} \mathsf{I} & \mathsf{A} \\ (0) & \mathsf{I} \end{pmatrix},$$

where

$$\mathsf{A} = \begin{pmatrix} 1/n & 0 \\ 0 & 1/n \end{pmatrix} = (1/n)\,\mathsf{I}.$$

Thus

$$\mathsf{T} = \begin{pmatrix} \mathsf{I} & (1/n)\,\mathsf{I} \\ (0) & \mathsf{I} \end{pmatrix}, \tag{XV-18}$$

where I is the 2×2 identity matrix.

As usual, the refraction matrix is much more difficult. Refer to Eqs. (XIV-37). The elements of B in Eq. (XIV-37b),

$$\mathsf{B} = \begin{pmatrix} \bar{x}/\bar{\beta} & \bar{y}/\bar{\beta} \\ -\bar{y}/\bar{\beta} & \bar{x}/\bar{\beta} \end{pmatrix},$$

being of degree zero, remain unchanged in the paraxial approximation. Therefore the matrix W in Eq. (XIV-37a),

$$\mathsf{W} = \begin{pmatrix} \mathsf{B} & (0) \\ (0) & \mathsf{B} \end{pmatrix},$$

also remains unchanged.

Now refer to the matrix L in Eq. (XIV-37c) and C and D in Eqs. (XIV-37d) and (XIV-37e), respectively:

$$\mathsf{L} = \begin{pmatrix} \mathsf{C} & (0) \\ \mathsf{D} & \mathsf{C}^{\mathrm{T}-1} \end{pmatrix}, \qquad \mathsf{C} = \begin{pmatrix} H/\zeta & 0 \\ -\bar{z}'\bar{p}/\bar{\beta}\zeta & 1 \end{pmatrix}, \qquad \mathsf{D} = -\begin{pmatrix} \bar{z}''\zeta^2/H & 0 \\ 0 & \bar{z}'\zeta/\bar{\beta} \end{pmatrix}.$$

First, recall that $\bar{z}$ is a function of $\bar{\beta}^2$ and that $\bar{z}'$ must then be essentially linear in $\bar{\beta}$. The symmetric function $\delta = (\bar{x}\xi + \bar{y}\eta)/\bar{\beta}$, Eq. (XIV-4), is also linear. The product $\delta\bar{z}'$ appearing in the definition of H in Eq. (XIV-11), $H = \zeta - \delta\bar{z}'$, being of degree *two*, must vanish. In the paraxial approximation, then, $H = n$. By the same type of reasoning, $\bar{z}'/\bar{\beta}$ must be a constant. The skewness, being a quadratic, vanishes so that $\mathsf{C} = \mathsf{I}$. The matrix D becomes

$$\mathsf{D} = -n\begin{pmatrix} \bar{z}'' & 0 \\ 0 & \bar{z}'/\bar{\beta} \end{pmatrix}.$$

Finally, to the first order, $\bar{z}' = \bar{\beta}\bar{z}''$, so that, paraxially speaking,

$$\mathsf{D} = -n\bar{z}''\,\mathsf{I},$$

where I is the 2×2 identity matrix. Thus

$$\mathsf{L} = \begin{pmatrix} \mathsf{I} & 0 \\ -n\bar{z}''\,\mathsf{I} & \mathsf{I} \end{pmatrix}. \tag{XV-19}$$

Now put it all together in

$$\mathsf{X}_{\mathrm{R}} = \mathsf{T}^{*-\bar{z}}\mathsf{W}^{-1}\mathsf{L}^{*}\mathsf{L}^{-1}\mathsf{W}\mathsf{T}^{*}.$$

Calculate $\mathsf{L}^{*}\mathsf{L}^{-1}$:

$$\mathsf{L}^{*}\mathsf{L}^{-1} = \begin{pmatrix} \mathsf{I} & (0) \\ -n^{*}\bar{z}''\,\mathsf{I} & \mathsf{I} \end{pmatrix}\begin{pmatrix} \mathsf{I} & (0) \\ n\bar{z}''\,\mathsf{I} & \mathsf{I} \end{pmatrix} = \begin{pmatrix} \mathsf{I} & (0) \\ -(n^{*} - n)\,\bar{z}''\,\mathsf{I} & \mathsf{I} \end{pmatrix}.$$

This commutes with W and W^{-1} so that

$$\mathsf{W}^{-1}\mathsf{L}^{*}\mathsf{L}^{-1}\mathsf{W} = \mathsf{L}^{*}\mathsf{L}^{-1}$$

and therefore

$$\begin{aligned} \mathsf{X}_{\mathrm{R}} &= \begin{pmatrix} \mathsf{I} & -(\bar{z}/n^{*})\,\mathsf{I} \\ (0) & \mathsf{I} \end{pmatrix}\begin{pmatrix} \mathsf{I} & (0) \\ -(n^{*} - n)\,\bar{z}''\,\mathsf{I} & \mathsf{I} \end{pmatrix}\begin{pmatrix} \mathsf{I} & (\bar{z}/n)\,\mathsf{I} \\ (0) & \mathsf{I} \end{pmatrix} \\ &= \begin{pmatrix} \mathsf{I} & -(\bar{z}/n^{*})\,\mathsf{I} \\ (0) & \mathsf{I} \end{pmatrix}\begin{pmatrix} (0) & -\mathsf{I} \\ +\mathsf{I} & (0) \end{pmatrix}\begin{pmatrix} \mathsf{I} & -(n^{*} - n)\,\bar{z}''\,\mathsf{I} \\ (0) & \mathsf{I} \end{pmatrix}\begin{pmatrix} (0) & \mathsf{I} \\ -\mathsf{I} & (0) \end{pmatrix}\begin{pmatrix} \mathsf{I} & (\bar{z}/n)\,\mathsf{I} \\ (0) & \mathsf{I} \end{pmatrix}, \end{aligned}$$

a product of three translations and two semi-involutions. These matrices for transfer and refraction generate a subgroup of the group of lenslike transformations that in turn is a proper subgroup of $Sp(4)$. This subgroup can be seen to be homomorphic to a subgroup of the 2×2 symplectic group $Sp(2)$, whose generators were given in Eq. (XV-14). The nature of the homomorphism can be seen by comparing the above expression for X_{R} with

$$\begin{pmatrix} 1 & -\bar{z}/n^{*} \\ 0 & 1 \end{pmatrix}\begin{pmatrix} 0 & -1 \\ 1 & 0 \end{pmatrix}\begin{pmatrix} 1 & -(n^{*} - n)\,\bar{z}'' \\ 0 & 1 \end{pmatrix}\begin{pmatrix} 0 & 1 \\ -1 & 0 \end{pmatrix}\begin{pmatrix} 1 & \bar{z}/n \\ 0 & 1 \end{pmatrix},$$

a product of the translation and semi-involution generators.

The process of lens design can then be visualized as the problem of representing a given element of the symplectic group as a word in the generators of that group. Suppose we construct a matrix representing a perfect lens, specifying its characteristics.

An example would be the matrix for the perfect copy lens, X_{c}, in Eq. (XIV-42). The specific characteristics would be determined by numerical values for m, α, and β. As a rotationally symmetric symplectic matrix, it can be represented as a product of the generators of the group of rotationally symmetric lenslike transforms. In this form the given matrix is equal to a product of translations, rotations, and one type of semi-involution. These, in turn, can be interpreted as a sequence of refraction and transfer matrices. Such an interpretation constitutes a legitimate lens design.

Uniqueness poses a rather interesting problem. If this group of rotationally symmetric lenslike transformations is a free group, then the representation of an ideal lens as a product of the generators is indeed unique. Corresponding to any set of lens specifications there would be one and only one design configuration. On the other hand, if the group is not a free group, there would exist a set of relations. These, as representations of the unit element of the group, must consist of lens designs for a perfect copy lens. In this case, referring to Eq. (XIV-43), $m = 1$ and $\mathsf{K} = (0)$, and the perfect copy lens would be perfectly corrected for all conjugate planes. This is unreasonable. Therefore the existence of these relations, certainly as constructible lenses, is doubtful. It follows that there is a good chance that the group of rotationally symmetric lenslike transformations is a free group and that the design of a lens with a fixed and complete set of specifications is unique.

REFERENCES

Boerner, H. (1963). "Representations of Groups." North-Holland Publ., Amsterdam.

Campbell, J. E. (1903). "Introductory Treatise on Lie's Theory of Continuous Transformation Groups." Oxford Univ. Press, London and New York. (Reprinted Chelsea, Bronx, New York, 1966.)

Chevalley, C. (1946). "Theory of Lie Groups," Vol. I. Princeton Univ. Press, Princeton, New Jersey.

Dickson, L. E. (1900). "Linear Groups with an Exposition of the Galois Field Theory." Univ. of Chicago Press, Chicago. (Reprinted Dover, New York, 1958.)

Eisenhart, L. P. (1933). "Continuous Groups of Transformations." Princeton Univ. Press, Princeton, New Jersey. (Reprinted Dover, New York, 1961.)

Hua, L.-K., and Reiner, I. (1949). On the generators of the symplectic modular group. *Trans. Amer. Math. Soc.* **65**, 415–426.

Levi, H. (1961). "Elements of Algebra," 4th ed. Chelsea, Bronx, New York.

Magnus, W., Karrass, A., and Solitar, D. (1966). "Combinatorial Group Theory." Wiley (Interscience), New York.

Murnaghan, F. D. (1953). The parameterization and element of volume of the unitary symplectic group. *Proc. Nat. Acad. Sci. U.S.* **39**, 324–327.

Stanek, P. (1961). Two-element generation of the symplectic group. *Bull. Amer. Math. Soc.* **67**, 225–227.

Stanek, P. (1963). Concerning a theorem by L.-K. Hua and I. Reiner, *Proc. Amer. Math. Soc.* **14**, 571–573.

van der Waerden, B. L. (1949). "Modern Algebra," Vol. I (translated by F. Blum). Ungar, New York.

Weyl, H. (1946). "The Classical Groups." Princeton Univ. Press, Princeton, New Jersey.

XVI

Conclusion

Looking back over these fifteen chapters of manuscript, we see two distinctly different themes. One is concerned with the geometrical optics of the inhomogeneous medium, the other with the theory of lens design. Both spring from the same source, Fermat's principal, and both progress side by side, from the Introduction through the chapter on the Hilbert integral. At that point there is a parting of the ways. For lens optics, the Hilbert integral leads to Snell's law, the eikonal equation, ray tracing, classical aberration theory, and all of the subsequent material. For the inhomogeneous medium, the Hilbert integral provides a differential equation for an orthogonal surface. From the condition for the existence of a solution to this equation comes the criterion for the definition of an orthotomic system of rays. Several examples are studied, and we are led ultimately to a system of equations that mimic the Maxwell equations. For about five chapters the reader is dragged back and forth between these two rather different points of view.

Each theme has its own terminus. Each, unfortunately, ends not with firm declaratory statements summing up what is known and what has been proven but rather on a tentative, quizzical note. More questions have been asked than have been answered. In each of these themes a global theory has been the goal. In each case the arrow has fallen just a bit short of its mark.

In the case of the inhomogeneous medium the necessary mathematical background and the optics to which it is applied are developed together. Much of the mathematics is no longer (if it ever was) a part of the mainstream course material to which a fledgling opticist is exposed. The development itself is rather meticulous and extensive, yet applications to optics are never far in the background. Plausibility rather than strict mathematical rigor is the rule. All this along with the extensive use of the vector notation has the satisfactory smell of success about it. The mathematics and the optics make exceptionally good companions, one complementing the other, so that together they add up to a greater whole.

On the other hand, little of the material presented here is new. Certainly the ray equation was well known to Hamilton, as was the eikonal equation. (In this context, "eikonal" *is* a little strange.) There are hints of the application to optics of classical differential geometry in an early text by Forsyth on the subject. The definition of an orthotomic system of rays by means of an existence theorem for a total differential equation cannot be claimed. Herzberger made specific use of this principle in many of his earlier works. The only thing of any novelty at all is the reasoning and the sequence of steps that lead to what I have called (somewhat presumptuously) the pseudo-Maxwell equations. And this offering should be accepted with a grain of salt. Not all of the details of their derivation have been entirely worked out, and a mapping of their ultimate consequences is speculative and incomplete.

The other theme, geometrical optics and lens design, fares a little better. The material on generalized ray tracing, although not new, is exciting and promising. No credit can be taken for Herzberger's fundamental optical invariant, nor for most of the material in the chapter on classical aberration theory. On the other hand, although the idea of optical image formation as a transformation has persisted for many years, the idea of the use of the Jacobian of such a transformation appears to have been missed. The key to its application is of course the fundamental optical invariant of Herzberger. Once this is stated in terms of the Jacobian matrix, the application becomes clear.

Notwithstanding the clarity of the goal of such an application, the techniques that must be used all but obscure it in an avalanche of incredibly difficult calculations. To duplicate, for example, the results of the relatively trivial calculations in the chapter on classical aberration theory, using the X matrix formalism of the chapter on the lens equation, is a formidable task.

Why then should we bother? must we emulate the mountain climber and scale every peak just because it is there? The answer is simplicity itself. Our peaks must be scaled not only because they offer us a challenge but also because—and this should be our principal motivation—they may be concealing something useful or valuable to us. We are prospectors and not sportsmen.

Generalized ray tracing is some gleanings from a well-worked but not worked-out mine. The idea of lenses as an algebraic group is a novel find in virgin territory. Neither of these claims could possibly qualify as an *El Dorado*, yet each might yield a lode to satisfy the avarice of even the greediest forty-niner.

There have been many serious omissions. Among these is the complete absence of any mention of the Lagrange invariant or of the Petzval sum in the chapter on classical aberration theory. Both are cornerstones of any practical use of these results to optical design. My excuse is that these topics were not vital to the argument and therefore could be dispensed with, albeit with some reluctance, in an already overcrowded chapter. The point of the chapter was to show from beginning to end the derivation of the third order aberration coefficients, perhaps for the first time, from the eikonal functions. This allows one to see best the structure of these aberrations from the point of view of shifts in the image and pupil planes.

At any rate, Gentle Reader, at long last the task is finished. We are not sure of what we have accomplished. Perhaps we have done nothing more than provide geometrical optics, that frowzy old frump, with a bath and a new dress and made her presentable enough to be palatable to yet another generation.

Index

A

D

E

J

K

L

Q

R

T

PURE AND APPLIED PHYSICS

A Series of Monographs and Textbooks

Consulting Editors

H. S. W. Massey
University College, London, England

Keith A. Brueckner
University of California, San Diego
La Jolla, California

1. F. H. Field and J. L. Franklin, Electron Impact Phenomena and the Properties of Gaseous Ions. (Revised edition, 1970.)
2. H. Kopfermann, Nuclear Moments. English Version Prepared from the Second German Edition by E. E. Schneider.
3. Walter E. Thirring, Principles of Quantum Electrodynamics. Translated from the German by J. Bernstein. With Corrections and Additions by Walter E. Thirring.
4. U. Fano and G. Racah, Irreducible Tensorial Sets.
5. E. P. Wigner, Group Theory and Its Application to the Quantum Mechanics of Atomic Spectra. Expanded and Improved Edition. Translated from the German by J. J. Griffin.
6. J. Irving and N. Mullineux, Mathematics in Physics and Engineering.
7. Karl F. Herzfeld and Theodore A. Litovitz, Absorption and Dispersion of Ultrasonic Waves.
8. Leon Brillouin, Wave Propagation and Group Velocity.
9. Fay Ajzenberg-Selove (ed.), Nuclear Spectroscopy. Parts A and B.
10. D. R. Bates (ed.), Quantum Theory. In three volumes.
11. D. J. Thouless, The Quantum Mechanics of Many-Body Systems. (Second edition, 1972.)
12. W. S. C. Williams, An Introduction to Elementary Particles. (Second edition, 1971.)
13. D. R. Bates (ed.), Atomic and Molecular Processes.
14. Amos de-Shalit and Igal Talmi, Nuclear Shell Theory.
15. Walter H. Barkas. Nuclear Research Emulsions. Volume I.
    Nuclear Research Emulsions. Volume II. *In preparation*
16. Joseph Callaway, Energy Band Theory.
17. John M. Blatt, Theory of Superconductivity.
18. F. A. Kaempffer, Concepts in Quantum Mechanics.
19. R. E. Burgess (ed.), Fluctuation Phenomena in Solids.
20. J. M. Daniels, Oriented Nuclei: Polarized Targets and Beams.
21. R. H. Huddlestone and S. L. Leonard (eds.), Plasma Diagnostic Techniques.
22. Amnon Katz, Classical Mechanics, Quantum Mechanics, Field Theory.
23. Warren P. Mason, Crystal Physics in Interaction Processes.
24. F. A. Berezin, The Method of Second Quantization.
25. E. H. S. Burhop (ed.), High Energy Physics. In five volumes.

26. L. S. Rodberg and R. M. Thaler, Introduction to the Quantum Theory of Scattering.

27. R. P. Shutt (ed.), Bubble and Spark Chambers. In two volumes.

28. Geoffrey V. Marr, Photoionization Processes in Gases.

29. J. P. Davidson, Collective Models of the Nucleus.

30. Sydney Geltman, Topics in Atomic Collision Theory.

31. Eugene Feenberg, Theory of Quantum Fluids.

32. Robert T. Beyer and Stephen V. Letcher, Physical Ultrasonics.

33. S. Sugano, Y. Tanabe, and H. Kamimura, Multiplets of Transition-Metal Ions in Crystals.

34. Walter T. Grandy, Jr., Introduction to Electrodynamics and Radiation.

35. J. Killingbeck and G. H. A. Cole, Mathematical Techniques and Physical Applications.

36. Herbert Überall, Electron Scattering from Complex Nuclei. Parts A and B.

37. Ronald C. Davidson, Methods in Nonlinear Plasma Theory.

38. O. N. Stavroudis, The Optics of Rays, Wavefronts, and Caustics.